Space Science and Technologies

Series Editor

Peijian Ye, China Academy of Space Technology, Beijing, China

Space Science and Technologies publishes a host of recent advances and achievements in the field – quickly and informally. It covers a wide range of disciplines and specialties, with a focus on three main aspects: key theories, basic implementation methods, and practical engineering applications. It includes, but is not limited to, theoretical and applied overall system design, subsystem design, major space-vehicle supporting technologies, and the management of related engineering implementations.

Within the scopes of the series are monographs, professional books or graduate textbooks, edited volumes, and reference works purposely devoted to support education in related areas at the graduate and post-graduate levels.

More information about this series at http://www.springer.com/series/16385

Jianyin Miao · Qi Zhong · Qiwei Zhao ·
Xin Zhao

Spacecraft Thermal Control Technologies

版权专有　侵权必究

图书在版编目（CIP）数据

航天器热控制技术 = Spacecraft Thermal Control Technologies：英文/苗建印等编著. —北京：北京理工大学出版社，2020.9

ISBN 978-7-5682-9010-4

Ⅰ. ①航… Ⅱ. ①苗… Ⅲ. ①航天器-热控制-英文 Ⅳ. ①V448.2

中国版本图书馆 CIP 数据核字（2020）第 170247 号

Sales in Mainland China Only
本书仅限在中国大陆地区发行销售

出版发行 /	北京理工大学出版社有限责任公司
社　　址 /	北京市海淀区中关村南大街 5 号
邮　　编 /	100081
电　　话 /	（010）68914775（总编室）
	（010）82562903（教材售后服务热线）
	（010）68948351（其他图书服务热线）
网　　址 /	http：//www.bitpress.com.cn
经　　销 /	全国各地新华书店
印　　刷 /	三河市华骏印务包装有限公司
开　　本 /	710 毫米 × 1000 毫米　1/16
印　　张 /	23.5
字　　数 /	445 千字
版　　次 /	2020 年 9 月第 1 版　2020 年 9 月第 1 次印刷
定　　价 /	132.00 元

丛书策划 / 李炳泉
策划编辑 / 孙　澍
张海丽
责任编辑 / 梁铜华
责任校对 / 周瑞红
责任印制 / 李志强

图书出现印装质量问题，请拨打售后服务热线，本社负责调换

Series Editor's Preface

China's space technology and science research have earned a place in the world, but have not been compiled into a series of systematic publications yet. In 2018, the series *Space Science and Technology* edited mainly by me and coauthored by the leading figures in China's space industry was published in China, when China Academy of Space Technology was celebrating the 50th anniversary of its founding. This collection contains 23 volumes in Chinese, only 10 of which have been selected, recreated and translated into English. In addition, each English volume has been recreated at the suggestion of the Springer, by deleting the contents similar to Springer's existing publications and adding the contents that are internationally advanced and even leading, and bear both Chinese characteristics and worldwide universality. This series fully reflects the professionalism and engineering experience recently accumulated by Chinese scientists and engineers in space technology and science research.

As the Editor-in-Chief of this series, I always insist that this collection must be of high quality, either in Chinese version or English version. First, the contents of this series must be condensed and sublimated based on the combination of theory and practice, so as to provide both a theoretical value and engineering guidance. Second, the relationships between the past knowledge and present achievement and between other people's work and our own new findings should be properly balanced in the book contents to ensure the knowledge systematicness and continuity and to highlight new achievements and insights. Each volume intends to introduce something new to the readers. Third, the English version should be customized for international exposure and play a solid supporting role for China to contribute to the world's space field.

This collection consists of 10 volumes, including *Spacecraft Thermal Control Technologies, Spacecraft Power System Technologies, Spacecraft Electromagnetic Compatibility Technologies, Technologies for Spacecraft Antennas Engineering Design, Satellite Navigation Systems and Technologies, Satellite Remote Sensing Technologies, Spacecraft Autonomous Navigation Technologies Based on Multi-source Information Fusion, Technologies for Deep Space Exploration, Space Robotics, Manned Spacecraft Technologies*.

Spacecraft Thermal Control Technologies covers systematic and specialized spacecraft thermal control techniques, emphasizes their systematic engineering applications and the development of space thermophysics and summarizes the relevant knowledge on thermal control techniques and the design requirements of thermal control system. This volume has two distinct features:

1. Wide coverage. It covers all the aspects of spacecraft thermal control design, including requirement analysis, space environment overview, system-level thermal design, different thermal control techniques for different requirements, thermal analysis simulation technique and thermal test technique, and especially highlights the thermal control methods and techniques adopted by a number of Chinese spacecraft.
2. Strong practicability. This volume not only addresses the relevant technical principles, but also presents the application rules, taboos and typical cases of each technique. Based on more than 50 years of experience in the development of Chinese spacecraft, this volume introduces the system-level thermal design cases and typical part-level thermal design cases of Chinese spacecraft (including lunar probes, manned spacecraft and other spacecraft), which are of great engineering guidance significance.

The publication of this series adds a new member to the international family of space technology and science publications and intends to play an important role in promoting academic exchanges and space business cooperation. It provides comprehensive, authentic and rich information for international space scientists and engineers, enterprises and institutions as well as government sectors to have a deeper understanding of China's space industry. Of course, I believe that this series will also be of great reference value to the researchers, engineers, graduate students and university students in the related fields.

Peijian Ye
Academician
Chinese Academy of Sciences
Beijing, China

Preface

When it comes to China's spatial thermal control technology, one person must be mentioned. He is Prof. Guirong Min, an academician of both the Chinese Academy of Sciences and the Chinese Academy of Engineering, who is also the pioneer and founder of China's space technology, especially spatial thermal control technology. He has made systematic and creative achievements in spacecraft thermal control. Focusing on the research and development of China's first man-made Earth satellite "Dong Fang Hong 1," he led his team to develop a series of spacecraft thermal control techniques, satellite-specific thermal analysis and calculation methods, spatial thermal environment simulation theories and techniques, and heat pipe and superinsulation techniques. In particular, the theory and method of satellite orbit period-integral average heat flow created by him have become the theoretical basis of the thermal analysis and calculation and heat balance test/external heat flow simulation of Chinese spacecraft and are still guiding the thermal tests of Chinese spacecraft. The development of China's spatial thermal control technology is completely attributed to "standing on the shoulders of giants." Therefore, we would like to express our heartfelt gratitude to Prof. Guirong Min through this book.

About how to compile a professional book on spacecraft thermal control technology, we, the authors of this book, did not agree with each other at the beginning. After all, there has already been the *Spacecraft Thermal Control Handbook* written by Prof. David G. Gilmore et al. With rich contents, high authority and great international influence, the handbook has been published in 1994 and 2002. Through the exchange of views, we gradually reached a consensus that we should take 50 years of development and experience accumulation of China's space industry into consideration to write a practical professional book reflecting the current situation and progress of China's spacecraft thermal control technology. This is the original intention of this book.

Based on the above considerations, this book is divided into seven chapters. Chapter 1 "Introduction" includes the contents such as spacecraft thermal control tasks and their requirements, as well as spacecraft thermal characteristics. Chapter 2 "Space Environment" includes the contents such as launch environment, space environment on the Earth orbit, and lunar and planetary space environments.

Chapter 3 "Design of Spacecraft Thermal Control System" includes the contents such as mission characteristics, design principles, design methods and design stages. Chapter 4 "Common Thermal Control Technologies for Spacecrafts" includes the contents such as heat transfer technology, heat insulation technology, heating technology, cooling technology as well as temperature measurement and control technology. Chapter 5 "Typical Design Cases of Spacecraft Thermal Control" includes the design cases of thermal control system and components and other contents. Chapter 6 "Spacecraft Thermal Analysis Technology" includes the contents such as external heat flow analysis, radiation analysis, simulation of specific problems and thermal model modification. Chapter 7 "Ground-Based Thermal Simulation Test for Spacecrafts" includes the contents such as spatial thermal environment simulation method, external heat flow simulation device and heat flow measurement, and heat balance test method. The authors of this book have rich engineering experience, as they have been engaged in the development of spacecraft thermal control technology for many years. Among them, Qi Zhong is responsible for writing the Chaps. 1 and 6; Xin Zhao is responsible for writing the Chaps. 2 and 3; Jianyin Miao is responsible for writing the Chap. 4; and Qiwei Zhao is responsible for writing the Chaps. 5 and 7.

In addition, Yanchao Xiang (for Chaps. 2 and 7), Weichun Fu (for Chap. 4), Hongxing Zhang (for Chap. 4) and Hai Jiang (for Chap. 6) also participated in the compilation. Aside from the above participants, Liang Zhao, Yifan Li, Jianxin Chen, Jialin Sun, Lei Yu, Xianwen Ning, Yuying Wang, Changpeng Yang, Haiying Han, Shuyan Xue, Jianfeng Zhao, Ting Ding and Wenjun Li et al also provided strong support to the compilation of this book. Jiang He, Yawei Xu, Chang Liu, Qiang Zhou, Sixue Liu, Xiangheng Li and Qi Wu also made great efforts to complete this book. Meanwhile, several other spacecraft thermal control experts provided a wealth of technical information during the compilation process. We would also like to acknowledge their help to the book.

Yaopu Wen, Jingang Hu, Hanlin Fan and Wei Yao carefully reviewed the whole book and put forward many insights. The compilation of this book has received close attention and careful guidance from Peijian Ye, an academician of the Chinese Academy of Sciences, and a lot of care and support from many experts in the China Academy of Space Technology (CAST) and the Institute of Spacecraft System Engineering (ISSE). Yongfu Wang, Xiaoheng Liang and Xiujuan Liang from the Science and Technology Commission, ISSE, have also done a lot of work for the publication of this book. Here, the authors would like to express sincere gratitude to them.

Jianyin Miao
Qi Zhong
Qiwei Zhao
Xin Zhao
Institute of Spacecraft System Engineering
Beijing, China

Contents

1 **Introduction** 1
   1.1 Mission of Spacecraft Thermal Control 1
   1.2 Demand for Thermal Control 2
       1.2.1 Temperature Level 2
       1.2.2 Temperature Uniformity and Stability 4
       1.2.3 Wind Speed and Humidity 4
   1.3 Thermal Characteristics 5
       1.3.1 Heat Source 5
       1.3.2 Magnitude and Fluctuation of Heat Dissipation 6
       1.3.3 Heat Flux 7
       1.3.4 Thermal Capacity 8
   1.4 Main Constraints 8
   1.5 Main Technology of Thermal Control 10
   1.6 Main Tasks 12
   References 12

2 **Space Environment** 15
   2.1 Overview 15
   2.2 Environment at Launching Phase 16
   2.3 Earth Orbital Space Environment 19
       2.3.1 Earth Orbital Thermal Environment 19
       2.3.2 Other Earth Orbit Space Environment 29
   2.4 Moon and Planetary Space Environment 35
       2.4.1 Lunar Environment 36
       2.4.2 Mercury Environment 37
       2.4.3 Venus Environment 37
       2.4.4 Mars Environment 38
   2.5 Thermal Environment at Re-entry or Entry Phase 38
   2.6 Inductive Environment 39

		2.6.1	Inductive Environment Caused by Engine Operation	39
		2.6.2	Inductive Environment for Spinning Spacecraft or Equipment	41
	References			42

3 Design of Spacecraft Thermal Control Subsystem ... 45
3.1 Overview ... 45
3.2 Mission Characteristics ... 45
3.2.1 Ground Phase ... 46
3.2.2 Launch and Ascent Phase ... 46
3.2.3 Orbiting Phase ... 47
3.2.4 Reentry or Entry Phase ... 47
3.2.5 Landing Phase ... 48
3.3 Basic Principles of Thermal Control Design ... 49
3.4 Design Method of Thermal Control System ... 50
3.4.1 Thermal Control Design Requirements and Conditions ... 50
3.4.2 Selection of Thermal Design Cases ... 52
3.4.3 Selection of System Design Methods ... 55
3.4.4 Selection of Thermal Control Technologies ... 60
3.5 Thermal Control Design Stages and Key Points ... 61
3.5.1 Concept Phase ... 62
3.5.2 Initial Prototype Phase ... 62
3.5.3 Formal Prototype Phase ... 63
3.5.4 Operation Improvement Phase ... 64
References ... 64

4 Typical Thermal Control Technologies for Spacecraft ... 65
4.1 Overview ... 65
4.2 Heat Transfer Technology ... 66
4.2.1 Introduction ... 66
4.2.2 Thermal Conductive Materials ... 67
4.2.3 Heat Pipe ... 72
4.2.4 Thermal Interface Fillers ... 101
4.2.5 Thermal Control Coating ... 105
4.2.6 Fluid Loop ... 115
4.2.7 Convection Ventilation Device ... 127
4.2.8 Radiator ... 130
4.2.9 Consumable Heat Dissipating Device ... 136
4.2.10 Phase Change Material (PCM) Device ... 143
4.2.11 Thermal Switch ... 147
4.3 Thermal Insulation Technology ... 151
4.3.1 Introduction ... 151
4.3.2 Radiation Insulation ... 153

		4.3.3	Thermal Insulation of Heat Conductance	183
		4.3.4	Thermal Insulation Under Gaseous Environment	188
	4.4	Heating Technology		192
		4.4.1	Introduction	192
		4.4.2	Electrical Heating	193
		4.4.3	Radioisotope Heating Technology	201
	4.5	Temperature Measurement and Control Technology		205
		4.5.1	Introduction	205
		4.5.2	Thermometry Technology	206
		4.5.3	Temperature Control Technology	218
	References			224
5	**Typical Thermal Control Design Cases of Spacecraft**			227
	5.1	Overview		227
	5.2	Design Cases of Spacecraft Thermal Control System		228
		5.2.1	Thermal Control System Design of Remote Sensing Satellite	228
		5.2.2	Thermal Control Design of Communication Satellite	234
		5.2.3	Thermal Control System Design of Lunar Probe	245
		5.2.4	Thermal Control System Design of Manned Spacecraft	256
	5.3	Thermal Design Cases of Spacecraft Assembly		262
		5.3.1	Thermal Design of Propulsion System	262
		5.3.2	Thermal Design of Battery	269
		5.3.3	Thermal Design of Electrical Equipment	271
		5.3.4	Thermal Design of Camera	278
		5.3.5	Thermal Design of Antenna	282
		5.3.6	Thermal Design of Drive Mechanism	287
6	**Thermal Analysis Technology**			289
	6.1	Overview		289
	6.2	Space Energy Conservation Equation		290
		6.2.1	Thermal Network Equation	290
		6.2.2	Computational Domain and Boundary Conditions	292
		6.2.3	Discretization	294
		6.2.4	Thermal Model Construction and Solution Process	296
	6.3	External Heat Flux Analysis		298
		6.3.1	Sun Position	300
		6.3.2	Orbital Parameters	303
		6.3.3	Thermal Environment Parameters	303
		6.3.4	Staying on Celestial Body	306
	6.4	Radiation Computing		307
		6.4.1	View Factor	308
		6.4.2	Radiative Absorption Factor	314

	6.4.3	Radiative Heat	315
	6.4.4	Non-diffusive Radiation	317
	6.4.5	Spatial Decomposition Method for Radiation Calculation	318
	6.4.6	Residual Processing	318
6.5	Simulation of Specific Problems		319
	6.5.1	Flow and Heat Transfer in Pressurized Cabin	319
	6.5.2	Flow and Heat Transfer in Ducts	320
	6.5.3	Heat Transfer of Heat Pipe	320
	6.5.4	Low Pressure Gas Heat Conduction	321
	6.5.5	Thermal Behavior of Solid–Liquid Phase Change	323
	6.5.6	Thermal Behavior of Semiconductor Cooling	324
	6.5.7	Junction-Case Heat Transfer of Electronic Components	325
6.6	Equivalent Transformation of Radiation Term of Thermal Network		328
	6.6.1	Equivalent Heating	328
	6.6.2	Equivalent Heat Sink	329
6.7	Thermal Model Correlation		331
	6.7.1	Basic Knowledge of Thermal Model Correlation	331
	6.7.2	Parameter Analysis	334
	6.7.3	Correlation Method	338
References			340

7 Spacecraft Thermal Testing ... 341
- 7.1 Overview ... 341
- 7.2 Simulation Methods for Space Thermal Environment ... 341
  - 7.2.1 Vacuum ... 342
  - 7.2.2 Cold and Dark Background ... 343
  - 7.2.3 Orbital Heat Flux ... 344
- 7.3 Environmental Heat Flux Simulator and Heat Flux Measurement ... 345
  - 7.3.1 Environmental Heat Flux Simulator ... 345
  - 7.3.2 Environmental Heat Flux Measurement ... 348
- 7.4 Thermal Balance Test ... 351
  - 7.4.1 Thermal Test Model ... 351
  - 7.4.2 Determination of Test Cases ... 352
  - 7.4.3 Test Process and Method ... 353
- 7.5 Atmospheric Thermal Test ... 354
- 7.6 Low-Pressure Test ... 356
  - 7.6.1 Overview ... 356
  - 7.6.2 Selection of Test Gas ... 356
  - 7.6.3 Gas Temperature Simulation ... 358
  - 7.6.4 Flow Field Simulation ... 358
  - 7.6.5 Measurement ... 359

About the Authors

Jianyin Miao is a professor, Ph.D. supervisor and visiting professor of MIT. Jianyin Miao is a director of "Beijing Key Laboratory of Space Thermal Control Technology," a head scientist of heat pipes of "China Academy of Space Technology," an academic leader of space thermal control technology of "China Aerospace Science and Technology Corporation" and a member of editorial committee of "Spacecraft Engineering." Dr. Miao has carried out innovative work on spacecraft thermal control and made outstanding contributions to the successful flight of Chinese Chang'e 3 lunar missions.

Qi Zhong a research fellow, experts in aerospace thermal control, serves in Institute of Spacecraft System Engineering, China Academy of Space Technology, technical leader of a project, one of the academic and technological leaders of China Aerospace Science and Technology Corporation, awarded the State Special Bonus, involved in spacecraft thermal control sub-system design and thermal analysis research and takes part in thermal control sub-system development of the manned spacecraft and navigation constellation. Awarded six times for scientific and technological achievements, owns 20 invention patents and published more than 40 papers.

Qiwei Zhao, Ph.D. of engineering thermophysics, a professor of space thermophysics at China Academy of Space Technology and a member of editorial committee of "Spacecraft Engineering." Dr. Zhao was responsible for the thermal control design and testing of Tracking and Data Relay Satellites of China. He is advisor in thermal control design for the development of various communication satellite platforms of China, such as DFH-4S, DFH-4E and DFH-5.

Xin Zhao, professor, is a member of the professional committee of Science and Technology Commission of China Academy Space Technology. He has been actively engaged in the research of technologies in thermal control designing and thermal analysis. For the past decades, he has served as the chief designer of thermal control sub-system for many satellites. He was the recipient of several National or Ministerial-level Science and Technology Awards including the Second Class Prize of the National Science and Technology Progress Award (2014).

Chapter 1
Introduction

1.1 Mission of Spacecraft Thermal Control

As an essential part, thermal control system, together with the attitude and orbit control system, the structure and mechanism system, the power supply system, the TT&C system, the data management system and the payload system, constitutes the spacecraft. The thermal control technology serves for the whole spacecraft and other systems. Therefore, it is the generic technology of spacecraft engineering.

The missions of thermal control are to analyze and identify the external space environment, mission features and own aspects of spacecraft from the prelaunch phase to the ending of mission; by using the reasonable thermal control technology comprehensively on the premise that meeting the constraints from external environment and spacecraft, to adjust the absorption, transfer and dissipation of heat, so as to ensure that the thermal parameters satisfying reliable operation and expected performance of spacecraft.

In most cases, thermal parameters refer to the temperature (including the level, uniformity and stability) of spacecraft equipment, structure and air in pressurized cabin, also including fluid speed and gas humidity, etc. Sometimes, the control of some parameters requires the combination of thermal control technology and other technologies. For example, air velocity or humidity in pressurized cabin is controlled under the cooperation of thermal control and Environmental Control and Life Support (ECLS) technology.

To satisfy the aerospace mission's requirements, it is necessary to identify the demand of spacecraft for the thermal control, to understand the main constraints, to analyze the external space environment, mission features and own aspects of spacecraft and to determine and implement the relevant thermal control technology. The space environment will be described in Chap. 2. This chapter briefly introduces the demand of spacecraft for thermal control, thermal characteristics of spacecraft, the main constraints faced by thermal control, the common thermal control technology and the main tasks.

1.2 Demand for Thermal Control

The demand of spacecraft for thermal control mainly includes temperature level, temperature uniformity, temperature stability, wind speed and humidity, etc.

1.2.1 Temperature Level

The demand for thermal control is mostly represented by the requirement on temperature range to be kept within, which has a significant influence on the function, performance, reliability and service life of equipments and units. The temperature range requirements for thermal control generally include operating, non-operating and start-up temperature. The operating temperature refers to the temperature condition required to ensure the proper operation of the equipments or units. The non-operating temperature is the temperature condition that has to be guaranteed when equipment or units are powered off (also called hibernation state). This temperature range must be safe for equipment to avoid physical or chemical damage that could cause it to fail. That is, the equipment can survive under non-operating temperature and can resume to normal operating mode. The non-operating temperature is an important guarantee for the survival of equipment in hibernation. Its range is usually greater than that of operating temperature, which depends on the characteristics of the object. The start-up temperature refers to the temperature at which the equipment is switched on. At this temperature, the equipment is in a transitional state between non-operating and operating states.

Temperature range requirements vary with types of equipments or units, which are firstly originated from the physical mechanism for the realization of their functions or performances, and secondly result from the significant influence of temperature on reliability. General requirements of the commonly used equipments on operating temperature range are:

(1) Electronic equipment: For most electronic equipment, the interface temperature to be maintained by thermal control system ranges from −15°C to 50°C. In addition, the unit supplier shall be responsible for ensuring that the component temperature meets the derating requirements, for example, the maximum allowable junction temperature of diode should be no more than 175 °C, the temperature derating of level I should not exceed 100 °C and that of level II not exceed 125 °C.

(2) Special devices: take the CCD in camera as an example. For every 7 °C rise in the operating temperature, the dark electric current of imaging will be doubled [1]. Therefore, the CCD is required to be controlled at a lower temperature, e.g., between −15°C and 5°C. As for the solid-state laser, when its operating temperature rises from 27°C to 127°C, its light-emitting efficiency will reduce to 10% of original value. The HgCdTe detector should work below 80 K to ensure its quantum efficiency and detection rate.

(3) Battery: The operating temperature range of aerospace lithium battery is related to the spacecraft type, the service life and the number of charge and discharge cycles, etc. For example, NASA specifies the following temperature range requirements of lithium battery: range from −40°C to 40°C for deep space exploration mission, range from −5°C to 30°C for Earth-synchronous and sun-synchronous orbit spacecraft.

(4) Antenna: The operating temperature range requirement of antennas varies greatly with its specific types. For example, the TR components of phased array antenna have to be controlled within the range from −10°C to 60°C, that of most low-orbit spacecraft antenna within the range from −100°C to 100°C, and that of high-orbit fixed shaped parabolic reflector within the range from −150°C to 100°C.

(5) Engine/thruster: The temperature of bi-propellant thruster should be above 15 °C and that of catalyst bed of mono-propellant thruster should be generally not less than 120 °C, so as to ensure its efficiency and service life.

(6) Electromechanical products: The operating temperature of electromechanical products is usually restricted by the Hall elements and drive parts. The temperature of motor surface is usually controlled within the range from −50°C to 85°C. The operating temperature of mechanical arm joint drive component is usually controlled within −30°C to 65°C and that of joint reducer usually within −25°C to 50 °C.

(7) Structural parts: Taking into account the temperature requirements of adhesive used, honeycomb panel temperature range requirement is generally specified within range from −100°C to 100°C, thereby ensuring the structural stability and mechanical properties.

According to the corresponding requirements, the temperature range requirement of spacecraft can be classified by cryogenic, normal and high-temperature regions, between which there is no particularly precise boundary. In most cases, the normal temperature range is −15°C to 50°C, which is suitable for most electronic equipment. The cryogenic requirement is mainly applicable for infrared detectors. For example, for the JWST (James Webb Space Telescope), the temperature of near infrared devices has to be kept at not more than 37 K and that of middle infrared not more than 7 K [2, 3]. As a high-temperature requirement, the catalyst bed temperature of mono-propellant thruster shall be maintained beyond 120 °C. The most common requirements are in the normal temperature range, for example, the temperature of general structure is required to be kept in the range from −100°C to 100°C, and the electronic equipment must be kept within the range from −15°C to 50°C. Within the normal temperature range, temperatures of battery, gyroscope, accelerometer, optical remote sensor and atomic clock are usually specified in a narrower or even extremely narrow range.

It is noteworthy that to acquire the knowledge of the influence of temperature on reliability, service life or long-term performance, substantial samples should be investigated. Therefore, it is impractical to attempt to specify the requirement for thermal control based solely on the effect of temperature. The feasible approach in engineering is to determine the temperature requirement based on experience and cost trade-off. On the one hand, the temperature effect data accumulated during the research of materials, components and manufacturing can be used as the basis for determining the approximate range of temperature requirement. On the other hand, the experience and lessons learnt from experiments and application of similar products are also the references for determining such requirement. The implementation cost also has to be evaluated. If the temperature requirement roughly determined based on the above approaches is more costly or even impossible to achieve, the more precise temperature effect experiment has to be performed for investigating the applicability of relaxing the temperature requirement.

1.2.2 Temperature Uniformity and Stability

In addition, the temperature uniformity and stability are also required, where the temperature uniformity requirement is generally specified in the form of temperature difference or temperature gradient. For example, for the sake of performance, the temperature differences between propellant tanks and among battery cells are required to be within ± 5 °C. Most temperature uniformity requirements arise from the need to suppress thermal deformation. In the Gravity Recovery and Climate Experiment (GRACE) mission jointly developed by National Aeronautics and Space Administration (NASA) and Deutsches Zentrum für Luft und Raumfahrt (DLR), the temperature difference of sensor unit DSS baseplate in SuperSTAR accelerometer is even restricted to be not more than 0.1 K [4].

The temperature stability refers to the limitation to the fluctuation or changing rate of temperature over time. The narrow temperature range requirements of optical remote sensor and gyroscope mentioned above are also the requirements on temperature stability. Extremely stringent temperature stability requirement is not uncommon, for example, the temperature of physical parts of rubidium clock should be stabilized at (10 ± 0.1) °C in their whole life cycle; the temperature of SU Sensor DSS baseplate and ultra-stable oscillator of GRACE is required to be stabilized at ± 0.1 °C/orbit [4]; the temperature stability of mK level is specified by the Space Interferometry Mission (SIM); and the hydrogen clock even requires a temperature stability of sub-mK level (0.1 mK/day) [5].

1.2.3 Wind Speed and Humidity

In addition to the temperature requirement, manned spacecraft is also subjected to the wind speed, humidity, etc., in the astronaut activity area. For example, the wind speed should be neither too large nor too small; generally, 0.2–0.8 m/s is appropriate, which is mainly due to the requirement of human comfort. The relative humidity is also mainly based on this requirement, generally 30%–70%.

1.3 Thermal Characteristics

Spacecraft has many thermal characteristics, including heat dissipation, thermal conductivity of materials, specific heat of materials or thermal capacity of units, surface thermal optical property and other aspects closely related to thermal control. This section describes some thermal characteristics that cannot be determined by thermal control systems, including heat dissipation (the source, the magnitude and fluctuation), the heat flux density and the thermal capacity, etc.

1.3.1 Heat Source

Heat dissipation refers to the heat generated by spacecraft equipments, instruments and crews onboard. The heat is generated by transferring or transforming of process of electricity, chemistry, machinery, microwave, nuclear, biological/human metabolism, etc. Different types of heat dissipation come from different physical/chemical processes and have different heat dissipation magnitudes. The main source of heat dissipation can be summarized as follows:

(1) Electrical–thermal energy conversion: The most common example is the electronics devices. Most of the electrical energy of internal components will be directly dissipated into heat. Besides, as a widely used thermal control means, the electric heater also falls into this category.
(2) Chemical–thermal energy conversion: Some short-term vehicles, such as recoverable satellites, are not equipped solar battery cells but primary batteries instead; most spacecraft uses solar battery cells and rechargeable batteries for power supply during eclipse. In the discharge process, the chemical energy of both batteries will be converted into thermal energy. When the engine is fired, the propellant combustion is also a process in which chemical energy is converted into thermal energy.
(3) Mechanical–thermal energy conversion: Part of mechanical energy will be converted into heat by the mechanical motion friction in solar array drive assembly. In the thermodynamic cycles of refrigerator, the same transformation also happens in the process of working fluid being compressed.
(4) Microwave–thermal energy conversion: Some electromagnetic wave will be converted into heat directly when the microwave devices (e.g., microwave switch and traveling-wave tube) operate.
(5) Nuclear–thermal energy conversion: In some vehicles, nuclear energy is used to provide thermal or electrical energy. Nuclear energy is converted into heat in both radioactive decay of atoms and controlled fission or fusion of nuclear reactors.
(6) Biological/human metabolism-thermal energy conversion: occurred in manned spacecraft.

1.3.2 Magnitude and Fluctuation of Heat Dissipation

Magnitude of heat dissipated by spacecraft varies dramatically with types of spacecraft: Total heat dissipation of most vehicles is of hectowatt or kilowatt magnitude, that of pico- and nanosatellites may be as low as a few watts or dozens of watts, while that of certain large-capacity communication satellites can reach myriawatt magnitude and that of nuclear-powered spacecraft can even reach megawatt magnitude. Heat generated by different units or equipment varies greatly, mainly as: The CCD usually dissipates a few watts, the star tracker more than ten watts, the repeater dozens of watts, the power control unit in hectowatt magnitude and SAR antenna up to thousands of watts.

The overall heat dissipation of spacecraft may vary greatly with the mission phase and operation mode. Generally, during the phases of prelaunch, launching, orbit transfer, descent and landing, the total heat dissipation is relatively low, because only the attitude and orbit control system, the TT&C system, the thermal control system and the data management (or onboard data handling/integrated electronics) system operate, while the payload system is in standby mode. After entering the specified mission orbit, the payload system is switched on. The heat dissipation in HEO communication satellites and navigation satellites generally maintains the maximum level, while the payload system of LEO spacecraft usually operates intermittently. Therefore, the heat dissipation will fluctuate greatly. The total heat dissipation of manned spacecraft (e.g., space station) will increase greatly during the docking or separation of segments. For the spacecraft landing on the moon, Mars and other extraterrestrial bodies, its heat dissipation is highly related to the mission target. Cruising or exploring on celestial bodies can increase the heat, while its "sleep" can result in relatively low heat dissipation.

Furthermore, in the long run, the slow degradation of batteries and electronic components will lead to the growth of heat dissipation; even in the short run, the on–off switch of electric heating elements, the ripple of spacecraft bus voltage, the operation of engine or thruster, etc., will cause the fluctuation of the total heat dissipation.

Regarding the spacecraft as an object, the on and off of the equipment will change the heat dissipation, which means that the heat dissipation distribution over the spacecraft will also change. Sometimes, such change may be caused by the displacement of heat source. For example, after the docking of manned spacecraft, in addition to the change in operating scenario of equipments onboard, the transfer of astronauts between cabin segments or their extravehicular activities means that the heat source is redistributed at the same time.

The fluctuation of equipment heat dissipation can be roughly divided into three categories: quasi-constant, time-varying and changing with temperature. Most electronic equipment generates heat approximately constantly. The heat dissipation of detectors, data storage/transmission devices, engines/thrusters and radio transmitters is often intermittent with time. Heat generated by electronic thermostat often changes with temperature. Besides, when a battery is discharged, or is at the end

1.3 Thermal Characteristics

of charging, the heat dissipation is also affected by its temperature. There are other factors concerned to the fluctuation of heat dissipation. For example, the heat generated during discharging is closely related to the magnitude of discharging current; the heat dissipation of the power control unit is affected by the amount of current load and shunt; heat dissipated on the hot end of semiconductor cooler is not only related to the temperature of hot end, but also related to the temperature difference between hot and cold end, as well as to the electric current.

1.3.3 Heat Flux

The heat flux herein generally refers to the heat flow rate per unit area cross the installation interface of equipment or devices. Heat flux is another key thermal characteristic, which greatly influences the final heat transfer temperature difference and the temperature distribution. The heat dissipation and heat flux are usually considered together in choice of relevant thermal control technologies. Similar to the heat dissipation, heat fluxes vary because of diversity of equipment or units; and different parts of equipment are subjected to different heat fluxes.

The heat flux of equipment or units spans from a few tenths of W/cm^2 to several hundreds of W/cm^2. The heat flux cross the installation interface of most ordinary electronic equipments is less than 1 W/cm^2, that of power control unit is of W/cm^2 magnitude and that of high-power microwave devices and laser pumping source can reach several hundreds of W/cm^2.

In particular, although the average heat flux cross the mounting side of most electronic equipment is not high, it is usually non-uniform. Heat flux components located at certain regions may be very high. It is common that the local heat flux at the mounting plate of electronic components inside an equipment reaches at the magnitude of W/cm^2. Moreover, the case of more than 10 W/cm^2 or even 100 W/cm^2 is not rare. Therefore, thermal control designers should pay more attention to the local heat flux than the average heat flux. Firstly, the thermal control design should focus on the hot spots. For example, the heat pipe under the baseplate of equipment should be arranged right below the high heat flux regions; secondly, it is necessary to consider the adaptability of the thermal control technology itself to the high heat flux, such as whether the heat flux to which the heat pipe is subjected exceeds its limit. In short, the higher the heat flux, the more difficult it is to diffuse, transfer and reject the heat. Therefore, more dedicated efforts should be paid in the thermal control design, and appropriate thermal control measures should be taken.

In addition, when the spacecraft engine is operating, the thermal radiation of the nozzle and the flame, and the rarefied gas plume heating will jointly impose high heat flux upon surrounding surface. For example, the theoretical heat flux of the 490 N engine of Chinese high-orbit communication satellite impacting on the launch vehicle adapter (LVA) ring surface can reach a few tenths of W/cm^2; during the landing of Chang'e 3, the theoretical heat flux on the surface of landing legs reaches dozens of W/cm^2 in the mode of touchdown shutdown. Although these values are not more stringent than the local heat fluxes of some components, the involved field of this kind of heat flux is much larger. Heat shields must be used to prevent the spacecraft structure and equipment from being damaged by overheating. Besides, it is also necessary to consider whether the thermal control products can withstand the high heat flux.

1.3.4 Thermal Capacity

The thermal capacities of equipment or units onboard are usually not determined by the thermal control design. However, it is an important factor affecting the range and rate of temperature change.

For spacecraft interiors, the influence of equipments or units with large thermal capacity should be paid more attentions to in the thermal control design. For example, for the thermal balance test with relative large thermal mass, general stabilization criteria may result in "assumed" stabilization temperatures quite different from actual stabilization temperatures. As another example, for the large-capacity storage tank, if the conventional on–off automatic thermostat is used, the temperature overshoot during control will be more severe.

For spacecraft exteriors, hardware with small thermal capacity should be emphasized in the thermal control design. It is adequate for most thermal designs to use only the global annual average values of the earth-emitted infrared radiation and albedo, ignoring their variations with the geographical region, the season, the solar-elevation angle, the orbit inclination, and distance from subsolar point. For the internal parts of spacecraft, this does not result in the large fluctuation or deviation of temperature. However, for the external parts, the external thermal environment, component thermal capacity and heat rejection ability jointly determine the sensitivity of their temperature to the external thermal environment. The smaller the thermal capacity of hardware outside the cabin of LEO spacecraft is, the greater the influence of above factors is. If the temperature of units outside cabin needs to be kept in a narrow range, a wider range of thermal environment parameters should be generally selected as the design basis according to the external thermal environment-sensitive factors of units (i.e., more sensitive to the Earth IR or albedo? or controlled by both) and their thermal inertia (thermal capacity).

1.4 Main Constraints

The main constraints on spacecraft thermal control system include vacuum environment, thermal environment, microgravity or non-1g gravity, space radiation, etc.

1.4 Main Constraints

All spacecraft have to undergo the ground testing, the launch and the running in extra-atmospheric space segment. Some even have to land on extraterrestrial body or return to the Earth again. The main operating phase of most spacecraft is the extra-atmospheric space segment. Even the spacecraft which finally lands on a celestial body with atmosphere (e.g., Mars) will also experience a long period of vacuum state before reaching the destination. The vacuum is a key different environment between space mission and ground program. Specifically, convection cooling is impossible for the final heat rejection/removal of spacecraft due to the vacuum. Although local heat rejection of a few short-term missions can be accomplished via evaporation, sublimation or other forms of expendable heat rejecting technologies, and the rarefied gas flowing on Mars surface can reject heat. In most cases, the thermal radiation is the unique means for heat removal. The vacuum will accelerate the outgassing of materials, and contaminant may result in the performance degradation or failure of spacecraft. Hence, the selection of thermal control product or material shall meet the restrictions related to vacuum, weight loss, volatile condensable materials, etc. Furthermore, the outgassing and condensation of materials may also contaminate the thermal control coating or other materials, causing the unexpected evolution of performance. Therefore, during the thermal control design, cold welding due to vacuum must be avoided.

The thermal environment is another factor which is quite different from the ground thermal engineering. In most cases, the thermal environment of ground mainly refers to the convective heating or cooling of equipment from the atmosphere. Sometimes, the solar radiation heating after atmospheric attenuation may be considered, but the Earth albedo and infrared radiation can be ignored almost in any cases. In space, all the thermal environment factors in the majority of cases, including the direct solar radiation, celestial albedo and celestial infrared radiation, should not be ignored. These factors are often called external heat load or orbital heating and affected by many factors, such as the distances between spacecraft and the sun, between spacecraft and celestial body, the orbit, attitude, geometric configuration of the vehicle and thermal radiation characteristics of its surface material. Sometimes, it also involves aerothermodynamic or free molecular flow heating, nozzle or high-temperature gas thermal radiation and rarefied gas plume heating from fired engine or thruster. For the spacecraft landing on the celestial body with aerosphere, the influence of atmosphere on the spacecraft is similar to that of the Earth. However, the basic knowledge (e.g., temperature and wind speed) of celestial aerosphere recognized by human beings is far less adequate than that of Earth atmosphere. In short, compared with the ground engineering, the thermal environment of spacecraft is more complicated. Some of them can be adjusted through design. For example, the absorbed external heat load can be adjusted by thermal coatings; while some need to be complied, and some have to be considered adequate margin during design due to the large uncertainty. All the above imposes extra restrictions on the thermal control of spacecraft.

In most cases, spacecraft is in the microgravity state, which also brings some constraints. For example, the capture and collection of liquid in gas–liquid two-phase heat transfer device cannot be accomplished by means of gravity. The vibration of moving parts may disturb the attitude and pointing accuracy of spacecraft, or limit the required microgravity level. Therefore, the use of moving parts such as mechanical pump and compressor may also be restricted. After landing on the moon, or Mars, the gravity acceleration is different from the Earth, which may lead to the change in the operation performance of some thermal control products. Furthermore, the centrifugal acceleration of some spin-stabilized spacecraft and the short-term overload during launch, orbit maneuver or descent and landing may also result in the same effect like affecting the pressure load of thermal control fluid loop. Stated above, the changes of acceleration conditions that should be accommodated or considered are often more complicated than those on the ground.

The ultraviolet, the proton, the electron, the atomic oxygen, the space debris or the micrometer are all not negligible factors for the selection of spacecraft products and materials. There is no doubt that many of the above factors should be considered for thermal control. Because thermal control coatings are widely used on the surface of spacecraft, the above space environment might lead to degradation or even failure of material performance. Considering these factors, more constraints will be brought to the selection of thermal control materials.

The atmospheric pressure during the launch phase of spacecraft is also a factor to be considered for thermal control. Thermal control is not the solo, but almost the most involved subsystem to consider and adapt to the rapid de-pressure during the ascent. For example, fixation of multi-layer thermal insulation blankets must undergo this situation.

In addition, the spacecraft thermal control is also restrained by spacecraft geometric configuration and layout. Besides, thermal control must meet the requirements on reliability, safety, service life, EMC/ESD, and anti-radiation, and be compliant with the environmental requirements related to the vibration/shock/acoustics and aerodynamic impinging during launch/separation/entry/landing, or even the requirement in aspects of non-toxicity, flame retardancy and ergonomics. These requirements are not unique to aerospace engineering but often more rigorous compared to those on the ground.

1.5 Main Technology of Thermal Control

Technology can be regarded as methodology, which includes not only the theoretical knowledge of analysis and design, but also the methods of manufacturing, the know-how of combination of products or skills about tools. According to control modes, thermal control technology can be generally classified as the passive and active thermal control technology, and the former mainly features open-loop control. Parameters (e.g., temperature) to be controlled are not used for feedback. The inherent physical characteristics (e.g., thermal radiation property and thermal conductivity) of hardware are usually used to control the heat entering into or leaving out of a system, thus controlling the temperature of spacecraft equipment within the specified range. In the operating of active thermal control technology, target parameters such as temperatures are used for feedback. The method can be electronic power

1.5 Main Technology of Thermal Control

consuming or not. All in all, the active thermal control technology has strong adaptability and high adjustment precision, while the passive technology is simple, reliable and cost effective. Hence, the passive thermal control technology is a prior consideration in design. Supplemented by active thermal control, most spacecraft thermal control systems uses passive thermal control as the main method.

Thermal coating, heat pipe, multi-layer insulation blanket, conduction enhancing/suppressing material (like thermal filler/low-conductivity-washer), and the phase change meterial are generally classified as passive thermal control technology, while electric heater, pumped fluid loop, fan, variable-conductance heat pipe, loop heat pipe, thermal switch, louver, refrigerator as active thermal control technology. This is just a classification that is traditionally applicable to most cases. It is not absolute but subjective. Because the performance of a product depends not only on its main functions, but also on the way it is used. For example, if the electric heater is simply powered on, its function is passive thermal control. However, if it is combined with the temperature sensor as automatic thermostat, its function is active thermal control. Take the variable-conductance heat pipe as another example. If the reservoir (accumulator) temperature is not subjected to be closed-loop controlled, its heat transfer performance is then only adjusted automatically according to the heat load carried, without direct feedback parameter. In this case, it is also acceptable to be regarded as passive thermal control. The thermal radiation property of most thermal control coatings is approximately constant, but the electrochromic smart thermal control coating, with the participation of other controllers, can adjust the applied electric field according to the temperature and change its thermal radiation property, thus yielding the temperature control function. In this case, this coating used belongs to the typical active thermal control.

According to the purpose, the main technologies used in spacecraft thermal control can also be roughly divided into heat transfer, heat insulation, heating, refrigeration, temperature TT&C, etc. Chap. 4 describes these technologies according to the classification. This kind of classification is mainly used for narration. Because one technology may have more than one features. For example, dedicated thermal control coatings on exposed surface can regulate both the heat absorbed and emitted simultaneously.

Although the space environment brings constraints to the spacecraft thermal control, it also gives birth to some thermal control technologies which are not popular on the ground, but widely used in the aerospace field. For example, the multi-layer thermal insulation blankets are almost absolutely necessary for every spacecraft. Thermal control coating which spectrally selects the absorbed fraction of the imposed radiation heating is another example. In most cases, the condensed liquid backflow of heat pipe does not need to overcome the gravity because of spacecraft runs in the state of microgravity. So, it is widely used in the aerospace field. The water sublimator cannot work in the ground atmosphere, but it can be used on the spacecraft with vacuum or very low surrounding pressure.

In most cases, the temperature of spacecraft thermal control has to be maintained near the level at which it is assembled, i.e., around room temperature. However, with the development of deep space or scientific exploration, lower temperature level, such as dozens of K, is increasing demanded. The appealing for long-term compliance with the heat source at thousands of K also appears. Thermal control technology applicable to such temperature range requirement is far from mature. Besides, spacecraft with larger heat dissipation (e.g., dozens of kW) and higher heat flux (e.g., hundreds of W/cm^2) is also gradually increasing, thus putting forward more challenges for thermal control technology.

1.6 Main Tasks

The main works of thermal control oriented to spacecraft or aerospace equipments are involved with design, analysis, development, testing and on-orbit support and include determining the thermal configuration (configuration means functional and physical characteristics). That is, determining thermal configuration based on analysis, proposing requirements for the lower- (thermal control products) or upper-level (the system, i.e. the spacecraft), integrating the thermal control system, and demonstrating performance compliance with technical specification of the system through test. The works of spacecraft thermal control, as a part of the entire spacecraft development system engineering, are closely bundled to the procedure of spacecraft development.

In China, the spacecraft development is usually divided into concept, engineering model/prototype development, flight model development and refining/improving phases following first flight. Sometimes, the feasibility should be illustrated before the concept phase. Once the technical route is determined, the key technical breakthrough must be completed in the concept phase. Configuration must be clearly defined after the prototype development phase. During this phase, the compliance of thermal control system as well as thermal products with system requirement should be qualified. The final configuration is determined and demonstrated during the flight model development phase, in which all design, manufacture, assemble and test should be done on the flight model; based on assessment of flight performance, the proposals about subsequent design improvements are paid attention to in the refining/improving phase. Thermal control analysis goes along through all phases. The items and scheme of verification/demonstration campaigns mainly depend on the readiness of technology or product adopted. For example, for the high readiness level technology or product, the items to be verified by test can be reduced.

References

1. L. Zhitao, X. Shuyan, C. Liheng, Thermal control of high-power focal plane apparatus. Opt. Precis. Eng. 16(11), 2187–2192 (2008)
2. K. Parrish, S. Glazer, S. Thomson, The cryogenic thermal system design of NASA's James Webb Space Telescope (JWST) integrated science instrument module (ISIM), in *35th International Conference on Environmental Systems and 8th European Symposium on Space Environmental Control Systems*, Roma (2005)

3. P.E. Cleveland, K.A. Parrish, Thermal system verification and model validation for NASA's cryogenic passively Cooled James Webb Space Telescope (JWST), in *35th International Conference on Environmental Systems and 8th European Symposium on Space Environmental Control Systems*, Roma (2005)
4. R.P. Kornfeld, B.W. Arnold, M.A. Gross, N.T. Dahya, W.M. Klipstein, GRACE-FO: the gravity recovery and climate experiment follow-on mission. J. Spacecr. Rockets 56(3), 931–951 (2009)
5. D.G. Gilmore, *Spacecraft Thermal Control Handbook, Volume I: Fundamental Technologies*, 2nd edn. (The Aerospace Press, El Sequndo, 2002)

Chapter 2
Space Environment

2.1 Overview

Generally, space environment refers to the astrospace environment outside the Earth's atmosphere. In the research oriented to spacecraft, the space environment scope includes the environment at special stages of space exploration such as spacecraft launching, return and re-entry, in addition to the environment of other celestial bodies like the moon, Mars, etc. The space environment is particularly complicated and even unpredicted and is quite different from ground environment. For example, the typical space environment on the Earth orbit mainly consists of vacuum, microgravity, particle, solar radiation, Earth albedo and Earth infrared radiation and so forth. However, the Mars environment consists of Mars gravitational field, the atmosphere environment based on CO_2, storm and other space environments.

The space environment has an important effect on the thermal control design and ground test verification of the spacecraft, mainly including the effect of thermal condition, space particles and radiation as well as special gravitational field. For example, the thermal effects of space environment, which includes the aerodynamic heating, solar radiation, Mars windstorm and engine plume, have an influence on the environmental heat flux absorbed by spacecraft. The space particles (such as proton and electron) and solar UV radiation would affect spaceborne thermal control materials and components through the atomic oxygen's erosion on spacecraft surface material and the solar UV radiation's damage to the thermal control coating of spacecraft surface, thereby affecting the thermal control material's absorption of space radiation and its own thermal radiation properties. The space microgravity, the gravitational fields of the moon and other celestial bodies, the acceleration during orbital maneuver, and the accelerations during launching, return and re-entry would affect the two-phase heat transfer technique usually used for thermal control. Usually, traditional aluminum–ammonia heat pipe with axial groove is merely applicable to microgravity environment in which the heat transfer capacity of the heat pipe would be significantly affected by gravitational field and acceleration. The statistical data have revealed that most of the abnormal spacecraft phenomena are related to space environment. Hence, during the thermal control design of the spacecraft, the influence and restrictions of space environment shall be closely concerned, and feasibility of ground test verification shall be considered as well.

This chapter introduces the launch environment, Earth orbit environment, interplanetary missions' environment and spacecraft-induced environment. The influence of these environments on thermal control design of spacecraft is also introduced briefly.

2.2 Environment at Launching Phase

The design of thermal control system is mainly adapted for a variety of space environments on orbit. Furthermore, ground environments including spacecraft transportation, test on launchpad, preparation before launching and ascent phase that the spacecraft has undergone shall be also taken into consideration in order that the spacecraft temperature would not exceed the allowable range at the initial stage of mission.

A spacecraft is usually power-off in the process of spacecraft transport from technical zone to launching zone (on the launchpad). During transportation, the ambient temperature and humidity should be controlled to an allowable storage temperature without condensation.

At the test and storage stages in launching zone, a special AC system could be designed to control environment temperature. In the rocket fairing, the ambient temperature would be adjusted by blowing in air or nitrogen with certain temperature (10 °C–25 °C), humidity (35%–55%) and lower speed (less than 2 m/s), avoiding direct blowing onto the spacecraft body. The specific range of gas temperature control depends on the temperature requirements, operation status and mode of all devices before launching. A specialized air-convection or fluid loop apparatus shall be designed to cool the devices which cannot be sufficiently cooled merely with cold air of AC system for the large-scaled or complicated spacecraft such as the manned spacecraft. The choice of the apparatus would lead to the increase of complexity and cost. As for the spacecraft without a fairing, temperature rising caused by aerodynamic heating at ascent phase shall be thoroughly considered as well. Generally, the range of temperature rising is closely up to the trajectory, flight attitude and so forth. In this case, it may be considered to provide cold air with lower temperature as far as possible under the condition of causing no condensation, thus reducing the initial temperature at the launch moment of the spacecraft as far as possible and adapting to the influence of pneumatic heating in the ascent phase.

Thermal environment becomes much worse from the launching phase to the injection phase. In the first few minutes, environment temperature depends on density of radiation heat flux on the inner side of the fairing surface. Then, the aerodynamic heating would rapidly increase the heat flux density. Density of radiation heat flux on the inner surface of each section of CZ-3A fairing is shown in Fig. 2.1. It is seen in this figure that at 200 s, the maximum heat flux density on the surface of the fairing

2.2 Environment at Launching Phase

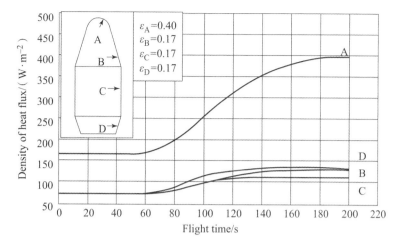

Fig. 2.1 Density of radiation heat flux on surface of each section of CZ-3A fairing [1]

is as high as 400 W/m². The rise of the faring temperature mainly leads to the rise of temperature of the external equipment, such as solar array, MLI, antenna and other light parts.

When the rocket flies high enough in 2–5 min after launching, the influence of aerodynamic heating would not exist anymore and the fairing could be separated. Influence of free molecular heating (FMH), which is caused by impact of a single molecular on the spacecraft, shall be considered after the fairing is separated. FMH flux is very high, but it lasts shortly. The change of FMH flux density after CZ-3A fairing has been separated at the ascent phase, is provided in Fig. 2.2. It is seen in

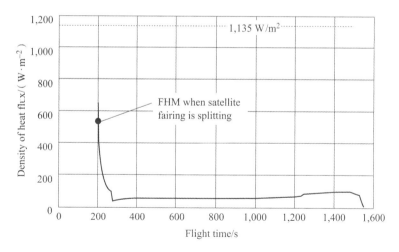

Fig. 2.2 Change of FMH flux density after CZ-3A fairing separation [1]

this figure that FMH flux density could be as high as 1,135 W/m², and it could drop to about 100 W/m² in 2 min.

Starting from fairing separation, the spacecraft is exposed to thermal environments of free molecular flow, solar radiation, Earth infrared radiation, Earth albedo radiation, etc., and sometimes, the thermal environments also include plume heating of rocket, upper stager or spacecraft engines. For example, when the third-stage rocket engine of CZ-3A is operating, the heat flux density on the splitting surface of satellite and rocket could reach 700 W/m² at most. Therefore, the influence of FMH, orbit shadow time, rocket or upper stage attitude shall be fully considered in the thermal design, and the initial temperature before launching shall be adjusted to ensure the required temperature if necessary.

The ascent phase could last for 20–45 min, and then, the spacecraft would enter the transfer orbit or directly enter the final mission orbit. Generally, only the spacecraft flying on LEO or large elliptical orbit could enter the final mission orbit directly. For spacecraft operating on GEO, most of them would operate on the transfer orbit firstly. At this time, the spacecraft would be exposed to the thermal environments such as solar radiation, Earth infrared radiation and Earth albedo radiation. Since the orbit shadow of the transfer orbit lasts for as long as 3.5 h, which is almost three times that of GEO, the temperature of some devices in the spacecraft would be lower than the lower limit of the allowable temperature. Hence, effective measures shall be taken in thermal control design. The spacecraft often takes a few hours or even weeks to set up normal operation attitude and keep stabilized after entry into the operation orbit. The payload would start to work when the operation attitude has been set up and platform equipment test has been successfully completed. Electrical heaters could be used to keep warm according to operation procedures at this phase.

Most spacecraft would be affected by FMH only at the ascent phase. However, for spacecraft of which perigee altitude is lower than 180 km, FMH on its operation orbit shall also be assessed. Generally, FMH flux shall be considered according to the following formula:

$$Q_{\text{FMH}} = \frac{1}{2}\alpha\rho u^3 \tag{2.1}$$

where ρ refers to the local atmospheric density in kg/m³; u refers to velocity (in m/s) of heating surface parallel to the spacecraft; α refers to dimensionless adjustment coefficient (which is 0.6–0.8 and 1.0 in conservation condition).

What is more, the Earth gravitational field and acceleration at ascent phase have some influence on thermal control technology as well. The acceleration would vary with missions. For example, heat pipes with axial groove may fail to operate normally due to the influence of Earth gravitational field and acceleration.

2.3 Earth Orbital Space Environment

The thermal control of spacecraft is a procedure to manage and control heat. The orbital environment, especially the thermal environment, plays an important role in this procedure. In addition to vacuum and microgravity environments, the Earth orbital thermal environments also include solar radiation, solar radiation (reflected radiation) reflected by Earth and infrared radiation sent by Earth, as shown in Fig. 2.3.

2.3.1 Earth Orbital Thermal Environment

2.3.1.1 Direct Solar Radiation

The sun is a huge thermal irradiator with a diameter of 1.393×10^6 km, a surface temperature of about 5,800 K and a total radiation power of 3.85×10^{26} W. The sun is the external heat source that has the greatest influence on the spacecraft when it flies in the solar system, which could not only directly provide thermal radiation energy to spacecraft, but also indirectly heat the spacecraft through albedo solar radiation parameters, such as spectrum distribution, intensity and parallelism of light, shall be concerned by the thermal design engineer since solar radiation is an important factor in the thermal control design of spacecraft.

The spectrum distribution in the entire solar system could be deemed as constant. The spectral distribution of solar radiation outside the Earth's atmosphere and on the surface of Earth is very close to that of 5,900 K blackbody radiation (there are deviations between spectral distributions on some wave bands and the blackbody because of the solar atmosphere), as shown in Fig. 2.4. Hence, the sun would be

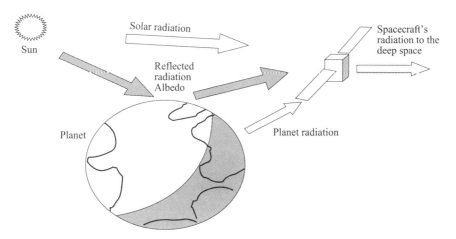

Fig. 2.3 Typical spacecraft thermal environment

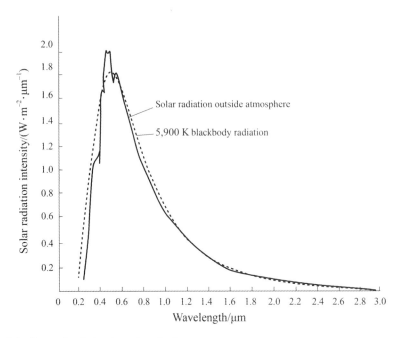

Fig. 2.4 Curve of radiation intensity of solar spectrum

assumed as a blackbody in the thermal control design. According to the Wein's displacement law, the maximum radiation wavelength of the blackbody is inversely proportional to its absolute temperature, and the peak wavelength in the power spectrum of solar electromagnetic radiation is about 0.48 μm. Of the total solar electromagnetic radiation energy, visible light (0.38–0.76 μm), infrared part (0.76–2 μm for the near-infrared ray, above 2–1,000 μm for the middle and far-infrared spectrum),and ultraviolet part (0.20–0.38 μm for the near ulatraviolet ray, below 0.2 μm for the far ultraviolet ray) account for 46%, 47% and less than 7%, respectively, and these three parts together account for more than 99% of the total radiation. Spectrum energy distribution values are indicated in Table 2.1. Because the solar radiation wavelength is much smaller than that of atmospheric radiation (3–120 μm), the solar radiation is called shortwave radiation, and the ground and atmospheric radiations are known as long-wave radiation.

When solar radiation passes through the atmosphere, one part of it (mainly the visible light) arrives on the ground and the other part is absorbed, scattered and reflected by the atmospheric molecular and moisture. For example, ozone has strong absorption of UV radiation, resulting in significant attenuation of radiation below 0.4 μm and zero attenuation of radiation below 0.3 μm, while some IR is absorbed by CO_2, vapor and other gases in atmosphere. A part of the scattered solar radiation returns to the space (becoming a part of Earth albedo radiation), and the other part reaches the ground. Therefore, after the solar radiation has passed through the atmosphere, the radiation intensity and spectrum energy distribution reaching the

2.3 Earth Orbital Space Environment

Table 2.1 Radiation density of solar spectrum on mean Earth–sun distance outside Earth's atmosphere

$\lambda/\mu m$	$S_\lambda/(W \cdot cm^{-2} \cdot \mu m^{-1})$	$F_{(0 \sim \lambda)}/\%$	$\lambda/\mu m$	$S_\lambda/(W \cdot cm^{-2} \cdot \mu m^{-1})$	$F_{(0 \sim \lambda)}/\%$
0.180	0.000,13	0.002	0.355	0.108,30	4.920
0.190	0.000,27	0.003	0.360	0.106,80	5.320
0.200	0.001,07	0.008	0.365	0.113,20	5.720
0.210	0.002,29	0.021	0.370	0.118,10	6.150
0.220	0.005,75	0.050	0.375	0.115,70	6.520
0.225	0.006,49	0.073	0.380	0.112,00	7.000
0.230	0.006,67	0.097	0.385	0.109,80	7.410
0.235	0.005,93	0.120	0.390	0.109,80	7.820
0.240	0.006,30	0.143	0.395	0.118,90	8.240
0.245	0.007,23	0.168	0.400	0.142,90	8.730
0.250	0.007,04	0.194	0.405	0.164,40	9.250
0.255	0.010,40	0.230	0.410	0.175,10	9.920
0.260	0.013,00	0.270	0.415	0.177,40	10.570
0.265	0.018,50	0.330	0.420	0.174,70	11.220
0.270	0.023,20	0.410	0.425	0.169,30	11.860
0.275	0.020,40	0.490	0.430	0.163,90	12.470
0.280	0.022,20	0.560	0.435	0.156,30	13.080
0.285	0.031,50	0.660	0.440	0.181,00	13.730
0.290	0.048,20	0.810	0.445	0.192,20	14.420
0.295	0.058,40	1.010	0.450	0.200,60	15.140
0.300	0.051,40	1.210	0.455	0.205,70	15.890
0.305	0.060,30	1.120	0.460	0.206,80	16.650
0.310	0.068,90	1.660	0.465	0.204,80	17.410
0.315	0.076,40	1.920	0.470	0.203,30	18.170
0.320	0.083,00	2.220	0.475	0.204,40	18.920
0.325	0.097,50	2.550	0.480	0.207,40	19.680
0.330	0.105,90	2.930	0.485	0.197,60	20.430
0.335	0.108,10	3.320	0.490	0.195,00	21.160
0.340	0.107,40	3.720	0.495	0.196,00	21.880
0.345	0.106,90	1.120	0.500	0.194,20	22.600
0.350	0.109,30	4.520	0.505	0.192,00	23.310
0.510	0.188,20	24.020	0.720	0.131,40	48.860
0.515	0.183,30	24.700	0.730	0.129,00	49.830
0.520	0.183,30	25.380	0.740	0.126,00	50.770
0.525	0.185,20	26.060	0.750	0.123,50	51.690

(continued)

Table 2.1 (continued)

$\lambda/\mu m$	$S_\lambda/(W\cdot cm^{-2}\cdot \mu m^{-1})$	$F_{(0\sim\lambda)}$ /%	$\lambda/\mu m$	$S_\lambda/(W\cdot cm^{-2}\cdot \mu m^{-1})$	$F_{(0\sim\lambda)}$ /%
0.530	0.184,20	26.740	0.800	0.110,70	56.020
0.535	0.181,80	27.420	0.850	0.098,80	59.890
0.540	0.178,30	28.080	0.900	0.088,90	63.360
0.545	0.175,40	28.740	0.950	0.083,50	66.540
0.550	0.172,50	29.180	1.000	0.074,60	59.460
0.555	0.172,00	30.020	1.100	0.059,20	74.410
0.560	0.169,50	30.650	1.200	0.048,40	78.390
0.565	0.170,50	31.280	1.300	0.039,60	81.640
0.570	0.171,20	31.910	1.400	0.033,60	84.340
0.575	0.171,90	32.540	1.500	0.026,70	86.650
0.580	0.171,50	33.180	1.600	0.024,40	88.610
0.585	0.171,20	33.810	1.700	0.020,20	90.260
0.590	0.170,00	34.440	1.800	0.015,90	91.590
0.595	0.168,20	35.060	1.900	0.012,60	91.640
0.600	0.166,60	35.680	2.000	0.010,30	93.490
0.605	0.164,70	36.800	2.100	0.009,00	94.200
0.610	0.163,50	36.900	2.200	0.007,90	94.830
0.620	0.160,20	38.100	2.300	0.006,80	95.370
0.630	0.157,00	39.270	2.400	0.006,40	95.860
0.640	0.154,40	40.420	2.500	0.005,40	96.290
0.650	0.151,10	41.550	2.600	0.004,80	96.670
0.660	0.148,60	42.660	2.700	0.004,30	97.010
0.670	0.145,60	43.740	2.800	0.003,90	97.310
0.680	0.142,70	44.810	2.900	0.003,50	97.580
0.690	0.140,20	45.860	3.000	0.003,10	97.830
0.700	0.136,90	46.880	3.100	0.002,60	98.040
0.710	0.134,40	47.880	3.200	0.002,30	98.220
3.300	0.001,90	98.370	5.000	0.000,383	99.512
3.400	0.001,66	98.505	6.000	0.000,175	99.718
3.500	0.001,46	98.620	7.000	0.000,099	99.819
3.600	0.001,35	98.724	8.000	0.000,060	99.878
3.700	0.001,23	98.819	9.000	0.000,038	99.914
3.800	0.001,11	98.906	10.000	0.000,025	99.937
3.900	0.001,03	98.985	20.000	0.000,001,6	99.991
4.000	0.000,95	99.058	40.000	0.000,000,1	99.998
4.500	0.000,59	99.331			

Note $F_{(0\sim\lambda)}$ refers to the function of solar radiation, indicating percentage of solar radiation with wavelength of $0 \sim \lambda$ in the entire wavelength [2]

2.3 Earth Orbital Space Environment

ground are changed. Then again, for spacecraft that operates outside the atmosphere, these complicated factors are not considered.

Even if solar UV radiation takes a small proportion of the total solar radiation energy (UV radiation smaller than 0.3 μm only accounts for 1% of the total solar radiation energy), it could damage the thermal control coating on the surface of spacecraft. Under long-term UV radiation, solar absorptance α_s of the surface of thermal control coating obviously increases. Hence, thermal design engineers should take the influence of UV radiation on the thermal control coating into consideration. Influence of the newly developed thermal coating shall be researched by vacuum UV radiation test [3].

The sun is a very stable heat source. 11-year solar cycle has little influence on the intensity of solar radiation, and its change could be always kept within 1%. At the mean distance between Earth and sun outside the Earth's atmosphere, the total power of solar radiation received per unit area (which is perpendicular to the solar radiation direction) is called a solar constant. The current widely accepted solar constant is 1 367 W/m². Since the Earth orbit is elliptical, the solar radiation intensity reaching the Earth varies by ±3.5% as Earth–sun distance changes. At apogee, the solar radiation intensity would reach the minimum value of 1,322 W/m²; while at perigee, it would reach the maximum value of 1,414 W/m². The solar radiation intensity is inversely proportional to the square of the distance from the Earth to the sun. Solar radiation intensity J_s at different distances d from the sun could be calculated by the following formula:

$$J_s = \frac{P}{4\pi d^2} \tag{2.2}$$

where P refers to the total power of sun, which is 3.85×10^{26} W. The solar radiation intensities at the mean distance from the sun to the celestial bodies in the solar system are listed in Table 2.2.

Table 2.2 Solar radiation intensities of celestial bodies

Celestial body	Solar radiation intensity J_s
Mercury	667
Venus	191
Earth	100
Moon	100
Mars	43.1
Jupiter	3.69
Saturn	1.1
Uranus	0.27
Neptune	0.11
Pluto	0.064

Note: The mean distance between Earth and sun is defined as 1 AU (astronomical unit)

The solar radiation intensity near Mercury is about 6.7 times as high as that at sun–Earth; the solar radiation intensity near Saturn is only 1% of that at sun–Earth distance; while the solar radiation intensity near Neptune is only 1‰ of that at sun–Earth distance. Some outer planet detectors flying outside the Earth orbit would be in a deep cold environment with little heat source, which is difficult to generate power with solar energy, while the inner planet detectors flying within the Earth orbit would encounter high environmental heat flux.

The divergence angle of solar beam near the Earth (1 AU from sun) is about 0.5°, which could be regarded as parallel beam in the thermal control design. However, for spacecraft flying very close to sun, the influence of the divergence angle shall be considered.

2.3.1.2 Earth Albedo Radiation

The Earth albedo mainly includes two parts: solar radiation scattered by moleculars and micronic dust in the atmosphere and solar radiation reaching the Earth ground and reflected by it. The spectrum distribution of Earth albedo radiation is inconsistent with the direct solar radiation due to the absorption effect of the atmosphere, but from the perspective of thermal control engineering, this difference is small enough to be ignored for most spacecraft.

Earth albedo is the most complex external heat source, which varies not only with seasons, day and night time, but also with geographic latitude and longitude. Judging from the type of radiation, the Earth albedo is more complicated than the other two heat sources. Solar radiation is directional radiation. The infrared radiation on the surface of the Earth is diffuse radiation, which follows the Lambert's cosine law, while the distribution of albedo on the surface of Earth does not follow Lambert's cosine law strictly, since some of the Earth's surface is diffuse reflection, some of the surface is mirror reflection, and quite a few of them are in between. In order to simplify the analysis, the Earth albedo is generally assumed to be diffuse reflection distribution in the thermal control design of most spacecraft, i.e., it is assumed to follow Lambert's cosine law distribution. In other words, the albedo intensity per unit area of the Earth is the same in all directions in the hemisphere.

The proportion of solar radiation reflected by the Earth's surface and the atmosphere is called the Earth albedo ratio, which is closely related to the characteristics of the Earth's surface and the atmosphere. For example, the albedo ratio of clouds is as high as 0.8, while that of surfaces such as water and forest is as low as 0.05. At the same time, the Earth albedo ratio changes with the geographic latitude, and its specific change curve is shown in Fig. 2.5. The choice of Earth albedo ratio is a tough job in the thermal analysis of spacecraft. However, taking the thermal inertia of most spacecraft into account, the orbit average albedo ratio can be used in the thermal control design. As far as the Earth is concerned, the average albedo ratio in the general thermal analysis is 0.30–0.35.

As shown in Fig. 2.5, the increase of geographic latitude increases the albedo ratio. Therefore, in order to calculate the Earth albedo ratio more accurately, the average

2.3 Earth Orbital Space Environment

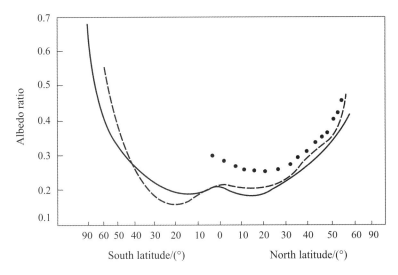

Fig. 2.5 Variation curve of Earth albedo ratio with latitude [4]

albedo ratio can be corrected according to the characteristics of orbital inclination. The specific correction method can refer to the data recommended in Table 2.3.

The Earth albedo radiation intensity J_a incident on the surface of spacecraft is a complex function related to the Earth albedo characteristics, the height of the spacecraft and the angle between the local vertical line and the solar rays and so on, which can be expressed by the following formula:

$$J_a = \rho F S \tag{2.3}$$

Table 2.3 Albedo ratio and orbit inclination

Orbit inclination angle/(°)	Earth albedo ratio ρ		
	Minimum value	Average value	Maximum value
±90	0.38	0.42	0.46
±80	0.34	0.38	0.42
±70	0.30	0.34	0.38
±60	0.26	0.30	0.34
±50	0.22	0.27	0.32
±40	0.19	0.24	0.29
±30	0.20	0.24	0.28
±20	0.20	0.24	0.28
±10	0.20	0.24	0.28

where F refers to the view factor between the surface of spacecraft and the Earth. The albedo radiation intensity of GEO satellites is only 5‰–8‰ of the solar constant, so the influence of albedo radiation can often be ignored when analyzing the environment heat flux of GEO satellites.

It is worth noting that the above is an approximate treatment of the issue. For complex spacecraft, especially those operating on LEO, the albedo heat flow on each outer surface needs to be accurately calculated according to the orbital position. At present, such complicated operations can be handled by professional software.

2.3.1.3 Earth's Infrared Radiation

Due to the relatively low temperature, the Earth's surface radiates all the heat as infrared wavelength ranging from 2 to 50 μm and with a peak wavelength of about 10 μm. The spectrum distribution is shown in Fig. 2.6, and its radiation is equivalent to that of 288 K blackbody. The Earth's infrared radiation seen on the spacecraft is actually a combination of radiation from the upper atmosphere (which has an effective blackbody temperature of 218 K) and radiation from the Earth's surface passing through the infrared window (transparent window), and the radiation intensity is equivalent to about 255 K of blackbody radiation which is about 237 W/m², because the atmosphere is basically opaque in infrared, and there is a water vapor absorption band at 4.8–8.0 μm, a carbon dioxide absorption band at 13.5–17.0 μm, and only a transparent window (ozone absorption band around 9.5 μm) in the spectral range of 8.0–13.0 μm.

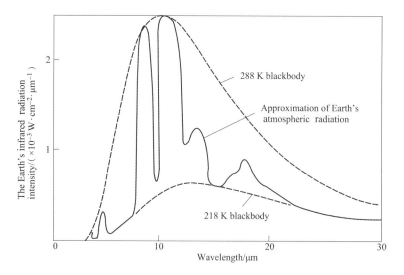

Fig. 2.6 Distribution of infrared radiation intensity on Earth [5]

2.3 Earth Orbital Space Environment

Since the temperature of the Earth's surface varies with time and geographic locations, the intensity of thermal radiation received by the spacecraft will also vary with time and its orbital position. In fact, due to the large thermal inertia of the Earth, diurnal fluctuation and seasonal changes have little influence on the temperature of the ground and the atmosphere. Furthermore, considering the thermal inertia of the spacecraft, error caused by using the average Earth radiation is usually small. In the actual thermal control design, it can be assumed that the Earth has the same intensity of thermal radiation on the whole surface area, i.e., the Earth and its atmosphere are assumed to be isothermal bodies with constant temperature. According to the energy balance of the Earth, the energy E_{IR} (which serves as the average Earth radiation mentioned above) radiated by the Earth and its atmosphere shall be equal to the solar energy absorbed by it:

$$4\pi R_{rad}^2 E_{IR} = \pi R_{rad}^2 (1-\rho) S \tag{2.4}$$

where R_{rad} refers to the radius of the Earth and its atmosphere, with a value of 6 408 km; ρ refers to the albedo ratio to sunlight, which is usually 0.3.

It is indicated in Formula (2.4) that:

$$E_{IR} = \frac{1-\rho}{4} S \tag{2.5}$$

If one considers more carefully, the maximum Earth infrared radiation intensity can be 247 W/m² (at perihelion, it is equivalent to the Earth's surface temperature of about −16 °C), and the minimum Earth infrared radiation intensity can be 214 W/m² (at aphelion, it is equivalent to the Earth's surface temperature of about −25 °C).

Since the radiation intensity decreases in inverse proportion to the square of the orbital height, the approximate value of the Earth's radiation intensity E_p (in W/m²) at the specified orbital height can be calculated according to the following formula:

$$E_p = 237 \left(\frac{R_{rad}}{R_{orbit}} \right)^2 \tag{2.6}$$

where R_{orbit} refers to the orbit radius. For other planets, it needs to be verified according to specific conditions. For example, on Mercury, a sidereal day and a Mercury year are of the same order of magnitude, equivalent to 59 Earth days and 88 Earth days, respectively. Terminator line moves very slowly. Temperature difference between the under sunlight part of Mercury and the under shadow part of it is kept at more than a hundred of degrees Celsius. The eccentricity of Mercury's orbit is also very large. The solar radiation intensities at perihelion and aphelion differ by more than 2 times and vary largely with the changes of seasons.

The infrared radiation intensity of the Earth incident on a certain surface of the spacecraft is also related to the relative position of the surface and the Earth surface, i.e., related to the view factor between this surface and the Earth surface. Most spacecraft have a surface facing the ground (which can be called a parallel surface with its normal pointing to the center of the Earth, generally referring to the $+Z$ surface) and several vertical surfaces (i.e., the surface normal is perpendicular to the line connecting the center of the surface and the center of the Earth, generally referring to the $+Y/-Y$ surface and the $+X/-X$ surface). The view factors between these surfaces and the Earth's surface are shown in Fig. 2.7.

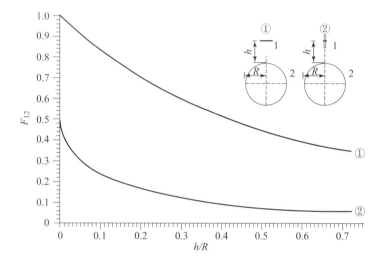

Fig. 2.7 View factors between parallel surface, vertical surface and Earth surface of spacecraft

The infrared radiation intensity of the Earth incident on these surfaces of the spacecraft can be calculated according to the following formula:

$$E_p^1 = 237 F_{12}^1 \quad E_p^2 = 237 F_{12}^2 \tag{2.7}$$

Taking the Earth as a blackbody is actually an idealized assumption, while the radiant energy emitted from the surface of an actual object is generally smaller than that of a blackbody at the same temperature. In order to characterize the radiation properties of actual surfaces, a concept of emissivity is introduced. In the thermal control design of spacecraft, the temperature requirement involved is generally below 2,000 K, at which the energy in the visible spectrum (wavelength less than 0.76 μm) is less than 1.4%, and more than 98% of the radiant energy is concentrated in the infrared spectrum. In addition, the concept of infrared hemisphere emissivity is often used in spacecraft thermal control design, which is directly referred to as infrared emissivity in some chapters of this book.

2.3.2 Other Earth Orbit Space Environment

In addition to the Earth orbit thermal environment, there are several space factors that shall be considered in the thermal control design; otherwise, it will affect the thermal control design and even lead to failure of the thermal control task.

2.3.2.1 Neutral Atmosphere

1. **Atmosphere distribution**

Aviation activities are mainly carried out in the lower atmosphere, while space activities are mainly carried out in the upper atmosphere and its outer space. The operation of spacecraft will be affected by the upper atmosphere.

The density, temperature and composition of the Earth's atmosphere vary greatly with the altitude. According to the thermal characteristics, the Earth's atmosphere includes five layers: troposphere, stratosphere, mesosphere, thermosphere and escape layer.

The tropospheric atmosphere is also called lower atmosphere, which is atmosphere from the ground to that below the tropopause, and the height of the tropopause is 8–18 km. In this layer, the atmospheric density is high, accounting for about 3/4 of the total mass of the Earth's atmosphere. The atmospheric pressure is 1 standard atmospheric pressure (1.013×10^5 Pa) on the ground, which decreases exponentially with the height and drops to 0.2 standard atmospheric pressure at the top of the layer. The free path of atmospheric molecules is much smaller than the dimension of aircraft or spacecraft, and it is completely in a viscous flow state of molecules. The vertical distribution of atmospheric temperature in the layer is characterized as follows: The temperature decreases with altitude; the average temperature at the top of the layer is −50°C to 55 °C; the average temperature decreases by about 6 °C every 1 km. The temperature at the bottom layer is high, and it is low at the upper layer, which forms convective motion of the atmosphere. Most weather phenomena occur in the troposphere. The proportion of various components, which includes nitrogen (about 78%), oxygen (about 21%), some inert gases and trace amounts of hydrogen and methane, etc., in the atmosphere at the layer, also does not change with the altitude. In addition, there are some components with varying contents, such as water vapor, carbon dioxide and ozone.

The area from the tropopause to altitude of about 50 km is the stratosphere. Ozone in the layer is heated up by absorbing solar UV radiation, which causes the atmospheric temperature in the layer to rise with the increase of vertical height, and makes the convection movement very weak, so the atmospheric movement is mainly horizontal. The temperature at the top of the layer is about 0 °C. Atmospheric mass in troposphere and stratosphere almost accounts for 99.9% of the total atmosphere, and this space is for activities of the aircraft.

The area with an altitude ranging from 50 km to 80–85 km is called the middle layer. The stimulated carbon dioxide in this layer could cool down by radiating infrared rays. The increase of altitude decreases the temperature in the layer by an average of about 3 °C per kilometer. The temperature at the top of the layer would drop to about 180 K, the lowest temperature of the Earth's atmosphere.

Above the middle layer is the thermal layer. Due to weak atmospheric turbulence and convection, the atmosphere begins to diffuse and separate, and the proportion of various components no longer remains constant. As the altitude increases, the proportion of heavy molecules and atomic components gradually decreases, and that of light atoms and molecules, such as hydrogen and helium, increases. Components such as oxygen atoms in the layer would absorb UV rays from the sun, which causes temperature rise with the altitude. The temperature at the top of the 300–400 km layer is about 600 K in the mean solar year and can reach 2,000 K in the solar maximum year. However, the atmospheric density is very thin, and the molecular free path is much longer than the dimension of spacecraft, and the molecules seldom collide with the spacecraft. Therefore, although thermal motion of atmospheric molecules is very acute, almost no heat energy is transferred to the spacecraft, which will not rise the temperature of the spacecraft.

The oxygen molecules would be dissociated by solar UV radiation at altitude over 80 km to oxygen atoms. Since oxygen atoms recombine slowly, there are abundant oxygen atoms in the thermosphere. At an altitude of about 100 km, the number density of oxygen atoms reaches the maximum. At a range of 100–700 km, the atomic oxygen becomes the chief ingredients of the atmosphere. Above the thermal layer is called escape layer where atmosphere temperature no longer changes with altitude and there are few collisions between particles. The particles may escape from the Earth's gravity and escape into space due to the high temperature and high velocity.

Sometimes, the atmosphere over 110 km is called the upper atmosphere, which is the space for spacecraft activities. The atmosphere in the range of 110–1,000 km plays an important role in the operation of spacecraft, and the increase of altitude decreases the influence of the atmosphere. Generally, the direct influence of the atmosphere over several thousand kilometers could be ignored.

2. **Influence of atomic oxygen on thermal control design**

The atomic oxygen's erosion toward materials is the main influence of the upper atmosphere on the spacecraft. Atomic oxygen can seriously erode hydrocarbon thermal control materials that are usually used outside the spacecraft, especially LEO spacecraft. The materials can be easily oxidized or react with the atomic oxygen. When the spacecraft is operating at 7–8 km/s on LEO, the energy of atomic oxygen hitting the spacecraft surface is 4–5 eV, and its flux density is about 10^{15} atoms/(cm^2·s) at the altitude of 200 km and about 10^{12} atoms/(cm^2·s) at the altitude of 600 km. Interacting with the atomic oxygen with strong oxidizing, large flux density and high energy, the materials on the surface of spacecraft would be eroded and degraded, thereby affecting the normal operation and service life of the spacecraft. The atomic oxygen flux in the collision direction is the product of atomic oxygen concentration (atoms/cm^3), spacecraft velocity (m/s) and mission time (s), wherein the product of atomic oxygen concentration and spacecraft velocity is the atomic oxygen flux

2.3 Earth Orbital Space Environment

density (atoms/(cm^2·s)). For example, at an altitude of 500 km, the maximum concentration of atomic oxygen is about 6×10^7 atoms/cm^3 and the spacecraft's speed is about 8,000 m/s, then the atomic oxygen flux density is 6×10^7 atoms/cm^3 × 8,000 m/s = 4.8×10^{13} atoms/(cm^2·s), and the atomic oxygen integral flux within one year is 4.8×10^{13} atoms/(cm^2·s) × 31.54×10^6 s = 1.51×10^{21} atoms/cm^2. For long-term spacecraft, its influence cannot be ignored.

The material mass loss caused by atomic oxygen per unit surface area per unit time can be expressed by the following formula:

$$dm = \rho_m R_{ef} \phi dA dt \tag{2.8}$$

where ρ_m refers to the density of the material (g/cm^3); R_{ef} refers to denudation rate or reaction efficiency (cm^3/atom), which is a proportional constant obtained from experiments; ϕ refers to atomic oxygen flux density (atoms/(cm^2·s)). The denudation rates of various materials are listed in Table 2.4 in detail.

Table 2.4 Atomic oxygen denudation rates of common space materials

Material	Denudation rate/($\times 10^{-24}$ cm^3·atom^{-1})	Material	Denudation rate/($\times 10^{-24}$ cm^3·atom^{-1})
Aluminum	0	Gold	0
Titanium	0	Magnesium	0
Silver	10.5	Fused silica	0
Polyimide	2.6–3.0	Polyester film	1.5–3.9
Carbon, graphite	1.2–1.4	Polyethylene	3.3–3.7

The loss rate of material thickness is:

$$\frac{dx}{dt} = R_{ef} \phi \tag{2.9}$$

For LEO spacecraft, especially the long-term operating LEO spacecraft, the atomic oxygen environment must be concerned. The total integral flux of atomic oxygen in the lifetime shall be calculated, and sensitive materials shall be evaluated according to the orbit of the space mission. The following methods are generally adopted to overcome the influence of atomic oxygen:

(1) Select materials with lower denudation rate to resist denudation of atomic oxygen. As indicated in Table 2.4, the atomic oxygen denudation rates of various materials vary greatly: The denudation rate of gold is almost zero, while that of silver could be as high as 10.5×10^{-24} cm^3/atom. The atomic oxygen denudation rate of the commonly used Kapton is 3.0×10^{-24} cm^3/atom. If it is directly applied on the windward surface of the spacecraft on a circular orbit of altitude of 200 km, the annual denudation thickness may reach 0.38 mm. Therefore, selecting materials with low denudation rate will be an effective thermal designing method.

(2) Use protective coating. Coating a thin layer of anti-atomic oxygen on the surface of the material with high denudation rate plays a significant role in reducing atomic oxygen denudation. In order to prevent the coating from changing thermal properties of the material surface and affecting thermal control performances of the spacecraft, the coating should be very thin. Coating a layer of less than 0.1 μm aluminum oxide (Al_2O_3) or silicon dioxide (SiO_2) on the surface of Kapton material can reduce the denudation rate by 1–2 orders of magnitude, thus not only improving the ability to resist atomic oxygen denudation, but also minimizing the influence on the original thermal properties of the material.
(3) Keep the surface sensitive to atomic oxygen away from the windward side. The spacecraft is operating at a speed close to 8 km/s. The flux of atomic oxygen on the windward side is greater than that on the leeward sides. Avoiding setting up the sensitive surfaces on the windward side will significantly reduce the damage caused by atomic oxygen.

2.3.2.2 High-energy Particles

There is high-energy particle radiation in the near-Earth space. These particles include protons, electrons, various heavy ions and neutrons, which mainly come from the Earth's radiation belt, solar cosmic rays and galactic cosmic rays. Different from the low-energy particles in the ionosphere and magnetosphere, high-energy particles have a strong ability to penetrate with large energy and high speed. Different from the thermal plasma and low-energy particles, the influence of high-energy particles which penetrates into the materials or devices of the spacecraft and human tissues is no longer limited to the surface of spacecraft; thus, they could affect the interior of the spacecraft. The depth of penetration depends on the energy and the type of the charged particles and the density of the material. For materials with low density, particles can penetrate through a greater thickness. Protons with energy of 10 MeV can penetrate aluminum plates about 0.7 mm thick, while protons with energy of 1 GeV can penetrate 1 m. The mass and volume of electrons are much smaller than those of protons, so electrons with the same energy have stronger penetration ability, and 500 keV electrons can penetrate aluminum plates with a thickness of 0.7 mm.

The penetration of high-energy particles into materials will do much harm to spacecraft. The degree of risk is related to the dose, dose rate, particle energy and particle properties. Electrons of keV magnitude can only penetrate into the shallow surface layer of 1 μm to cause surface charging and discharging. Electrons of MeV magnitude can penetrate the skin into the medium inside the spacecraft and cause hazards, such as total dose effect, single particle effect, charging and discharging inside the medium, wherein the total dose effect is closely related to the thermal control design.

The total amount of energy absorbed by a substance per unit mass over a period of time during which the material is radiated by particles is called the total dose. The effect or damage on materials and devices caused by the total dose is called the total dose effect. It includes ionization damage and displacement damage of high-energy particles on materials, resulting in material performance change, component performance decline or failure and biological body damage. Radiation damage is related to radiation type, radiation energy and the performance of radiated material. High-energy particles and solar proton events in the radiation zone are important sources of the total accumulated dose. The international unit of radiation dose is Gy, which represents 1 J (Joule) of energy deposited in 1 kg of material. Traditionally, the unit commonly used for semiconductor materials is rad silicon (rad (Si)), which deposits 0.01 J of energy in 1 kg of silicon material.

The radiation damage thresholds of some materials are listed in Table 2.5. After receiving a certain dose of radiation, some synthetic materials would suffer from performance degradation, and transparency of optical materials would reduce. Metal materials with high radiation resistance are often used to shield other radiation-sensitive materials or devices.

Table 2.5 Radiation damage thresholds

Material	Organism	Electron device	Lubricant	Ceramic glass	Polymeric material	Structural metal
Threshold /rad	10^1–10^2	10^2–10^6	10^5–10^7	10^6–10^8	10^7–10^9	10^9–10^{11}

Thermal properties of the thermal control coatings would be degraded under the combined action of solar UV and particle radiation, thereby leading to the increase of spacecraft temperature and even deviating from the designed value. In order to evaluate its influence, in addition to conducting experiments on orbit, it is also necessary to conduct simulation experiments on the ground. Therefore, it is necessary to determine the energy spectrum and orbital integral flux (particle number/(cm^2·d)) of charged particles. The calculation shall be carried out based on electronic model and proton model of the radiation belt. According to the calculation results, the tests shall be carried out in accordance with the requirements of GJB 2502.6 *Test Method for Thermal Control Coatings of Spacecraft Part 6: Proton Irradiation Test in Vacuum* and GJB 2502.7 *Test Method for Thermal Control Coatings of Spacecraft Part 7: Electron Irradiation Test in Vacuum*.

Charged particles can also bring issues about thermal effects. The thermal control design for room temperature level can be ignored since the heat flux generated by the charged particles is very small. Then again, it has an important influence on the cryogenic radiator. For example, if the temperature of the radiator is 70 K on GEO, the temperature of the radiator heated by the charged particles would be 72.9 K at 1.5 times the average radius height of the Earth. At a height of 2.0 times the average radius of the Earth, the radiator temperature is 74.4 K. It assumes that the balance temperature of the radiator is 0 K, and the radiator would be heated to 27.3 K on GEO (6.6 times the average radius height of the Earth).

2.3.2.3 Micrometeoroids and Space Debris

Meteoroids orbiting the sun in space are solid blocks or small particles with a speed of 11–72 km/s relative to the Earth. The quantity of meteoroids with large mass is relatively very small. Micrometeoroids are abundant in the solar system, and most large meteoroids can be observed and tracked by means of the ground radar and optical instruments, thus obtaining the motion laws of them. The dimension and the flux of the micrometeoroids can only be calculated by examining the impacted surface of the test sample on orbit. The diameter of micrometeoroid is mostly less than 1 mm, and the mass is less than 1 mg. The smaller the mass is, the higher the flux of micrometeoroids will be.

Generally, micrometeoroids occur sporadically and dispersedly. When the Earth moves near the orbits of broken comets or asteroids, it will encounter a relatively dense group of micrometeors, showing meteor showers in the Earth's atmosphere. Dense meteor swarms pose a threat to the safety of spacecraft on orbit, and there have been reports of spacecraft hit and damaged by the Perseid meteor shower. The spacecraft can adjust its attitude before passing through the meteor group to prevent the meteor group from attacking sensitive components.

All wastes left behind on orbit by human space activities are called space wastes, also known as space debris. Thousands of spacecraft have been launched into space. Each launch will send some other incidental objects into space at the same time. In addition, the failed spacecraft and spacecraft wastes have all been space debris.

These fragments vary widely in size, ranging from large ones such as last stage launch vehicles to small ones such as alumina particles ejected from the solid rockets. A large number of space debris are less than 1 cm, but at present, only debris larger than 10 cm can be tracked regularly by the radar or optical methods. As of April 2017, more than 184,000 space debris have been registered. It is estimated that the number of debris that can damage spacecraft amounts to millions. The level of damage on spacecraft by space debris depends on the speed, size, shape and other factors. The debris at a speed of 3 km/s may also do harm to the spacecraft. Space debris is mainly distributed on the orbits with plenty of spacecraft, such as the orbit below 2,000 km or near 20,000 km and the GEO. The height of debris on LEO gradually decreases due to atmospheric resistance, and it takes years or even hundreds of years for the debris to fall into the atmosphere and burn down. Debris near GEO would exist for a longer period of time. With frequent space activities, the amount of space debris would further increase.

The relative motion speed between micrometeoroid or space debris and spacecraft is extremely high. Even tiny particles would cause damage once they collide with the spacecraft. Small particles will erode the coating material on the surface of spacecraft, changing its thermophysical properties and affecting the thermal control design. At the same time, it would also damage the surface of optical devices and affect their optical performances. Larger particles can penetrate the outer shell of the spacecraft and penetrate the sealed cabin or radiator of the spacecraft, thus causing leakage of the gas or fluid inside. Secondary debris would be generated when penetrating the shell, which will further endanger the internal instruments and equipment. In serious cases, it would lead to catastrophic accidents. For example, the thermal insulation foam falling off the fuel tank of the space shuttle Columbia damaged the thermal protection material on its orbiter, resulting in disintegration of the space shuttle Columbia by the aerothermal and aerodynamic when re-entering the atmosphere.

In order to avoid the damage of micrometeoroids and space debris, especially to protect the manned space missions from threats or catastrophic accidents, avoidable and protective measures shall be adopted. For short-term manned space missions, the launch window of spacecraft shall be selected and the operation orbit shall be assessed carefully, and the long-term operation mission performs orbit maneuver according to the monitoring and early warning information of debris bigger than 10 cm, thereby avoiding collision with the registered space debris. Buffer shielding technology, such as sandwich structure or multi-layer structure with protective walls, shall be adopted to protect the sealed cabins or the sensitive parts of spacecraft.

2.4 Moon and Planetary Space Environment

When spacecraft fly in the interplanetary space, solar radiation is the main environmental heating. While flying over the planet, the spacecraft will receive heat generated by planet IR and albedo radiation. For spacecraft landing on the surface of planet, it will also be directly affected by the temperature of the planet surface. Geometric and orbital parameters of typical celestial bodies in the solar system are listed in Table 2.6.

The solar radiation intensity (W/m^2) at different distances from the sun can be calculated as follows:

$$\text{Solar radiation intensity} = \frac{1367}{L^2} \tag{2.10}$$

where L refers to the average distance between celestial bodies and the sun in astronomical units (AU), and its value can be referred to in Table 2.6. In order to understand the thermal environment encountered during interplanetary flight, the concept of "reference sphere" [6] is used here. The reference sphere is an isothermal sphere with an absorptance and emissivity of 1.0. The temperature of the reference

Table 2.6 Geometric parameters and orbit parameters of typical celestial bodies in solar system

Star	Orbital semimajor axis /AU	Perihelion distance /AU	Aphelion distance /AU
Earth	1.000,0	0.983,3	1.016,7
Moon	1.000,0	0.983,3	1.016,7
Mercury	0.387,1	0.307,5	0.466,7
Venus	0.723,3	0.718,4	0.728,2
Mars	1.524,0	1.381,0	1.666,0

sphere can roughly indicate the degree of its cold and heat. The sphere equilibrium temperatures of typical celestial bodies in the solar system are listed in Table 2.7.

Table 2.7 Balance temperatures of typical celestial bodies reference sphere

Star	Balance temperature of reference sphere /°C
Mercury	174
Venus	55
Earth	6
Mars	−47

2.4.1 Lunar Environment

The moon has no atmosphere, and the gravity field is about 1/6 of the Earth's gravity field. A "lunar day" is equivalent to 28 Earth days, i.e., the lunar day and the lunar night are equivalent to 14 Earth days, respectively. The lunar surface is characterized by high temperature in the day and low temperature in the night, which has the most significant impact on the lunar surface landing and patrol detectors. The lunar surface temperature at the subsolar point can reach 120 °C at lunar noon and drop to −180 °C at lunar night. The IR of the lunar surface is calculated by cosine function (to about 70°), i.e., it decreases with the increase of angle starting from the subsolar point.

The average albedo ratio of the moon is 0.073, and the albedo ratio of different geological areas is different. For example, it is 0.129 in Langrenus Crater, 0.126 in Copernican-type craters and 0.123 in Apennines Mountain. For example, in Langer, the Endeavour Pit is 0.129, Copernicus Pit is 0.126. Therefore, the lunar surface has an absorptance as high as that of the black paint, making the surface temperature in the sunshine area very high.

For spacecraft flying around the moon or landing on it, the thermal control designers should pay attention to some factors according to the aforesaid thermal environment characteristics:

IR intensity on the lunar orbit is very large, so its influence on the radiator surface of the spacecraft is more serious than that of the spacecraft on the Earth orbit. And the arrangement positions of radiators and the spacecraft attitude should be selected to keep the view factor between radiators and the lunar surface as small as possible. At present, since the solar absorptance of radiators of most spacecraft is low and the infrared emissivity of them is high, radiators could be designed to point to the sun and avoid the lunar surface to some extent.

Though the view factor between the radiator facing the zenith and the lunar surface is small, the temperature of the exposed devices located on the lunar landers may raise by at least 10 °C if the lander is close to some low mountains with high temperature. Therefore, the existence of mountains or the inclination angle of the spacecraft cannot be ignored.

Another important factor is the dust on the lunar surface. Because the solar absorptance of lunar dust is very high, a small amount of lunar dust falling on the radiator can significantly improve the solar absorptance. It is easy to be thrown up when a lunar rover or a person walks on the lunar surface, so the crew of Apollo Lunar Rover often had to stop and wipe off the lunar dust on the surface of the radiator.

Another problem that needs to be noticed is the influence of the extremely low thermal conductivity of lunar soil. The low thermal conductivity leads to quick dropping of the surface temperature of -180 °C in the shadowed area quickly, and these shadowed spots reduce IR heating on the equipment nearby. For example, if the tire on one side of Apollo 14 transporter shields the tire on the other side, temperature of the blocked tire will be lower than the minimum temperature limit.

2.4.2 Mercury Environment

Mercury is surrounded by a thin layer of atmosphere, which is composed of hydrogen, helium, oxygen, sodium, calcium and potassium. The atmospheric pressure is about 10^{-15} Pa, so it is often regarded that there is no atmosphere on the surface of Mercury. Since Mercury is the closest planet to the sun, it is also the hottest planet with an average surface temperature of 179 °C and a maximum temperature of 427 °C. Mercury's orbital period is about 88 Earth days, and its rotation period is approximately 58 Earth days. Mercury's "day" is as long as 176 Earth days. Such slow rotation enables the surface facing the sun to reach a thermal balance state under the function of solar flow, and while the surface facing against the sun is very cold and with a minimum temperature of -173 °C. Mercury's surface temperature from subsolar point to the terminator can be expressed by the Hanson's cosine function.

The calculated planetary radiation is very considerable and with the radiation intensity ranging from 6 to 12,700 W/m^2.

2.4.3 Venus Environment

Venus' atmosphere is composed of dense mixtures of gases and 96% of them is CO_2, which brings severe greenhouse effect to the Venus. The atmospheric temperature on the surface of Venus is as high as 460 °C, while the atmospheric pressure is 93 standard atmospheres. The gravity field on the surface of Venus is 8.87 m/s^2, which is about 90.5% of the Earth's gravity field. Venus has an orbital period of 224.7 Earth days and a rotation period of 243 Earth days.

The orbital thermal environment of Venus is colder than the environment around Mercury because Venus is farther from the sun. Albedo ratio of Mercury is very small, so most of the incident solar radiation is absorbed by the surface of Mercury and then dissipated as IR. Venus is completely covered by clouds, so its albedo ratio is very high (about 0.8). The high albedo ratio makes the temperature at the top of the cloud low, so Venus IR radiation is even smaller than that of the Earth.

2.4.4 Mars Environment

Mars, with an average distance of 2.279×10^8 km (1.523 7 AU) to the sun and orbital period of 687 Earth days and 1.88 Earth years, is the closest outer planet to the Earth. The gravity field on the surface of Mars is about 38% (i.e., 3.72 m/s^2) of the Earth. The rotation period of Mars is 24.6 h, which is slightly longer than the Earth day. The Mars surface has thin atmosphere with an average atmospheric pressure of 0.63 kPa. The main composition of the atmosphere is CO_2 (about 95.3%) and a small amount of N_2, Ar, O_2, H_2O, etc. When the sun keeps heating the atmosphere, the pressure gradient and temperature gradient caused by it would generate wind on the surface of Mars. The wind direction and speed are always changing during the day. The wind speed below 20 m/s on the Mars could be an acceptable designing consideration for thermal engineer.

Mars atmosphere contains obvious suspended dust which will block sunlight to some extent. The concept of light depth is usually used to describe the blocking effect of Mars atmosphere (including dust) on light. Light depth (τ) is used to indicate the attenuation degree of light when it passes through the atmosphere at a certain incident angle.

Strong winds on Mars will roll up dust to form dust storms (the average wind speed is 50 m/s, and the maximum wind speed is 150 m/s). When a dust storm becomes most severely, it will directly affect the intensity of solar radiation incident on the surface of Mars. In addition, the dust storms on the Mars will also roll up Mars dust which would finally fall on the surface of the thermal control coating and change its thermophysical parameter.

2.5 Thermal Environment at Re-entry or Entry Phase

The thermal environment at re-entry or entry phase refers to the aerodynamic heating environment generated during high-speed re-entry or entry in the atmospheric environment. When re-entering from LEO, the general re-entry speed is about 7.82 km/s. As for the deep space probe returning to the Earth's ground, the re-entry speed could even exceed 11 km/s, and the surface of the deep space probe will bear the aerodynamic heating load with high heat flux density. For the purpose of preventing the aerodynamic heating from affecting the internal equipment of the returner (or return capsule), the ablation layer material with sufficient thickness shall be coated over the surface of such spacecraft for thermal protection. The generated heating could be effectively taken away through a series of complex physical processes (such as melting, evaporation and radiation) and chemical (such as decomposition and depolymerization) processes of the ablation material at high-temperature environment. Even so, some heat will still be transferred into spacecraft, which is an environmental factor that must be taken into consideration. Considering temperature rise of the internal equipment caused by aerodynamic heating during re-entry, it is an effective measure to reduce initial temperature of the returner or return capsule before re-entry by precooling it in advance, thereby ensuring the temperature rise of equipment within the range of temperature. For example, before the returning of Shijian 10, the inner equipment temperature of the return capsule shall be reduced to 10–15 °C by regulating the temperature of the working fluid in the single-phase fluid loop. Considering the increase of temperature caused by both aerodynamic heating and heat dissipation of equipment during the return process, the maximum temperature of scientific experiment apparatus in the whole process of landing will not exceed 25 °C, which maintains the success of space scientific experiment.

For the landing probe in deep space exploration mission, similar aerodynamic heating effect exists due to the existence of thin atmosphere when it lands on the surface of other planets with atmospheric environment, such as Mars lander.

In addition, the acceleration and gravity field of re-entering the Earth or entering other celestial bodies are similar to those of the launch phase and will also have certain influence on the choice of thermal control technology.

2.6 Inductive Environment

The space environment exists objectively and independently without depending on the spacecraft or equipment itself. However, some environments are induced by special working modes and special space equipment of spacecraft. Hence, additional or extra environmental factors must be considered in the design. As different subjects may have different inductive environments, the induction environment discussed in this section mainly includes the induction or additional thermal environment generated by engine ignition, spin motion of spacecraft or equipment, etc.

2.6.1 Inductive Environment Caused by Engine Operation

When the engine works, it will produce high-speed and high-temperature plume. The temperature of the engine nozzle will rise and balance at a very high temperature level within tens of seconds. Meanwhile, the engine also has a large temperature gradient along the axis, as shown in Fig. 2.8. The high-temperature plume and high-temperature nozzle wall during the operation of engine have an impact on the thermal control of spacecraft, which requires the attention of thermal control engineers. Generally, the plume heat flow of the engine shall be analyzed and obtained by professional software, and the analysis results shall be verified by engine plume test if necessary. The influence of engine plume on spacecraft thermal control is mainly indicated in two aspects: thermal effect and pollution.

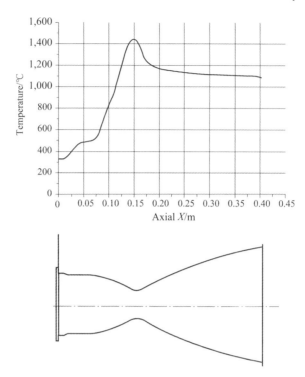

Fig. 2.8 Temperature distribution of nozzle along axial direction during operation of 7,500 N engine

The thermal effect of engine plume should be primarily concerned in the thermal control design. The thermal effect of plume varies from different types of engines, which shall be analyzed specifically aiming at specific engine types, operation status of engine and layout status. The distribution of plume heat flux on panels caused by operation of 7,500 N engine on Chang'e 3 is shown in Fig. 2.9. When landing on the lunar surface, the density of plume heat flux on the landing buffer mechanism was as high as 4.2×10^5 W/m², so it was very important to carry out thermal protection design for the landing buffer mechanism.

The other influence of the engine plume is its pollution to the thermal control coating. The influence of engine plume on the surrounding thermal control coating, especially on optical solar reflector (OSR, which is often called secondary surface mirror), shall be considered in the thermal control design. Some research has shown that some particles in the engine plume would deposit on OSR nearby the engine, with more tend to deposit on both sides of the engine and relatively less on the back of engine. The process of particle deposition is a long-term continuously accumulating process. The deposition of particles will increase the solar absorptance of OSR coating, thus increasing the temperature of the radiating surface and reducing the radiating efficiency. Furthermore, the deposition quantity of plume pollutants is closely related to the surface temperature of the coating. The temperature increase of the surface of coating can effectively reduce the deposition quantity of pollutants [7].

2.6 Inductive Environment 41

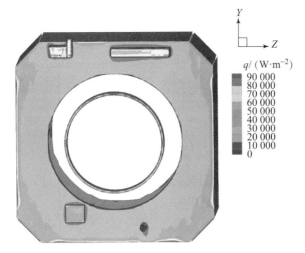

Fig. 2.9 Plume heat flow distribution on deck during operation of Chang'e 3 7,500 N engine

2.6.2 Inductive Environment for Spinning Spacecraft or Equipment

The centrifugal force will be generated by high-speed rotation of spacecraft or equipment, which will affect the movement of working medium inside the heat pipe and further affect the operation of heat pipe that drives the working medium to flow with a very small capillary force, since ensuring that the working medium can flow from the condensation section back to the evaporation section is a priority for normal operation of the heat pipe. For spinning spacecraft, the centrifugal force on the backflow of heat pipe working medium shall be thoroughly considered. During design and application of heat pipes, it is very important to prevent centrifugal force from blocking the backflow of working medium. There are two ways to prevent or reduce the influence of centrifugal force: ① The layout of heat pipes shall be in favor of promoting backflow of working medium with help of centrifugal force; ② reducing rotation speed of spacecraft could weaken the influence of centrifugal force.

If the centrifugal force effect caused by spacecraft spin is not handled properly, the heat pipe will not operate normally, and such failure once occurred to Dongfanghong 2 satellite. The application case of heat pipes on this satellite is shown in Fig. 2.10: The heat pipe transfers heat of TWT to the despinning assembly shield, and the spin speed of satellite is 50 r/min. The original design is shown in Fig. 2.10, where an inward bending part in the middle of heat pipe makes the working medium in the condensation section of heat pipe fail to flow back to the evaporation section under the centrifugal force produced by high-speed rotating despin assembly. The operation failure of heat pipes leads temperature of TWT to rise to 60.7 °C. The specific reason is illustrated in the curved part of dotted line as shown in Fig. 2.10. Since the satellite is spinning, the working medium in heat pipe would evaporate after absorbing heat

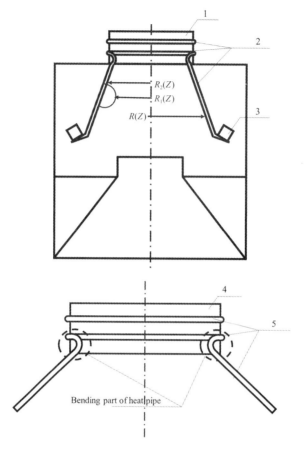

Fig. 2.10 Layout of heat pipe for spin satellite. 1, 4—Despin assembly shield; 2, 5—Heat pipe; 3—TWT

at TWT and condense to liquid at the cover of the despin assembly. Then, under the combined action of centrifugal force and capillary force, the liquid flowed back to TWT to keep evaporating. If local reverse bending happens (as indicated in the dotted line of Fig. 2.10), the liquid may flow from $R_2(Z)$ to $R_1(Z)$ under the function of centrifugal force, where the liquid is blocked and unable to flow to the evaporation section of heat pipe.

References

1. CZ-3A Series Rockets Handbook, 2003 version
2. S.M. Yang, W.Q. Tao, *Heat Transfer*, 3rd edn. (Higher Education Press, Beijing, 1998)
3. Test method for thermal control coatings of spacecraft Part 5: Ultraviolet irradiation test in vacuum, GJB 2502.5-2015

References

4. T.H. Vonder Haar, V.E. Suomi, Satellite observation of the Earth's radiation budget. Science 163(3868) (1967)
5. G.R. Ming, S. Guo, *Spacecraft Thermal Control*, 2nd edn (Science Press, Beijing, 1998)
6. D.G. Gilmore, *Spacecraft Thermal Control Handbook, Volume I: Fundamental Technologies* (The Aerospace Press, EI Sequndo, 2003)
7. X.X. Lin, Z. Wen, J.S. Tao, Simulation analysis of deposition effect of molybdenum atoms from ion thruster plume on satellite OSR thermal properties. Spacecr. Eng. 25(3), 52–56 (2016)

Chapter 3
Design of Spacecraft Thermal Control Subsystem

3.1 Overview

The purpose of spacecraft thermal control is, according to the requirements of a flight mission and the condition of internal and external thermal loads to which the spacecraft is exposed in the whole life cycle, to systematically take thermal control measures to organize the heat exchange processes inside and outside the spacecraft so as to keep the temperature of onboard equipments, structures and living environment within the prescribed limits.

In general, in order to meet the temperature requirements of the spacecraft, various thermal control solutions can be taken. The designer should make a good trade-off among all the temperature indicators from the perspective of system design to decide the optimum solution.

In this chapter, the mission characteristics, design principles and system design methods of spacecraft thermal control subsystem are presented. On this basis, the development phases of thermal control subsystem and their corresponding focuses are briefly introduced.

3.2 Mission Characteristics

The design of thermal control subsystem should fully consider and ensure that the spacecraft requirements for temperature and other indicators can be met in various stages from ground phase to orbiting and then returning to the ground or landing on other celestial bodies, that is, under the environmental conditions in the whole life cycle. The targeted measures in whatever nature must take into account the specific mission characteristics of the spacecraft and adapt to various environmental conditions the spacecraft might encounter.

3.2.1 Ground Phase

Because of the presence of air on the ground, the spacecraft and its internal devices exchange heat mainly through convection. The purpose of spacecraft temperature control on the ground is mainly to ensure the spacecraft's need for ground power-on testing. This can be usually guaranteed by a ground-based air-conditioning system. Even when sitting on the launchpad, the spacecraft can be supplied with suitable ambient temperature through an air-conditioning system specially configured on the pad. However, the designers of ground support system must fully consider the local climate characteristics, launchpad features, spacecraft launch time, spacecraft temperature requirements and other factors to ensure that the entire air-conditioning system has a sufficient cooling or heating capacity.

For some special devices, such as the batteries and atomic clocks with strict temperature requirements and the solar array driving apparatus (SADA) with strict humidity requirements, enhanced heating or convection cooling measures should be additionally taken. In order to protect the thermal coating with high cleanliness requirement from contamination, the cleanliness requirements of the ground facilities, launchpad and fairing interior must be considered.

Considering the influence of local temperature, the possible temperature changes during the ascent phase and the temperature changes caused by spacecraft powering, it is necessary to control the initial temperatures of some devices, especially the payloads with stringent temperature requirements, within the specified ranges before launch. In some cases, special ground temperature regulation might be necessary to accurately control the initial temperature before launch.

3.2.2 Launch and Ascent Phase

The launch and ascent phase is the period from the blastoff and launch to flying through the dense atmosphere at a high speed and finally the separation of the spacecraft (or the upper stage, if any) from the launch vehicle. During this period, the condition (or mode) of spacecraft operation and the external environment conditions are complex and completely different from those in the orbiting phase. On the one hand, in order to ensure adequate power supply during orbit insertion, payloads and some platform equipments are often in a power-off state. On the other hand, the ascending spacecraft undergoes aerodynamic heating and free molecular heating. Especially for those spacecraft without a fairing, the direct high-speed friction between their surfaces and the atmosphere produces aerodynamic heating.

For most spacecraft with a fairing, their temperature generally tends to decrease during this period, especially after the fairing separation. However, due to the short duration of launch and ascent phase, the temperature of the devices inside the spacecraft normally does not change distinctly. The remarkable temperature changes usually occur on the external devices or components, thus demanding the analysis of their temperature changes during this phase in some cases.

For the spacecraft without a fairing, aerodynamic heating would cause an abrupt rise of the surface temperature of the spacecraft. For these spacecraft, the temperature rise of the equipments will be more significant even if effective insulation measures are taken.

3.2.3 Orbiting Phase

During the on-orbit phase, many factors need to be considered, including space environment, spacecraft mission characteristics, operation modes and so on. These factors would lead to the changing of on-orbit external heat radiation and internal heat dissipation with time and space. Thermal designers need to fully consider various boundary conditions and internal thermal condition to ensure that the temperature of each part is within the specified range.

Low Earth orbit (LEO) spacecraft are usually inserted directly into their mission orbits. Other spacecraft such as LEO constellations launched by one vehicle, GEO satellites and interplanetary probes are generally inserted into a temporary parking orbit or a transfer orbit before formally entering their mission orbits. Therefore, the thermal environment characteristics of these intermediate orbits should be carefully analyzed. Before the commencement of formal operation, the spacecraft usually needs to pass the on-orbit test for the performance of some platform equipments and loads. In this case, the temperature level inside the spacecraft is generally maintained by means of additional heating.

For a spacecraft with stably operating payloads and relatively simple attitude changes, several typical working conditions are generally considered to cover all the normal working conditions on the orbit. For example, for geosynchronous orbit (GEO) communication satellites, only three conditions in winter solstice, summer solstice and spring equinox need to consider. For a spacecraft with non-continuously operating payloads, various attitudes and even multiple mission orbits, the combination of multiple factors must be considered. The combined extreme condition must have the largest envelope. Thus for such spacecraft, the choice of working conditions is the focus of consideration.

3.2.4 Reentry or Entry Phase

For the spacecraft that needs to reenter the Earth or another celestial body with atmosphere, its surface temperature will generally rise sharply due to the effect of aerodynamic heating when reentering or entering the atmosphere. The aerodynamic heating effect or temperature increase will be related to the atmosphere density, spacecraft velocity and attitude, aerodynamic spacecraft configuration and so on.

The key of this phase is to consider thermal protection and insulation in order to reduce the effect of aerodynamic heating on the temperature rise of internal equipments. In addition, the environmental pressure around the spacecraft will change from vacuum to the atmospheric pressure of the Earth or an extraterrestrial object. The convective heat exchange will develop from nothing, and the performance of multilayer insulation material will also degrade under vacuum conditions. All these changes need to be considered in the thermal control design.

Furthermore, before reentry or entry, the configuration or composition of the spacecraft might change greatly to adapt to high-speed reentry or entry. For example, the Earth-returning satellite needs to be separated from the service cabin before returning. The reentry capsule of a manned spacecraft needs to be separated from the propulsion capsule. In the process of reentry or entry, the operating condition and heat dissipation of internal payloads also change according to the task requirements. Meanwhile, the long working period of propulsion subsystem would generally result in the high temperature of engine jet pipe and the plume effect, which needs special consideration in the thermal control design.

Therefore, for a spacecraft demanding reentry or entry, the spacecraft configuration or composition as well as the condition of heat exchange between its internal and external environments and between internal equipments are much different from those at the orbiting stage. These are the significant features of this stage and the focus of thermal control design.

3.2.5 Landing Phase

After reentering and landing on the surface of Earth, the spacecraft still needs to wait for a period of time before being recovered. Due to the thermal inertia effect, the spacecraft temperature will continue to rise for a while. At this time, the outside is already atmospheric, and the temperature varies with region and season. Therefore, the temperature control of the spacecraft is required under such conditions.

For the spacecraft expected to land on the surface of another planet, full consideration must be given to the conditions of the planet surface, including gravity conditions, atmospheric conditions, topography, ground surface temperature and thermophysical properties, in addition to the operating condition of the spacecraft. Due to the uncertainty of the surface conditions of the planet, a sufficient design margin or a thermal control technology with high regulation capability generally needs to be considered for such a spacecraft.

3.3 Basic Principles of Thermal Control Design

Thermal control design is often a complex system design process involving many factors. Different designers may adopt different thermal control solutions. However, the function and performance of the thermal control subsystem should meet the design specifications and ensure an acceptable design margin. Thermal control design should also comply with the design requirements for reliability, safety, space environment adaptability, testability, maintainability and supportability (or REATMS for short). In addition, the following design principles should be considered [1].

- Manufacturability

The designer should notice the technical feasibility of thermal control measures and take into full account the various operations required in manufacturing, installation, testing and other steps in the prelaunch process as well as their impact on the performance of thermal control subsystem.

- Inheritability

Priority is given to standardized flight-proven technologies and products to shorten the development period, reduce the development cost and improve reliability.

- Verifiability

The thermal control solution or technology that can be verified by simulation or thermal testing should be chosen to avoid the design unable or hard to be verified.

- Adaptability

Since some parameters and interfaces will inevitably change during the spacecraft development, the lack of certain design adaptability will often lead to annoyingly frequent modification of the thermal control design. At the same time, for the spacecraft on orbit, the thermal control subsystem also needs to have certain adaptability when the condition of thermal environment or internal heat source deviates from the design value.

- Advancement

The proper use of new technologies and products that have been fully tested or qualified can improve the performance of thermal control subsystem and the overall performance and competitiveness of the spacecraft.

- Economy

Economy should be considered from two aspects. On the one hand, while ensuring the function and performance of thermal control subsystem, the development costs, including the costs of design, materials and devices, stand-alone equipments and test verification, should be controlled as much as possible. Meanwhile, it is necessary to optimize the development process, improve efficiency, shorten the development cycle, and reduce manpower and equipment costs. On the other hand, greatest efforts should be made to minimize the mass of thermal control subsystem and the onboard power consumption and create conditions for the improvement of spacecraft performance.

3.4 Design Method of Thermal Control System

With the diversification of space missions and the development of space technology, the quantity and types of spacecraft are unexpectedly increasing. From the perspective of mass, there are not only the MicroSats or CubeSats that weigh only a few dozens or even ten grams, but also many large-scale spacecraft that weigh more than ten tons or even hundreds of tons (such as International Space Station). From the perspective of orbit, there are the LEO spacecraft at the altitudes of 200–300 km, the geosynchronous orbit (GEO) spacecraft at the altitude of 36,000 km and the lunar probes orbiting around the moon and the deep space probes orbiting around the sun. From the perspective of mission, there are the spacecraft remotely sensing the surface of the Earth or other planets, the spacecraft for space exploration, and the spacecraft for soft landing and exploration (such as on the surface of the moon or the Mars). From the manned or unmanned perspective, there are unmanned spacecraft, near-Earth manned spacecraft (such as Shenzhou spacecraft series, Tiangong Space Lab, Space Station, etc.), as well as manned lunar landing vehicles (such as Apollo spacecraft series). Although different spacecraft have different masses, flight altitudes and mission characteristics, they usually share some thermal design methods and techniques. For example, most spacecraft adopt passive thermal control technologies, while manned spacecraft and deep space probes tend to rely on active thermal control technologies in order to improve their adaptability and self-regulation to mission diversity and complex environments.

3.4.1 Thermal Control Design Requirements and Conditions

1. **Design Requirements**

The thermal design requirements generally include:

(1) The operating temperature range, storage temperature range, start-up temperature range, temperature uniformity (including temperature difference and temperature gradient), temperature stability and temperature control accuracy of onboard equipments.
(2) The temperature and humidity of gaseous environment in the pressurized capsule.

3.4 Design Method of Thermal Control System

(3) The flow field in the pressurized capsule.
(4) Mass.
(5) Power consumption.
(6) Life.
(7) REATMS requirements.
(8) Ergonomics.
(9) The interfaces with the spacecraft and other subsystems and systems.

2. **Design Conditions**

To design the thermal control subsystem, the following basic conditions or information should be acquired and mastered:

(1) Mission.
(2) Space environment conditions.
(3) Orbital parameters.
(4) Flight procedures and attitudes.
(5) Spacecraft configuration and structure.
(6) Layout of instruments and devices.
(7) The material, size, mass, heat dissipation, thermophysical properties and operating modes of all onboard devices and components.
(8) Environmental conditions of testing and launching sites.
(9) Other technical specifications.

Fully understanding and analyzing the above requirements and conditions is of benefit to the design of various stages of thermal control subsystem. For example, the decision on the dominance of either passive or active thermal control technology should be made in the schematic design phase according to the characteristics of space missions and the complexity of space environment. However, it should be noted that, in the schematic design phase, although many details are still not defined, preliminary thermal balance calculation is still required. The purpose of such calculation is to preliminarily determine the basic strategy of thermal control, that is, to decide whether the module-specific or integrated thermal design, whether the insulation design or heat dissipation design and whether the radiation/conduction or convection/conduction should be adopted. To be specific, if heat dissipation design is adopted, it is required to estimate whether the radiation area on the exterior spacecraft surface is large enough to radiate all the heat in the spacecraft at the given temperature. In the prototyping phase, the thermal control design should be completed according to the design requirements and design constraints, and its compliance should be verified through analysis and testing. If necessary, some constraints and technical indicators can be adjusted. Then based on the thermal control design, the state of flight model can be defined. In the formal prototype phase, the thermal control design, development and test verification should be completed under the finally defined design requirements and conditions.

3.4.2 Selection of Thermal Design Cases

3.4.2.1 Thermal Design Considerations

In the whole life cycle from the launch phase to the end of the service, the spacecraft will undergo a variety of operating conditions and external environments. Thereby, the life cycle of the spacecraft can be generally divided into different mission phases. For example, the mission of a lunar landing rover can be divided into the phases such as launch, Earth–moon transfer, circumlunar, powered descending and lunar terrene exploration. The analysis of a "thermal case" is generally to analyze and evaluate the balance of internal heat consumption, external heat flow and heat dissipation of the spacecraft based on the thermal control requirements in different mission stages. In the whole operation period, the working condition is continuously changing for the following reasons:

(1) Thermal environment variation. The spacecraft will experience different thermal environments, generally including the Earth surface environment in the launch-awaiting phase, the Earth atmosphere environment in the launch and ascent phase, the planetary or lunar orbit environment in the orbiting phase, the planetary or lunar atmospheric environment in the atmosphere-entering phase, and the planetary or the lunar surface environment in the planetary (or lunar) surface roving phase and returning phase. The variation of these thermal environments will result in the fluctuation of the heat flow incident on the surface of the spacecraft. When the spacecraft orbits a planet (or the moon), the variation of such factors as the orbit altitude, the distance between the planet (or the moon) and the sun, the angle between sunlight and spacecraft surface and the time of Earth shadow and sunshine can lead to the fluctuation of the direct sunlight, albedo and planetary radiant heat flux incident on the surface of the spacecraft. In addition, when the spacecraft lands or cruises on the surface of a planet (or the moon), the factors such as season, the distance between the planet (or the moon) and the sun, the angle between sunlight and spacecraft surface, day and night, atmospheric environment and topographical features will also affect the environmental heat flux absorbed by the spacecraft.

(2) Changes in the attitude and configuration of the spacecraft. According to its mission characteristics, the spacecraft flying on orbit must maintain a regularly changing attitude such as spin-stabilized Earth orientation, three-axis stabilized Earth orientation or inertial orientation to a space target in order to ensure that the payload can point at, track and scan the target and the local orientation of some devices can be controlled (e.g., the sun-pointing control of solar wings). The change of attitude directly affects the variation of the heat flow reaching the surface of the spacecraft, which is closely related to the choice of radiators. In fact, it is also necessary to consider the on-orbit attitude maneuver, namely changing from one attitude to another, which will also result in the changes in absorbed orbital heat flux. In addition, in different operation phases, the configuration of the spacecraft or its combined body may change. For example, the combination

or separation of multiple cabins/modules, as well as the movement of some parts such as solar arrays, large antennas, robotic arms and deployable thermal radiator will result in the variation of the spacecraft configuration. These changes will inevitably affect the heat flow incident on the surface of the spacecraft.

(3) Degradation in the thermophysical parameters of the thermal control coatings on the spacecraft surface. Due to the complexity of space environment, these thermal control coatings can be affected by atomic oxygen, charged particles, ultraviolet radiation, high vacuum and contaminants. These factors will generally cause the increase in solar absorptivity, resulting in the increase in the heat absorbed by the spacecraft on orbit and the overall temperature rise.

(4) Changes in the heat dissipation of inboard heat source. The heat generated by most electronic equipments is stable and unconverted, while the fluctuations in the heat dissipation of some equipments fall into three cases. The heat dissipation of the devices such as detecting sensors, data storage/transmission devices, engines/thrusters, transmitters and electric heating elements is obviously intermittent. The heating power generated by the devices like batteries changes slowly due to the degradation of the power supply efficiency. The heating power generated by atomic clocks, gyros and other devices with internal temperature control changes with temperature.

Among the above points, the points 1, 2 and 3 are closely related to the absorption of external heat flux, while the point 4 is closely related to internal heat dissipation. It can be seen that the task of the thermal design is to ensure that the temperature of the spacecraft is always within the required range under various thermal conditions. However, the design of a thermal control model for so many thermal cases will lead to a long unnecessary design cycle. Usually, the selection of several typical thermal cases will suffice for the design.

3.4.2.2 Determination of Thermal Design Cases

Due to the ceaseless changing of internal heat dissipation and external thermal environment, most spacecraft always run in a typical transient thermal process throughout the life cycle, rather than in the proper "thermal equilibrium state." For this reason, the thermal designers have to face various challenges when carrying out the tasks of thermal design, thermal analysis and ground testing verification. Therefore, from the engineering perspective, some extreme cases, including the worst hot case and the worst cold case, are often selected in thermal design and used to determine the assumed engineering-oriented "thermal equilibrium states" at high and low temperature for design, analysis and testing verification. With the thermal design based on typical cases, it can be expected that, the largest analysis/test envelope derived from the extreme cases can nearly cover all the actual temperature conditions and ensure that the temperature requirements are met in all thermal cases. The above two extreme cases occur, respectively, when the sum of the absorbed orbital heat flux and the internal heat dissipation is the maximum or minimum. In fact, either the combination of maximum–maximum or that of minimum–minimum is rarely present, although it is unavoidable for the presence of the combination in some situations. The absorbed external heat flux generally reaches the peak value at the end of life of the spacecraft when the sunlight exposure factor (i.e., ratio of illumination time to orbital period) is at its maximum. Also, the absorbed external heat flux generally reaches the bottom at the beginning of the life when the factor decreases to the minimum. If the worst hot and worst cold cases are used as design cases, all other cases will be enclosed in between. Hence, thermal control design usually accounts for these two cases.

1) Determination of the extreme cases of internal heat dissipation

There are three main types of conventional internal heat dissipation including long-term stable heat dissipation (e.g., for most of platform equipments), regular short-term heat dissipation (e.g., for some detection payloads operating at high frequency) and random short-term heat dissipation (e.g., for some detection payloads operating only for a few times).

It is generally effortless to determine the worst case with the minimum heat dissipation, in which only the long-term stable heat dissipation (even in safe mode) needs to be considered. For the case with the maximum heat dissipation, the thermal designer should comprehensively investigate a variety of combinations of long-term stable heat dissipation and various short-term heat dissipations. For example, for a low Earth orbit spacecraft, the case of the maximum heat dissipation usually includes the combination of the condition with the longest continuous working period of regular short-term heat dissipation in an orbital period and the condition with most continuous working cycles in one day. In many cases, the random short-term heat dissipation cannot be ignored.

2) Determination of the extreme cases of absorbed orbital heat flux

The determination of the extreme cases of orbital heat flux is much more complicated. In addition to the factors described above, the designer needs to consider the thermal control design state of the spacecraft, in particular the layout and area of the radiators. For example, for a three-axis stabilized Earth-oriented hexahedral spacecraft with the β angle of $0°–90°$, the combination of radiators on different sides of the spacecraft will result in different extreme cases of absorbed orbital heat flux and different β angles accordingly [2, 3].

The said extreme cases can generally be determined by a theoretical method. An enumeration method can also be used to analyze the orbital heat flux with given various β angles and various attitudes and finally determine the extreme cases of absorbed orbital heat flux by comparison. In general, the following factors should be considered in the determination of the extreme orbital heat flux cases [4]:

(1) Maximum/minimum solar radiation intensity.
(2) Maximum/minimum $|\beta|$ angle.
(3) Critical β angle.
(4) Zero β angle.
(5) Beginning of life (BOL) and end of life (EOL).

3.4 Design Method of Thermal Control System 55

For the determination of the extreme cases, the following supplementary instructions need to be added:

Firstly, due to the influence of such factors as the operating characteristics of payload and the operating mode of power subsystem, the extreme orbital heat flux and the extreme internal heat dissipation may not occur at the same moment. Then, it is necessary to separately consider the two extreme cases in order to find out the extreme value of the combinations of absorbed external heat flux and internal heat dissipation.

Secondly, the extreme cases for some special equipments (such as battery, atomic clock, etc.) or cabins with independent thermal control design may not occur at the same moment as those for the entire spacecraft. In this case, the additional analysis and verification especially for those equipments should be carried out after accomplishing the thermal control design for the extreme cases of the entire satellite.

3.4.3 Selection of System Design Methods

Thermal control subsystem is an important part of the spacecraft and directly serves the entire spacecraft (including payload, equipment). Therefore, the design of thermal control subsystem cannot be separated from the integrated design of the spacecraft. To better complete the task, the thermal designer should fully understand the relevant information such as the mission of the spacecraft, the function of the payload and the feature of equipment layout. The system also has its own special features. While meeting the constraints rooted in external environment and spacecraft on thermal control technology, the designer should comprehensively apply reasonable thermal control techniques to regulate and control the process of heat collection, transmission, dissipation and storage. In some cases, the method of compensating heating or cryogenic refrigeration will be considered to achieve the thermal control target of the spacecraft.

3.4.3.1 Selection of Design Layout

The layout designs can be divided into two categories: insulation layout design and coupled layout design. The insulation layout designs include the thermal insulation between devices or components, between platform and payload, and between modules. The coupled layout designs include equipment–equipment, cabin–cabin and payload–platform thermal coupling.

Thermal insulation design is adopted for the equipments or stand-alone devices with high-temperature control accuracy, high-temperature stability or high-temperature level requirement to achieve thermal isolation between specific equipment and other equipments (or cabin plates). In addition, an exclusive thermal radiator is usually set for the equipment to directly remove its waste heat. For example, in order to avoid the freezing of propellant, the propulsion pipeline and correlative components (such as pressure sensors, engines, self-locking valves and filters) are thermally insulated from the mounting plate. In order to guarantee the temperature stability of a time–frequency reference device (rubidium clock, etc.), the thermal insulation between the device and other devices (or cabin plates) should also be considered. The thermal insulation design between payload and platform is similar to that between the components or stand-alone devices.

Based on the concept of independent module design adopted by many spacecraft, a spacecraft is generally composed of several independent modules including a platform module (or service module), a payload module and other modules (sometimes including a propulsion module or reentry module). The platform cabin is mainly provided with the equipments which are generally in long-term operation mode. The payload cabin is mainly provided with the payloads which are usually in intermittent operation mode. In order to facilitate the standardized, modularized and generalized design, thermal insulation measures are generally taken between the cabins. In this way, independent thermal design is adopted for each cabin, in which the heat is radiated through radiators.

The method of thermal coupling is usually to reduce the temperature difference between devices through isothermal design. Meanwhile the temperature fluctuation of short-term operating equipments can be reduced by coupling them together to increase the heat capacity. An orthogonal embedded heat pipe network consisting of multiple heat pipes perpendicular to each other in two directions is an effective method for thermal coupling. The orthogonal embedded heat pipes embedded in the honeycomb plates constitute a fully orthogonal embedded heat pipe network. In this method, the requirement for equipment layout is relatively relaxed, but the processing and implementation of heat pipe network are more difficult.

The design method of inter-cabin thermal coupling is to form effective thermal coupling between different cabin segments through various thermal control measures. Finally, all the heat in the entire spacecraft is dissipated to the space through some controllable thermal radiators and can be dynamically adjusted by the use of bypass valves or louvers. In addition to fluid loop technology, the coupling between two cabins can be achieved by other measures (such as convection/ventilation techniques) in some cases. However, the cabin equipments with relatively strict temperature requirement or large power consumption fluctuation can also be thermally coupled with the fluid loop through some specific thermal control techniques (such as loop heat pipe, heat switch). From the perspective of thermal control, inter-cabin thermal coupling breaks the physical isolation between different independent cabins, as in the case where the requirement for integrated layout is relatively low. From the macroperspective, it has realized the isothermal and controllable heat management of the whole spacecraft system.

3.4 Design Method of Thermal Control System

The integrated thermal coupling between the payload and platform is one of the current mainstream methods of thermal design. The payload is thermally coupled with the platform through various thermal control means, by which a portion of the platform heat is transferred to the main body or peripheral structure (such as the shading cover or barrel of the optical system) of the payload. The heat transferred to the payload can help maintain the temperature level of the payload. This coupling method could greatly reduce the compensated heat originally required in the independent thermal control design of the payload, and relieve the design burden on power supply and distribution subsystem, and thereby optimize and improve the performance of the whole spacecraft.

3.4.3.2 Selection of Thermal Regulation Method

1. **Heat dissipation and heat storage**

The selection and design of thermal radiators are very important, because the heat generated by most spacecraft is ultimately radiated to the deep space. The radiator conditions such as the installation position, orientation, surface area and temperature should be determined according to both the analysis results of the cases and the temperature limit requirements for equipment operation and storage. The type, quantity and thermal control coating of the radiators should also be determined.

In general, the panel surface on which the incident environmental heat flux is stable and as little as possible is the best choice for the radiator. At the same time, the field of view of the thermal radiator facing the deep space should be unobstructed. For example, the structural panel opposite to the sunlight direction is preferable to be the radiator for the sun-synchronous orbiting spacecraft, while the south and north structural panels are suitably selected as the radiator for the geosynchronous orbiting spacecraft. When it is difficult to avoid the influence of solar or planetary radiation on the radiators because of the mission characteristics, the use of additionally designed sunshade plates, umbrellas or large-sized antennas to reduce this influence should be considered.

The area and temperature of the radiating surface of the thermal radiator should be determined according to the orbital heat flux absorbed by the radiator, the heat dissipation, operating temperature range and storage temperature range of the spacecraft equipments, and the heat transfer relationship between the internal heat source and the radiator.

In addition to the structural panel radiator using spacecraft structural plate as the base, the body-mounted radiator mounted on the outer spacecraft surface as an independent structure, deployable radiator or other radiator types may also be selected for heat dissipation. In fact, it is also conceivable to design a radiator that can be extended, folded and adjusted to avoid the influence of orbital heat flux. For example, some high-power communication satellites have used deployable radiators for heat dissipation.

To use thermal control coatings on the thermal radiator to accommodate the absorption of environmental flux and dissipation of heat to outer space is an important method to control the temperature of the spacecraft. The solar radiant heat flux absorbed by the radiator surface should be minimized, and the heat flux dissipated into the space should be increased as much as possible. Thermal radiator coatings with low solar absorptivity and high emissivity, such as organic/inorganic white paint and second-surface mirrors (often referred to as optical solar reflectors), are commonly selected.

Furthermore, the heat emitted by the radiator can be adjusted by means of louvers, thermal switches, electric heat shields, variable conductance heat pipes, loop heat pipes, capillary pump loops, fluid loop and heat pump loop. In order to improve the heat efficiency of structural panel radiator, the heat pipes can be embedded in the spacecraft structure plate or mounted on the structure plate.

Besides, there is another way of heat rejection, that is, dissipation-type heat rejection, which evaporates or sublimates a kind of consumable substance into space. The latent heat of phase change of the substance is used to cool down the equipments or the crew. For example, this method has been applied to extravehicular activity.

In some cases, the temperature rise of high-power equipment in transient operation cannot be suppressed by conventional heat rejection methods, or the waste heat is intended to be stored and utilized to compensate for some equipments. At this time, it is possible to make use of phase change heat storage technique, which can use the latent heat of the phase change material to achieve heat storage. Solid–liquid phase change materials are commonly used for this purpose.

2. **Heat collection and transmission**

Usually, the high-emissivity coating on the surface of the spacecraft equipment can be used to directly or indirectly transfer the heat flux to thermal radiator through radiation.

If the heat generated by the operating equipment is hard to be effectively controlled by radiation alone, it can be directly transmitted to the thermal radiator by means of thermal contacting or through high thermal conductivity components. For example, most of the devices in the communication satellite cabin are directly mounted on the thermal radiators. In order to strengthen the contact heat transfer, thermal filler may be applied on the contact interfaces, or the heat transfer resistance may be adjusted by using heat-conducting switches.

If some devices have over high heating power and cannot be directly placed on the thermal radiators, their heat can be collected and transmitted by means of constant-conductance heat pipes, fluid loops, loop heat pipes, capillary pump loops, convection ventilation loops, heat pump loops and other means. If the heat flux from a device to the radiator is collected and transmitted through constant-conductance heat pipes, the amount of collected and transported heat will depend on the heat transfer capacity of heat pipes. The thermal resistance mainly depends on both the thermal contact resistance between heat pipes and the device and that between heat pipes and the radiator. By using the pump-driven fluid loops, the heat from the equipments installed on cold plates, microchannel heat sinks and jet or sprayer devices are collected, and transmitted to thermal radiators through the heat exchangers. The collected and transmitted heat flow can be adjusted by fluid loop devices such as pumps and valves. The heat dissipation of the equipment is collected by the evaporator on a loop heat pipe or a capillary pump loop and transmitted to the thermal radiator through gas tube pipe and condenser. In some pressurized cabins, the forced convection heat transfer between gas and equipment is used for heat collection and transfer. After being controlled by regulating the gas flow, the collected heat flux is finally transmitted to thermal radiator and then to space. The heat dissipation of the equipment is collected by the evaporator in the heat pump loop. The working fluid is compressed by the pump and transmitted to the thermal radiator to dissipate heat at the temperature higher than the heat source and then condense.

3. **Compensatory heating**

Normally, it is difficult to meet the minimum operation temperature requirements of the equipment by thermal control technology (such as insulation) alone. In this case, thermal compensation is necessary to ensure that the equipment can work normally at low temperature or meet the low-temperature survival demand (the equipment does not work). Compensatory heating is the most commonly used method in spacecraft thermal design. It is to convert electric or solar energy into thermal energy and provide thermal compensation for the equipment. The methods of compensatory heating include electric heating compensation, waste heat utilization, solar energy utilization, radioisotope heater source compensation and the like.

Among them, electric heating compensation is the most conventional, reliable and stable method. It realizes thermal compensation by using controllers and electric heaters. The thermal compensation can be achieved by uninterrupted constant current heating, or controlled by means of manual control command, tele-command, program control command and automatic control. This method is applied to most of the existing spacecraft. Radioisotope heater unit (RHU) compensation is generally applied to deep space exploration missions to solve the problem of insufficient energy due to too long distance from the sun or the lack of sunlight for a long period of time, for example, in the cases of Mars exploration or lunar landing and cruising. Some indispensable protective measures need to be used for this method.

4. **Cryogenic refrigeration**

Cryogenic refrigeration is generally applicable to the payloads with low-temperature requirements, such as infrared detection devices or spaceborne high energy particle detection instruments. For these payloads, the insulation between the payload and the platform, as described in the previous section, should be generally designed to minimize the thermal coupling between the payload and the platform. In addition, separate cryogenic refrigeration techniques are also required for the payloads, including radiative cooling, thermoelectric cooling, mechanical cooling and storage cooling. When using different cooling methods, consideration should be given to reliability, vibration interference, weight and volume, in addition to the requirements such as low-temperature level, cooling capacity and cooling efficiency.

Generally, the design of thermoelectric refrigeration and mechanical refrigeration should consider the redundant design; otherwise, it will become a single point of failure affecting the payload performance. For the active cooling methods involving motion, the anti-vibration design should be considered, or the thermally coupling cryogenic loop heat pipes should be used to reduce the impact of vibration sources (such as compressor) on the detector. The storage refrigeration device utilizes the evaporation or sublimation of the cryogenic liquid or solid carried by the spacecraft at a low temperature to absorb the waste heat of the equipment, which is finally discharged to the space along with the exhaust. In the practice of some other countries, superfluid helium is used for infrared satellites to maintain the low temperature required by the detector. However, this is a consumable refrigeration method with limited service life. The key to this refrigeration design is the need to consider the structure and thermal control of the refrigerant medium and the effect of liquid sloshing on the spacecraft attitude.

3.4.4 Selection of Thermal Control Technologies

There are also many types of spaceborne thermal control technologies. At present, thermal control technologies are conventionally divided into passive thermal control technology and active thermal control technology.

The passive thermal control technology has the advantages of simple implementation, reliable performance, lightweight, low cost, good universality and long service life. Nevertheless, due to the strong dependence on overall spacecraft layout and structure and the poor adaptability, this technology is unable to adapt to the large variations of heat dissipation, especially inside the spacecraft. In general, the passive thermal control technology can be employed to achieve the purpose of thermal control if the external thermal environment and internal heat dissipation of the spacecraft do not change much and there is no onboard equipment with strict temperature requirement. As the main basic thermal control technology in most common use, this technology is almost indispensable for all spacecraft. The passive thermal control can be generally achieved through thermal coating, heat conduction material, interfacial heat transfer enhancement, thermal insulation, heat pipe, phase change material and other techniques.

In spite of the advantages such as low requirement for the overall spacecraft layout and structure, self-adjustability and strong adaptability, the active thermal control technology has some shortcomings, such as complex implementation, low reliability, heavyweight, high cost and poor universality. When the external thermal environment and internal heat dissipation of the spacecraft fluctuate drastically with time or the temperature requirement of equipment is too strict, the use of passive thermal control technology alone is not enough to control the temperature of spacecraft or equipment within the specified range. In this case, active thermal control technology can be employed additionally. Active thermal control can be generally achieved by self-adjusted coating, louver, convection ventilation, fluid loop, cryogenic refrigeration and other techniques.

The general principle of selecting the thermal control technology is to give priority to passive thermal control and then consider active thermal control. Several factors such as task features and space environment characteristic, temperature requirement, lifetime requirement, development cost and energy condition are the key points to consider when selecting the passive-dominated or active-dominated thermal control technology. However, the demand for active thermal control technology will increase with the development of space technology. Meanwhile, with the advancement of science and technology, active thermal control technology will develop toward the directions such as structural simplification, weight reduction, reliability improvement, cost reduction and operational convenience. It will be more and more widely applied to the design of spaceborne thermal control subsystem.

3.5 Thermal Control Design Stages and Key Points

The thermal control design of a spacecraft is generally divided into concept phase, initial prototype phase, formal prototype phase and operation improvement phase [5]. For the flight missions with a certain degree of inheritance, the thermal control design sometimes directly commences with the initial prototype phase or even the formal prototype phase. The design phases are shown in Fig. 3.1.

- The key work of concept phase is the accomplishment of key technical breakthroughs and the development of engineering prototype.
- The key work of initial prototype phase is the completion of thermal control test verification and product identification.
- The key work of formal prototype phase is to control the technical conditions of thermal control design and products.
- The key work of operation improvement phase is to demonstrate and verify the improved thermal control design.

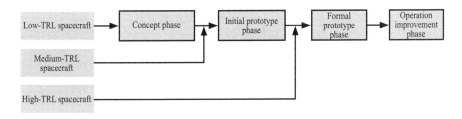

Fig. 3.1 Process of spacecraft thermal control design

3.5.1 Concept Phase

The concept phase of spacecraft thermal control design is defined as a development phase to verify the possibility of thermal control concept through a test, obtain necessary performance data and coordinate the interfaces.

The work in this phase mainly includes concept demonstration, simulation and verification, conceptual experimental verification, key technologies refining, key technology roadmap formulation, key technology breakthrough and concept phase summarization.

Thermal simulation analysis and key technology breakthrough are of the importance in the concept phase. Thermal simulation analysis is the main means of thermal control design and development in the concept phase. Key technology generally refers to the bottleneck technology that supports the completion of a spacecraft mission. Therefore, the completion of key technical breakthrough is a necessary condition for the entry into the next development phase. The technical development process of concept phase is shown in Fig. 3.2.

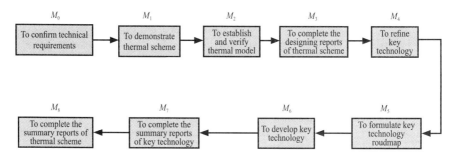

Fig. 3.2 Development process of concept phase

3.5.2 Initial Prototype Phase

The initial prototype phase of spacecraft thermal control design is defined as a development phase to verify the reasonability and correctness of thermal design and the coordination between systems, carry out various thermal simulation tests and assess the product performance indexes.

The tasks in the initial prototype phase mainly include the detailed thermal design, the final assembly design for thermal control, the development of initial prototype and qualification product, and the participation in an integrated spacecraft test, the thermal balance testing of initial prototype, the correction of thermal analysis model and the summarization of thermal control development.

The thermal balance testing of initial prototype is the focus of initial prototype phase. The thermal balance test of initial prototype, in which the test cases should at

3.5 Thermal Control Design Stages and Key Points

least cover the worst hot case, worst cold case and fault case, verifies the satisfaction of the thermal requirements comprehensively. Sometimes, some specific thermal testing conditions are set to amend thermal model or investigate the potential optimum modes of thermal control subsystem on orbit.

It should be noted that for a spacecraft with a certain degree of inheritance, the thermal control design and analysis can be carried out though simulation analysis and then the initial prototype phase can be directly entered. The correctness and conformity of thermal control design can be verified by the thermal balance test of formal prototype. For newly developed thermal control products, materials and coatings, their qualification should be completed in the initial prototype phase before entering the formal prototype phase. The technical process of initial prototype phase is shown in Fig. 3.3.

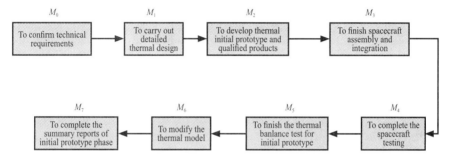

Fig. 3.3 Development process of initial prototype phase

3.5.3 Formal Prototype Phase

The formal prototype phase of spacecraft thermal control design is defined as a development phase where the improved thermal control subsystem or product can meet the overall technical requirements and can be available for space flight test or application.

The tasks in the formal prototype phase mainly include the detailed thermal control design, the final assembly design for thermal control, the development of formal prototype product, the participation in an integrated spacecraft test, the thermal balance test and thermal vacuum test of formal prototype, the summarization of formal prototype development, the prelaunch preparation, and the on-orbit flight control and assessment.

Among them, technical condition control and thermal balance testing are the focus of formal prototype phase. The thermal designers should take the final design condition of formal prototype as baseline, strictly control the technical state of thermal control subsystem and focus on the impact on the interfaces between thermal control subsystem and other subsystems. The focus of thermal balance test in this phase is on the completeness and adequacy of test items, the rationality and correctness of test methods, the accuracy of simulation test conditions and the conformity of test results.

3.5.4 Operation Improvement Phase

The operation improvement phase of spacecraft thermal control design is defined as a development phase where the spacecraft meets the predetermined thermal control performance requirements after passing the orbital flight test, and the thermal control design is improved after being qualified and reviewed. The focus of this phase is on the full demonstration and testing of the necessity and impact of the improved design.

The tasks in the operation improvement phase mainly include the improved thermal control design, the general final assembly design for thermal control, the development of approved thermal control product, the integrated testing, the thermal balance testing and thermal vacuum testing, the summarization of ex-works development, the prelaunch preparation and the on-orbit flight control.

It should be noted that for the spacecraft in series production, the thermal balance test for formal prototype can be carried out only on the first spacecraft. After the on-orbit flight verification of the first spacecraft, the formal thermal balance test for the subsequent spacecraft of the same kind whose technical status has no significant change can be canceled so that their thermal control design can be verified only by the modified thermal simulation analysis. Nevertheless, the technological consistency among thermal control product development, component assembly and final assembly should be under tougher control and can be indirectly verified by thermal vacuum testing.

References

1. General specification for thermal control system for spacecrafts, GJB 2703A-2006
2. X. Zhao, Study on change rule of β angle for various orbits in satellite thermal design. Spacecr. Eng. 17(3), 57–61
3. X. Ning, X. Zhao, C. Yang, Optimum design method of combined type radiator for inclined-orbit satellite. Spacecr. Eng. 21(5), 48–52
4. Modeling requirements for thermal analysis of spacecraft, Q/W 1626-2018
5. Guideline for Spacecraft Thermal Design, GJB 1029A

Chapter 4
Typical Thermal Control Technologies for Spacecraft

4.1 Overview

Common spacecraft thermal control technologies mainly include heat transfer technology, thermal insulation technology, heating technology, cooling technology, temperature measurement and thermal control technology, etc. Among them, heat transfer technology involves thermal conductive materials, heat pipes, thermal interface materials, thermal control coatings, fluid loops, convection ventilation devices, radiators, consumptive heat dissipating devices, phase change material devices, thermal switches, etc. Thermal insulation techniques include multi-layer insulation blanket, thermal insulation materials for conductive insulation, microporous materials for atmospheric environment insulation, etc. Heating techniques mainly include electric heating and isotope heating. Cooling techniques mainly include radiator, thermo-electrical cooler, cryogenic cooler and so on. Because some published books have presented the cooling technologies for space application, this book will not present them. Temperature measurement and control techniques mainly include the temperature measurement based on temperature sensors like a thermistor, and the temperature control based on temperature control unit and related control strategy.

In the spacecraft-oriented thermal design, the temperature control of onboard equipments, structural components and cabin air is usually realized by the comprehensive use of various thermal control technologies introduced in this chapter in accordance with the requirements for overall mission, flight orbit, spacecraft attitude, general spacecraft configuration, load heat dissipation and temperature.

This chapter will focus on introducing the functions and working principles, categories and characteristics as well as selection principles of each thermal control technology and finally illustrating the application of each thermal control technology in the form of application cases. It is expected that this chapter could provide guidance for spacecraft thermal control designers to select and use thermal control technologies.

4.2 Heat Transfer Technology

4.2.1 Introduction

Heat transfer technology is generally used in the places where heat dissipation is high. The heat generated by equipments and astronauts needs to be taken away to ensure that the temperature of the equipments and of the air inside the sealed cabin will not exceed the allowable upper limit. In the thermal design process of a spacecraft, heat transfer design is often adopted in the following situations:

(1) Additional enhanced heat transfer technology is necessary to intensify the heat dissipation capacity of some equipment when the equipment fails to achieve the expected temperature control goal by relying on its own radiating capacity. For example, heat transfer technology, i.e., heat conduction connection, could be used at the mounting interface of equipment by coating a thermal interface material (like thermal grease) over the interface. In addition, the equipment mounting plate shall also be made of a material with good thermal conductivity, such as aluminum alloy. High thermal conductive material and two-phase heat transfer technology (such as heat pipes) shall be used for heat expansion if necessary.

(2) When the instrument mounting plate is used as a fixed radiator, it is necessary to intensify heat transfer inside the mounting plate in most cases. For example, the temperature uniformity of the entire mounting plate could be realized by embedding heat pipes inside the plate or attaching the heat pipes to the plate surface, thus improving the radiating efficiency of the plate. And of course, the adoption and layout of the heat pipes depend on the needs of thermal design.

(3) Heat transfer technology shall be used when high-power equipment cannot be installed on the mounting plate which works as a radiator. For example, heat transfer techniques, such as heat pipe, loop heat pipe and fluid loop, are adopted by the high-power equipments inside the spacecraft cabin to collect and transfer waste heat to the radiator through which heat is dissipated to space. For the spacecraft with a sealed cabin structure, convection ventilation is also a choice to realize the thermal management inside the spacecraft and the temperature control of onboard equipments by carrying out heat transfer between different equipments inside the cabin, between equipment and heat dissipation surface, and between equipment and heat exchanger.

(4) The heat transfer technology with adjustment function is needed when the equipment has a narrow working temperature range in its intermittent operation or its temperature is greatly affected by the environmental heat flux alternation. For example, the phase change device could be used to restrain the influence caused by equipment heat loss or environmental heat flux fluctuation. Temperature control can also be realized by using the heat transfer techniques with regulating and controlling ability, such as variable-conductance heat pipe, heat switch, loop heat pipe and fluid loop.

(5) When the equipment is only used for one time or the spacecraft is on a short-term mission, the special heat transfer technology that supports the short-term mission could be adopted. For example, a consumption-type heat dissipation technology, such as sublimator or evaporator, could be used to reduce the demand for other resources like radiator, thus solving the heat dissipation problem in special cases when there is no radiator.

4.2.2 Thermal Conductive Materials

4.2.2.1 Functions and Working Principles

Thermal conductive materials are widely used in spacecraft thermal control to transfer heat rapidly and support the temperature control of relevant equipments. With the rapid development of aerospace technology, the electronic components have become lighter, thinner, shorter and smaller with constantly improved performance and significantly increased heat dissipation and power density. This makes it urgent to use the materials with high thermal conductivity to conquer the technical bottleneck, thus enabling high-efficiency materials with ultra-high thermal conductivity to be continuously used and verified in the spacecraft.

The functions of thermal conductive materials used in the spacecraft generally include two types: One is the heat expansion for the objects like high-power equipments, and the other is the point-to-point heat transfer. The working principles of the two typical applications are separately shown in Figs. 4.1 and 4.2.

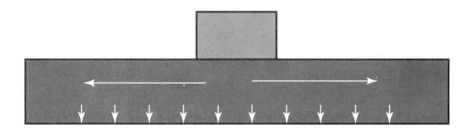

Fig. 4.1 Schematic diagram of heat expansion of thermal conductive materials

Fig. 4.2 Schematic diagram of point-to-point heat transfer of thermal conductive materials

4.2.2.2 Categories and Characteristics

The commonly used thermal conductive materials for spacecraft include two main categories, metallic materials and non-metallic materials. Metallic materials mainly include aluminum and aluminum alloy, copper and so on. Non-metallic materials mainly include aluminum-based silicon carbide, high thermal conductivity graphite, diamond, high thermal conductivity carbon fiber, aluminum-based diamond, etc. Among the above-mentioned materials, aluminum and aluminum alloy are traditional high-performance thermal conductive materials that are widely used in the spacecraft owing to their mature technology, low price, and good processability. With the development of material processing technology, ultra-high-performance thermal conductive materials, such as carbon fiber materials and diamonds, have developed rapidly. And they have been initially applied in the temperature control and thermal management system of spacecraft high-power high heat-flux equipment/device. These materials are preferred substitute materials for aluminum and copper due to their low coefficient of thermal expansion (CTE), ultra-high thermal conductivity and low density.

4.2.2.3 Selection Principles and Application Cases

1. **Selection principles**

Technically, the selection of thermal conductive materials depends on the heat transfer requirements and application constraints. The main considerations in the selection process are listed as follows:

(1) Whether the thermal conductivity of the material can meet the heat transfer requirements under the corresponding temperature conditions.
(2) Whether the density of the material can satisfy weight constraints.
(3) Dimension and accuracy requirements.
(4) Material processability.
(5) Expansion coefficient matching between a thermal conductive material and the controlled material.
(6) Whether the surface of the material will produce redundancies during its service life, such as shedding particles and falling powder.
(7) Whether the material has condensable volatile materials, and whether the total mass loss and the content of condensable volatile materials can meet the requirements.
(8) When bonding or welding is used, it is necessary to consider whether the surface characteristics of the thermal conductive material can meet the requirements and whether surface modification treatment is required.
(9) Whether the applications have the requirement for electrical insulation or conduction in terms of the electrical conductivity of the material.
(10) Whether the mechanical properties of the material, such as compressive strength, flexural strength, yield strength and elastic modulus, can meet the

requirements for installation stress and launching or landing mechanical environment.
(11) Whether the thermal conductivity, mechanical parameters, configuration and size of the material are stable under the service environment.

2. **Application cases**

Case 1: Application of aluminum and aluminum alloy as thermal conductive materials.

In general, aluminum and aluminum alloy components, e.g. aluminum honeycomb panels and cabinets in spacecraft, can be used as both the load-bearing structural material and the thermal functional material to realize the integrated design and use of structure and heat. In special cases, aluminum and aluminum alloy materials are only used as thermal functional materials, such as aluminum thermal doublers on the mounting surface of high-power equipment. For the aluminum honeycomb panels used for satellite structure, as shown in Fig. 4.3, their internal honeycomb structures and face panels are made of aluminum alloy. Figure 4.4 shows a typical electronic equipment cabinet whose frame structure is also made of aluminum alloy.

Case 2: Application of copper as thermal conductive material.

Copper has high thermal conductivity, favorable mechanical properties and good processability. In space missions, copper is mainly used in special parts or the places with special requirements due to its shortcoming of high density. Copper could be used as thermal conductive materials in the occasions where the heat power is high and the installation space is limited, such as heat dissipation of the antenna TR assembly array as shown in Fig. 4.5. When a flexible connection is required between the controlled object (such as the internal components of a camera that requires side swing) and the heat dissipation surface (such as the fixed heat dissipation surface for the camera), a copper guide rope can be used for flexible heat transfer. A typical copper guide rope for satellite is shown in Fig. 4.6.

Fig. 4.3 Aluminum honeycomb panels used for satellite structure

Fig. 4.4 Typical electronic equipment cabinet

Fig. 4.5 Copper strip used for heat dissipation of TR assembly array

Fig. 4.6 Typical copper guide rope for satellite

Case 3: Application of high thermal conductive graphite and high thermal conductive carbon fiber composites.

The application of high thermal conductive graphite and high thermal conductive carbon fiber composites can realize efficient heat transfer and structural weight reduction due to their small mass, high thermal conductivity and low thermal expansion coefficient. During application, the above-mentioned materials are prone to produce conductive redundancies, such as shedding particles, falling powder, etc. Therefore, their surface needs special protection treatments like metal encapsulation.

On a remote-sensing satellite made by China, three-line array mapping camera is the main load, and its imaging quality is the key to the success of the satellite flight mission. As a high-resolution space camera, its optical performance mainly depends on not only the design of optical system but also the temperature changes. The thermal deformation of components caused by the temperature level and temperature gradient is a non-negligible important factor that may affect the optical performance of the camera. The drastic changes of heat flux in the space environment where the camera is located have a great influence on the uniformity and stability of the camera temperature field. Although the MLI coated on the outer surface has good thermal insulation performance, the circumferential temperature difference of the lens barrel still exceeds the requirement. In order to solve this problem and improve the circumferential temperature uniformity of the barrel, a high thermal conductive graphite film (Fig. 4.7) shall be used to enhance the circumferential thermal conductivity of the camera barrel, thereby reducing the temperature difference and fluctuation of each part. In January 2012, this material was successfully applied to China's first 3D mapping satellite, achieving the maximum circumferential temperature difference lower than 0.2 °C and the radial temperature difference stability of <0.1 °C in the orbital period.

Case 4: Application of diamond materials.

Pure single-crystal diamond has ultra-high thermal conductivity and the maximum coefficient of 2,300 W/(m K), which is about 5 times that of copper and silver. It is an ideal thermal conductive material due to its extremely high hardness and good insulation. Diamond is generally used for thermal conduction or heat expansion in the areas where structural dimensions are limited. Before using it, the processing steps including cutting, grinding, surface modification (such as metallization, etc.) and so

Fig. 4.7 High thermal conductive graphite film dedicated to camera's structural temperature uniformity

forth are necessary for this material to meet the practical application requirements. In addition, diamond has a low expansion coefficient. The expansion coefficient of the single-crystal diamond is 1.1 ppm/K. When using diamond for bonding or welding, special attention needs to be paid to the expansion coefficient compatibility between single-crystal diamond and controlled object and the selection of bonding agents and welding materials so that diamond can adapt to the interfacial stress fatigue caused by temperature alternations.

Copper-based diamond radiator and laser diode diamond radiation fin are typical applications of diamond material.

4.2.3 Heat Pipe

4.2.3.1 Functions and Working Principles

The heat pipe, characterized by efficient heat transfer capacity, is the heat transfer device based on phase change process. It is mainly used for achieving the heat transfer between the heating device and the heat dissipation surface in the spacecraft, and realizing temperature uniformity in different parts of spacecraft or in the same component.

Heat pipe is a two-phase heat transfer device. The circulation of the working fluid is driven by the capillary force generated in the capillary structure [1]. The heat pipe is usually composed of a metal enclosure, a capillary wick and working fluid. The metal enclosure is a shell that encloses the capillary wick and the working fluid. The capillary wick is a structure that provides a capillary suction force and a flow passage for the backflow of liquid working fluid to transfer heat in the heat pipe.

1. **Axially grooved heat pipe**

According to the working temperature range, the axially grooved heat pipes usually include normal-temperature and low-temperature axially grooved heat pipes. Usually, the normal-temperature axially grooved heat pipe with ammonia as the working fluid is widely used in the spacecraft. Therefore, this section focuses on this type of heat pipe.

The axially grooved heat pipe mainly comprises a shell that is usually made of aluminum alloy or pure aluminum (6063 or 1060), a axial capillary channel (capillary wick) integrated with the shell structure, and working fluid. The capillary channels of axially grooved heat pipes are mainly shaped into rectangle, trapezoid or "Ω".

A complete heat transfer process in the axially grooved heat pipe is as follows. The evaporation section of the heat pipe is heated, and the liquid working fluid is heated and evaporated into gas. The gaseous working fluid flows to the condensation section and then condenses into liquid working fluid. The liquid working fluid flows back to the evaporation section under capillary pumping force. By repeating this cycle, efficient heat transfer can be achieved.

The working principle of a typical axially grooved heat pipe is shown in Fig. 4.8, and the cross section of this pipe is shown in Fig. 4.9.

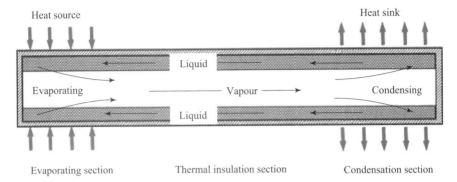

Fig. 4.8 Working principle diagram of a typical axially grooved heat pipe

2. **Loop heat pipe**

According to the working temperature range, the loop heat pipes are categorized into normal-temperature loop heat pipe and low-temperature/cryogenic loop heat pipe, but normal-temperature loop heat pipe is the most commonly used. The contents of this chapter are also limited to this pipe (hereinafter referred to as loop heat pipe).

The loop heat pipe is mainly composed of an evaporator, a condenser, an accumulator, a vapor line and a liquid line. The most prominent feature of the loop heat pipe is its localized capillary structural arrangement, that is, the capillary wick is only arranged in the heat absorption area of the evaporator. The pores of the capillary wick are micron-sized and can both increase the capillary force to get a larger heat transfer capability and effectively overcome the influence of gravity without negatively causing the increase in the pressure drop of the liquid backflow. The

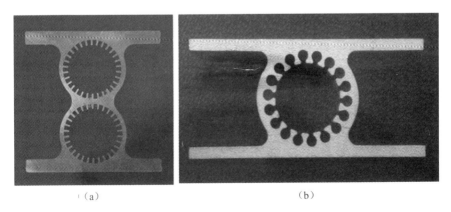

Fig. 4.9 Cross-sectional view of typical axially grooved heat pipe section. (a) rectangular axially grooved heat pipe; (b) "Ω"-shape axially grooved heat pipe

accumulator is connected to the evaporator to realize the discharging and suction of the liquid working fluid. The distribution of the vapor–liquid interface in the condenser shall be adjusted to adapt to the changes of the heat load and heat sink environment, and meet the requirements of temperature control. Compared with the conventional axially grooved heat pipe, the loop heat pipe has the characteristics of long heat transfer distance, large heat transfer capacity, flexible pipe arrangement, strong anti-gravity working ability and one-way heat transfer. Therefore, it is particularly suitable for carrying out long-distance heat transmission from the equipment (heat source) to the heat dissipation surface (heat sink), thereby achieving the heat dissipation and equipment temperature control.

A complete heat transfer process in the loop heat pipe is as follows. The liquid working fluid absorbs heat and evaporates on the outer surface of the capillary wick in the evaporator. The generated gaseous working fluid flows to the condenser through the vapor line, releases heat and condenses in the condenser, and becomes sub-cooled. The sub-cooled liquid working fluid flows through the liquid line and the accumulator into the interior of the capillary wick. The liquid working fluid flows through the capillary wick and finally reaches the surface of the capillary wick, where the liquid absorbs heat and evaporates again. Here, a heat transfer cycle is completed. The working principle of a loop heat pipe is shown in Fig. 4.10.

The main working modes of a loop heat pipe include:

1) Start-up mode [2]

After being started, the loop heat pipe operates and transfers the equipment heat to the heat dissipation surface, thus realizing thermal connection between the equipment and the heat dissipation surface. The auxiliary start-up heater separately installed on the evaporator is necessary for the start-up process, and the heat flux is generally no less than 1 W/cm^2. The smooth start-up of the loop heat pipe can also be accomplished by using the thermoelectric cooler (TEC) mounted on the accumulator to cool the accumulator. In order to speed up the start-up process, the auxiliary start-up heater

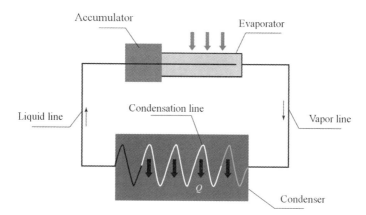

Fig. 4.10 Working principle of a loop heat pipe

4.2 Heat Transfer Technology 75

and the TEC can be simultaneously turned on. When the loop heat pipe is started, the auxiliary start-up heater can be turned off. If the sub-cooling degree of the return flow is sufficient, the TEC can also be turned off. The typical start-up process of the loop heat pipe is shown in Fig. 4.11.

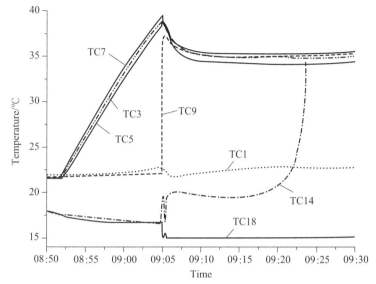

Fig. 4.11 Typical start-up curve of loop heat pipe

It can be seen from Fig. 4.11 that after the heat load is loaded on the evaporator of the loop heat pipe, the evaporator temperature (TC7) and the accumulator temperature (TC5) rise synchronously due to heat leak between the evaporator and the accumulator, but the evaporator outlet temperature (TC9) remains unchanged. In this case, no vapor exits from the outlet, which indicates that the loop heat pipe has not been started yet. After a while, when the temperature of the evaporator is higher than that of the accumulator to a certain extent, the working fluid in the loop heat pipe evaporator begins to evaporate and absorb heat. Then, the temperature of the evaporator and accumulator drops sharply, and the temperature of the evaporator outlet rises sharply, which means that the loop heat pipe is successfully started.

2) Cutoff mode

When the instrument is not working or is at its lower temperature limit, the heat transfer between the instrument and the heat dissipation surface can be cut off to realize thermal isolation between the instrument and the heat dissipation surface. The temperature of the accumulator could be increased above the temperature of the evaporator by turning off the TEC on the accumulator and simultaneously turning on the heater on the accumulator. The liquid working fluid is gradually discharged from the accumulator at this time. When it fills the condenser pipeline, vapor will not condense and release heat in the condenser and the loop heat pipe can stop operating gradually. The typical cutoff curve of a loop heat pipe is shown in Fig. 4.12.

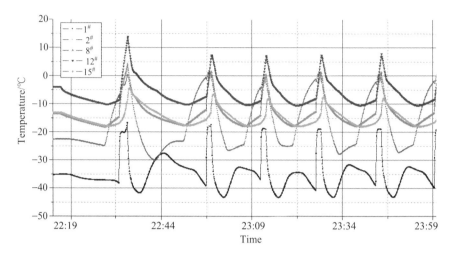

Fig. 4.12 Typical cutoff curve of loop heat pipe (see the color illustration)

It can be seen from the multi-blocking processes of heat pipe shown in Fig. 4.12 that the condenser temperature (1#) rapidly drops from around −20 °C to about −40 °C when the accumulator temperature (12#) is higher than the evaporator temperature (8#). This indicates that there is no significant heat transfer from the evaporator to the condenser, and the loop heat pipe is successfully cut off.

3. **Variable-conductance heat pipe (VCHP)**

VCHP is a kind of heat pipe that can basically keep the evaporator temperature constant by changing the size of the condensation area as the heat load changes [3]. The VCHPs can be mainly categorized into three types: the VCHP with cold gas storage chamber (which is controlled by heat sink temperature), the VCHP with hot gas storage chamber (which is controlled by heat source temperature) and the mechanical feedback VCHP [1]. This book will only introduce the VCHP with cold gas storage chamber. For other types of VCHP, please refer to professional heat pipe books.

The variable-conductance heat pipe with cold gas storage chamber (hereinafter referred to as VCHP) mainly comprises heat pipe body, gas storage chamber, working fluid, control gas, etc. The heat pipe body is generally an axially grooved heat pipe. The VCHP can be used for achieving the heat dissipation and temperature control of spaceborne equipments, and for manufacturing a radiator with variable radiating capacity. Different from the conventional heat pipe that can maintain good isothermality under working conditions, the VCHP is filled with the control gas that will not condense within the working temperature range of the heat pipe and can form partial pressure. Internally, the working VCHP can be divided into a working fluid section, a transition section and a control gas section. The controlled gas section can be regarded as an invalid section since it does not take part in the condensation heat exchange. The working process and principle of the VCHP are described below.

When the heat load of the heating section increases or the equipment temperature rises, the temperature of evaporation section of the VCHP will rise and the saturation pressure of the working fluid will go up accordingly to compress the control gas (according to the idea gas state equation). As a result, the transition section where the control gas and the working fluid vapor is mixed will move toward the cold end (control gas end), thereby increasing the effective heat transfer length of the VCHP. The coupling of VCHP and radiator can increase the radiation capacity, as shown in Fig. 4.13. Contrarily, when the heat load of heating section or temperature drops, the transition section will move to the hot end (working fluid end), thereby reducing the effective heat transfer length of the VCHP and the heat dissipation capacity of the radiator. With a reasonable design, the control gas section/condensation section of the VCHP can be completely covered by the radiator, thereby blocking the heat transfer between the radiator and the VCHP, as shown in Fig. 4.14. In theory, the heat dissipation capacity of the VCHP radiator can be adaptively adjusted from 0 to 100%.

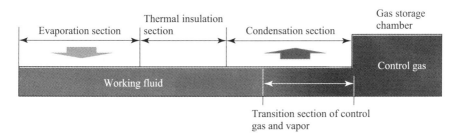

Fig. 4.13 Working principle of VCHP (1)

4. **Flexible heat pipe**

The flexible heat pipe is a constant-conductance heat pipe whose shell and capillary wick are flexible [1]. The flexible heat pipe mainly comprises a shell (including a

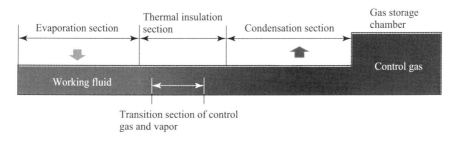

Fig. 4.14 Working principle of VCHP (2)

rigid section and a flexible section), a flexible capillary wick and working fluid. Its outstanding feature is the bending capacity that helps it swing or twist freely within a certain range, thereby realizing efficient flexible thermal connection between the heat source and the heat sink.

In the spacecraft engineering practice, flexible heat pipe is mainly used in the areas where there are relative vibrations or frequent movement between the cold end and the hot end, for example, in the camera that requires side swing and the sun-pointing rotating device. These devices require the flexible heat pipe to repeatedly swing or twist in a specified direction, which means that the heat pipe shall possess good fatigue resistance to work in the bending process.

The swing of the flexible heat pipe is shown in Figs. 4.15 and 4.16.

5. **PCM heat pipe of phase change material**

The phase change material (PCM) absorbs or releases a certain amount of latent phase change heat during the phase change process, without significant change of its temperature. The ammonia-containing axially grooved heat pipe could be combined with PCM according to their characteristics, i.e., the heat pipe with an additional PCM cavity is used to increase the heat capacity of heat pipe, thus dissipating the equipment waste heat and restricting the equipment temperature fluctuation. This type of heat pipe is referred as PCM heat pipe. A typical cross section of PCM heat pipe is shown in Fig. 4.17.

4.2.3.2 Categories and Characteristics

1. **Types of heat pipes**

The heat pipes commonly used in spacecraft are shown in Table 4.1, in which the recommended working fluids are for reference only and shall be selected as needed.

2. **Characteristics of heat pipe**

Typical characteristic indexes of the heat pipe mainly include heat transfer capacity, heat transfer temperature difference, critical heat flux density (including radial heat flux density and axial heat flux density, and generally referring to radial heat flux density unless otherwise stated), safety, reliability, service life, etc. Specific indexes vary with different specifications. For spacecraft-specific thermal design, it is obvious that special attention shall be paid to the characteristics related to heat transfer, such as heat transfer capacity, heat transfer temperature difference and critical heat flux.

1) Heat transfer capacity

As an efficient heat transfer device, the heat pipe has its limit of heat transfer capacity. The heat transfer capacity limits of different heat pipes vary with different working temperature conditions. The heat transfer capacity limits can be generally categorized into sound velocity-induced heat transfer limit, viscosity-induced heat transfer limit,

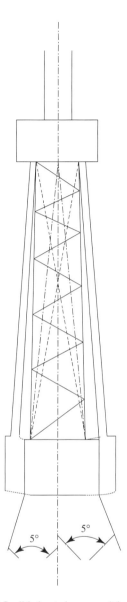

Fig. 4.15 ±5° reciprocal swing of flexible heat pipe around the center

carrying-induced heat transfer limit, capillary force-induced heat transfer limit and boiling-induced heat transfer limit. The ammonia-containing axially grooved heat pipe, whose heat transfer capacity limit is the capillary force-induced heat transfer limit, is commonly used in spacecraft. Circulating flow of the working fluid in heat pipe is driven by the capillary force (also known as capillary head) applied by the capillary wick. The capillary force is limited, because it depends on the surface

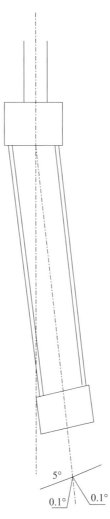

Fig. 4.16 ±0.1° reciprocal swing of flexible heat pipe centering on the 5° position

tension of the liquid working fluid, the contact angle and the size of the capillary groove. When the circulating velocity of the working fluid reaches a certain value, the heat transfer amount of the heat pipe also reaches a certain value. When the pressure drop of the circulating gaseous and liquid working fluid in the heat pipe is balanced with the maximum capillary force, the current heat transfer capacity will be the capillary force-induced heat transfer limit. The heat pipe featuring the capillary force-induced heat transfer limit has a temperature-dependent parabolic heat transfer limit curve (also called maximum heat transfer capacity), which is related to the change of physical properties, such as latent heat, surface tension, viscosity, density,

4.2 Heat Transfer Technology

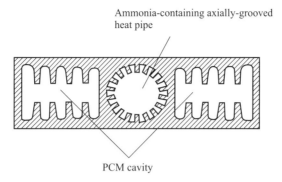

Fig. 4.17 Typical cross section of PCM heat pipe

etc. The maximum heat transfer capacity of ammonia-containing axially grooved heat pipe that changes with temperature is shown in Fig. 4.18.

The capillary structure dimension (groove width) of the axially grooved heat pipe is generally about several hundred micrometers. When ammonia is used as the heat transfer fluid, the capillary force provided by the capillary structure within the normal temperature range is generally in hectopascal order of magnitude. Although its anti-gravity performance is poor, it is very suitable for space microgravity environment. Figure 4.19 shows the variation of the maximum heat transfer capacity of a typical ammonia-containing axially grooved heat pipe when anti-gravity height gradually changes from 0 to 9 mm. It can be seen from the figure that the maximum heat transfer capacity shows a typical linear decline with the increase of anti-gravity height. The anti-gravity height refers to the height difference between the head of the condensation section and that of the evaporation section.

2) Heat transfer temperature difference

The ammonia-containing axially grooved heat pipe is generally considered to have a constant equivalent heat conductivity coefficient within the common intravehicular temperature range. Therefore, this type of heat pipe is sometimes called the constant-conductance heat pipe. In the thermal calculation, a fixed value is often taken regardless of the influence of temperature, and the heat transfer temperature difference of heat pipe is linearly related to the heat load. Since the heat pipe has high heat transfer efficiency in evaporation and condensation, its heat transfer temperature difference is small provided that the heat transfer capacity is within the limit. That is why the heat pipe has been widely used as a high-performance heat transfer technology.

The loop heat pipe has two working states, i.e., variable thermal conductance and constant thermal conductance. When a two-phase loop heat pipe accumulator is used, the location of vapor/liquid interface inside the condenser could be adjusted by increasing or decreasing the accumulator temperature so that the condensation heat resistance of the loop heat pipe can be changed (i.e., the thermal conductance is variable). In this case, the loop heat pipe can realize more accurate temperature control

Table 4.1 Types and characteristics of heat pipes commonly used in spacecraft

Name of heat pipe	Working fluid	Recommended working temperature range /°C	Remarks
Axially grooved heat pipe	Ammonia	−60 to 80	Suitable for microgravity environment, and having poor anti-gravity capacity
	Ethane	−125 to 0	
Loop heat pipe	Ammonia	−60 to 80	Suitable for long-distance heat transfer in microgravity environment, and having strong anti-gravity capacity and the ability to control the evaporator temperature
	Ethane	−125 to 0	
	Propylene	−125 to 45	
	Nitrogen	−200 to −160	Suitable for long-distance heat transfer, and having strong anti-gravity capacity. The internal working fluid is in a supercritical state at normal temperature. A special supercritical start-up process is required before use
VCHP	Ammonia/non-condensing control gas	−60 to 80	Suitable for microgravity environment, and having poor anti-gravity capacity (when using axial groove). Variable thermal conductivity can be designed based on the variations of environmental temperature and heat load power. Non-condensing control gas is generally nitrogen, helium, neon, etc.
Flexible heat pipe	Ammonia	−60 to 80	Suitable for microgravity environment, and having certain anti-gravity capacity. The fatigue resistance of the flexible section (including internal capillary core) generally needs to be verified in special tests

(continued)

Table 4.1 (continued)

Name of heat pipe	Working fluid	Recommended working temperature range /°C	Remarks
PCM heat pipe	Ammonia/phase change material	−60 to 80	Suitable for microgravity environment, and having poor anti-gravity capacity. The heat pipe is to meet the heat transfer demand, and the phase change material is to meet the demand for heat storage and release

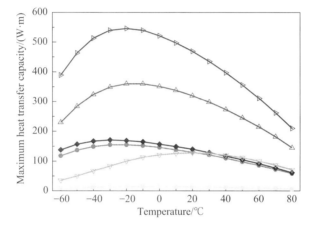

Fig. 4.18 Heat transfer capacity of different ammonia-containing axially grooved heat pipes

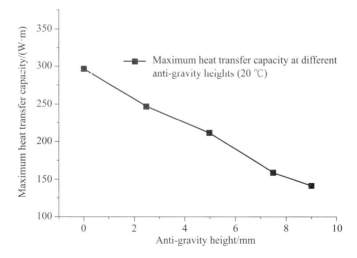

Fig. 4.19 Variation of maximum heat transfer capacity with anti-gravity height

under the condition of variable thermal conductivity, which generally corresponds to the condition of small heat load. With the increase of heat load or the cooling of accumulator, the accumulator in the loop heat pipe is gradually filled with liquid working fluid. Here, the position of vapor/liquid interface can no longer move toward the outlet of condenser (ignoring the cooling and contraction effect of the liquid working fluid), indicating that the loop heat pipe enters a constant-conductance status. The operating temperature of loop heat pipe will increase linearly with the heat load (regardless of the heating and expansion effect of the liquid working fluid). Typical temperature characteristics of the ammonia loop heat pipe under variable-conductance and constant-conductance conditions are shown in Fig. 4.20. No matter in the variable or constant-conductance mode, the heat transfer temperature difference is small. However, the increase of the heat flux on evaporator installation interface will enlarge the heat transfer temperature difference of the evaporator itself (including the saddle).

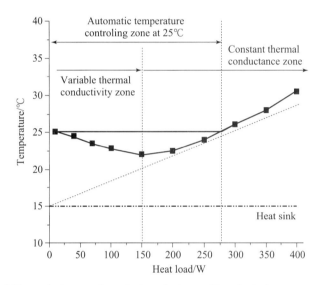

Fig. 4.20 Variable conductance and constant conductance of loop heat pipe

3) Critical heat flux

Another key characteristic of a heat pipe is critical heat flux, which refers to the heat flux along the radial direction of the heat pipe. When the radial heat flux loaded on the heat pipe reaches a higher level, the surface evaporation will no longer dominate the heat exchange of liquid working fluid in the capillary groove of heat pipe. Bubbles are generated in the liquid working fluid due to the increase of superheat, and in turn cause boiling and destroy the original working state. This is mainly manifested by the large temperature fluctuation or even a rapid rise of temperature. The radial heat flux in this case is called critical heat flux of the heat pipe.

4.2 Heat Transfer Technology

4.2.3.3 Selection Principles and Application Cases

1. **Selection principles**

Heat pipe shall be selected according to the characteristics of the heat pipe itself and the different service constraints. The common principles are as follows:

(1) Model selection: the heat pipe that has passed flight verification is preferred.
(2) Operating temperature range: the operating temperature range of heat pipe is determined by the nature of the working fluid, and shall cover all the service conditions. For the heat pipe that uses ammonia as working fluid, the operating temperature is generally −60 °C to 60 °C and the upper limit in special cases could be 80 °C.
(3) Heat transfer capacity and ultimate heat flux: The selected heat pipe shall meet the requirements of heat transfer capacity and heat flux under the service conditions, and the necessary engineering allowance shall be reserved.
(4) Manufacturability: The external structure of heat pipe products shall be designed separately according to the service constraints, and the technique reliability and stability shall be considered during its selection.
(5) Environmental adaptability: The heat pipe should meet the spacecraft requirements of mechanical stress, thermal cycling stress and other environmental stresses. The pressure inside the heat pipe is directly related to temperature, and the environmental stress caused by thermal cycling is distinctive.
(6) Backup: A proper backup shall be provided during application. The backup strategy shall be determined according to the specific thermal design and reliability design of the thermal control system.
(7) Service life: The service life shall meet the requirements for ground storage and on-orbit application in the total life cycle.

Different types of heat pipes have their particularities. The following individualized principles are recommended for heat pipe selection.

The following individualized principles shall be considered in the selection of ammonia-containing axially grooved heat pipe:

(1) Application scope: It is used for achieving the structural isothermality of instrument mounting plate and radiator and the heat transfer between instruments and their radiating surfaces.
(2) Installation interface and shape: The cross-sectional shape and size of the heat pipe that passes flight verification are fixed, and its length and shape could be customized as needed. The requirements for the flatness of heat pipe mounting surface and the straightness of heat pipe shall be put forward based on the technical requirement. During the design of non-straight heat pipe, it is necessary to consider whether the heat pipe shape could meet the requirements of plane bending or space bending. In general, heat pipes cannot be twisted for processing. If necessary, the testing verification of safety and heat transfer capacity shall be carried out for heat pipes, preferably pure aluminum heat pipes.

(3) Heat pipe installation: The installation modes such as pressurization connection, cementing, preembedding or insertion are adopted according to different applications.
(4) Heat transfer capacity: The heat transfer capacity of heat pipe is not only directly related to its type, but also related to temperature, pipe length, whether or not to bend, the number of bends and the mechanical environment in which the pipe is located. Attention should be paid to the aforesaid during selection. The heat transfer capacity of heat pipe is related to its operating temperature. Their relationship is a parabolic curve, with the peak value of heat transfer capacity in the temperature range of 0–30 °C. The heat transfer capacity is inversely proportional to the pipe length. The heat transfer capacity of a bending heat pipe has a certain degree of attenuation and needs to be tested and verified if necessary. The acceleration generated by spacecraft spin and orbital change will affect the heat transfer capacity of heat pipe. It is good for heat transfer if the acceleration direction is the same as the heat transfer direction of heat pipe; otherwise, the heat transfer capacity of heat pipe will be attenuated and finally lost. In gravitational fields (such as those of the Earth and moon), the heat pipe could work in horizontal mode or with assistance of gravity. In special circumstances, it could work under certain restriction of anti-gravity height, but its heat transfer capacity will be attenuated. The heat transfer capacity shall be derated during selection.
(5) Heat flux: The restriction of heat flux limit shall be considered. Since the heat flux limit of commonly used ammonia-containing axially grooved heat pipes is less than 4 W/cm^2, heat flux shall be derated during selection.
(6) Isothermality: The isothermality index under test conditions shall be determined in accordance with the operating requirements or relevant standards.
(7) Liquid slug effect: When the operating temperature of heat pipe is high, part of the head of its condensation section (under microgravity or in horizontal posture) will be blocked by liquid working fluid so that it will not take part in heat exchange and can be handled as an invalid section.

The following individualized principles shall be considered during the selection of ammonia-containing loop heat pipe:

(1) Application scope: It is used for the heat dissipation of a high-power load, the operation of a deployable radiator and the radiating surface isothermality through the coupling with ammonia-containing axially grooved heat pipe network.
(2) Installation interface and its shape: The evaporator length could be customized according to the size of thermal interface, the pipeline length could be determined according to heat transfer distance and layout, and the condenser could be designed and customized individually as needed. The flatness requirements of installation surfaces of loop heat pipe evaporator and condenser shall be proposed according to technical requirement. In general, vapor line and liquid line are made of annealed stainless steel, which could be flexibly bent according to the actual layout during application.

4.2 Heat Transfer Technology

(3) Heat transfer capacity: The heat transfer capacity of loop heat pipe is not only directly related to evaporator, but also related to temperature, pipe length, whether the line is bent and the number of bends, and the mechanical environment in which the pipe is located. However, the heat transfer capacity of loop heat pipe is less sensitive to pipe length and bending conditions than that of the ammonia-containing axially grooved heat pipe. The loop heat pipe could work against gravity on ground, i.e., the evaporator is located above the condenser. But, the increase of anti-gravity height will attenuate the heat transfer capacity. The specific degree of attenuation shall be determined by analysis and calculation or test according to the actual situation. The heat transfer capacity shall be derated during selection.

(4) Heat flux: The restriction of heat flux limit shall be considered. Generally, the heat flux limit of loop heat pipe is less than 8 W/cm^2, and the heat flux shall be derated during selection.

(5) Thermal resistance: The proper requirements for thermal resistance shall be put forward according to the thermal design requirements and the index characteristics of loop heat pipe.

(6) Temperature control characteristics: The loop heat pipe could control temperature, as a thermoelectric cooler and a thin-film heater are mounted on the loop heat pipe accumulator for temperature control. The accuracy of temperature control is generally better than ±2 °C.

(7) Adaptability to heat load: In general, loop heat pipes are not suitable for low-power applications. The specific power scope is related to the specific design of loop heat pipe and is generally determined through testing. Since the start-up operation of loop heat pipe takes time, the continuous working time for intermittent load thermal control shall be no shorter than 30 min. The heater can be used for heating compensation to maintain the continuity of heat load of the loop heat pipe if necessary.

The following individualized principles shall be considered during the selection of variable-conductance heat pipe (VCHP):

(1) Application scope: It applies to the cases where the heat dissipation of equipment or the external heat flux of radiating surface changes periodically.

(2) Selection principles: In general, ammonia-containing axially grooved heat pipes could be chosen as VCHP heat pipes. The gas storage chamber could be designed separately according to specific situation. The influence factors include the variation range of heat transfer capacity, the temperature change range of radiating surface, the operating temperature range, the requirements for temperature control, the installation interface location of gas storage chamber, the length of heat pipe condensation section and so forth.

(3) Control gas: The selection of control gas depends on the lowest temperature on the cold end of the VCHP and the corresponding pressure inside the heat pipe, i.e., the control gas is required not to condense at this moment.

(4) Installation interface and its shape: The length and shape of VCHP could be customized as needed. Attention should be paid to the particular requirements for installation interface of gas storage chamber.
(5) Thermal interface of gas storage chamber: Gas storage chamber shall work at proper temperature to better realize temperature control. Its thermal design and temperature control could be designed separately if necessary.
(6) Heat transfer capacity: Generally, it is slightly lower than that of ammonia-containing axially grooved heat pipe.
(7) Heat flux: It is generally as same as the heat flux of ammonia-containing axially grooved heat pipe.

The following individualized principles shall be considered during the selection of flexible heat pipe:

(1) Application scope: Flexible heat pipe could be used to solve the matching problem found in the installation of rigid heat pipe, or to realize the heat transfer between two parts in relative motion or vibration.
(2) Installation interface and its shape: The length of the rigid section (including evaporator and condenser) could be customized according to the size of heat interface, while that of the flexible section could be determined according to the requirements for heat transfer distance, layout and necessary bending, swing and vibration.
(3) Heat transfer capacity: Heat transfer capacity is not only directly related to heat pipe design, but also related to temperature and mechanical environment. Attention should be paid to the aforesaid during selection. The anti-gravity capability of flexible heat pipe is stronger than that of ammonia-containing axially grooved heat pipe, but is limited to some extent. The heat transfer capacity shall be derated during application.
(4) Heat flux: The restriction of heat flux limit of heat pipe shall be considered. Since the heat flux limit is less than 4 W/cm^2, the heat flux shall be derated during selection.
(5) Fatigue characteristics of flexible section: The flexible section is limited by the minimum allowable bending radius, the bending angle and the maximum allowable bending times, which shall be determined according to the actual situation. In general, the flexible section shall be tested for bending fatigue.

The following individualized principles shall be considered during the selection of PCM (phase change material) heat pipe:

(1) Application scope: It applies to the cases that require certain heat transfer capacity, higher temperature control precision, intermittent equipment operation or larger environmental heat flux fluctuation on the radiating surface.
(2) Installation interface and its shape: The length, shape and installation interface of the heat pipe shall be customized as needed.
(3) Selection of heat transfer and phase change energy storage: The heat pipe shall follow the principles of heat pipe selection, and the heat storage and dissipation of PCM shall follow the principles of PCM device selection.

4.2 Heat Transfer Technology

(4) Selection of PCM: Whether or not the latent heat of PCM meets requirements could be verified according to the load heat dissipation and its change, the environmental heat flux change on the radiating surface, and the temperature control precision. Generally, the point of solid–liquid transformation shall be in the range of temperature control threshold of the controlled object.

(5) Temperature scope: The temperature scope of this heat pipe is as same as that of ammonia-containing axially grooved heat pipe. The actual working point of PCM device is near the solid–liquid transformation point of PCM. The supercooling of solid PCM and the overheating of liquid PCM are allowed. In case of supercooling or overheating, not only the temperature constraints of ammonia heat pipe but also the safety issues caused by PCM expansion shall be considered. For example, when the heat pipe needs to be embedded and bonded, the influence of bonding solidification temperature of honeycomb panel on the safety must be considered.

2. **Application cases**

(1) Application cases of ammonia-containing axially grooved heat pipe

Ammonia-containing axially grooved heat pipes are widely applied in spacecraft-specific thermal control. They could be used for the thermal connection inside equipment, the heat transfer between equipment and radiating surface, the isothermality of equipment mounting plate and the isothermality between radiating surfaces in different locations of the satellite.

Case 1: Application in the thermal connection inside equipment.

Generally, when the heat pipes are used for thermal connection inside equipment, small ammonia-containing axially grooved heat pipes would be selected. An application case of CCD camera of a satellite is shown in Fig. 4.21. A small heat pipe is used to connect both ends of a copper plate in order to reduce the temperature differences between different CCD parts.

Case 2: Application in the heat transfer between equipment and radiating surface.

When the equipment cannot be directly coupled with and mounted on the radiating surface, the ammonia-containing axially grooved heat pipe could be adopted to achieve thermal connection between the equipment and the radiating surface. An application case of using ammonia-containing axially grooved heat pipe to achieve thermal connection between star sensor and radiator in a satellite is shown in Fig. 4.22. Heat inside the star sensor is transferred to the radiator through two heat pipes. By combining this measure with other thermal control measures, the temperature control of the star sensor is realized.

Case 3: Application in the isothermy of equipment mounting plate.

The majority of equipments and instruments inside the satellite are installed on the mounting plate (honeycomb panel). To achieve unified temperature control of

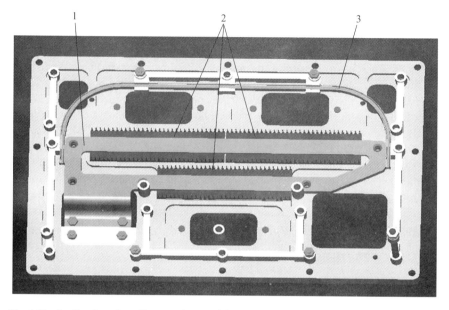

Fig. 4.21 Application of small ammonia-containing axially grooved heat pipe in thermal connection inside equipment. 1—Heat-conducting copper plate; 2—CCD part; 3—Small ammonia-containing axially grooved heat pipe

Fig. 4.22 Application of ammonia-containing axially grooved heat pipe in heat transfer between equipment and radiator. 1—Radiator; 2—Star sensor; 3—Heat pipe

4.2 Heat Transfer Technology

the equipments and improve the radiation capacity of the mounting plate (when the mounting plate is also used as a radiating surface), ammonia-containing axially grooved heat pipes are often adopted to achieve the isothermy of the honeycomb panel.

A schematic diagram of a typical honeycomb panel with isothermy design is shown in Fig. 4.23. Some heat pipes are embedded inside the honeycomb panel to achieve thermal control integration of heat pipes and honeycomb panel structure, while other heat pipes are externally bonded to the honeycomb panel and are orthogonal to the embedded heat pipes to form an orthogonal heat pipe network.

Case 4: Application in thermal coupling on different radiating surfaces.

When the imbalance of environmental heat flux and heat load between different radiating surfaces is obvious, ammonia-containing axially grooved heat pipes with large heat transfer capacity could be adopted to connect different radiators and to achieve heat transfer and control between different radiating surfaces. This measure could increase overall heat dissipation capacity of the satellite and reduce the heating compensation power of the low-temperature surface. An application case of ammonia-containing axially grooved heat pipes in Chang'e 1 is shown in Fig. 4.24. In this case, the ammonia-containing axially grooved heat pipes are adopted to connect two radiators in the $\pm Y$ directions of the satellite and to thermally couple with the embedded heat pipes inside radiators, thus improving overall heat dissipation capacity of the satellite and reducing the heating compensation power of the low-temperature surface.

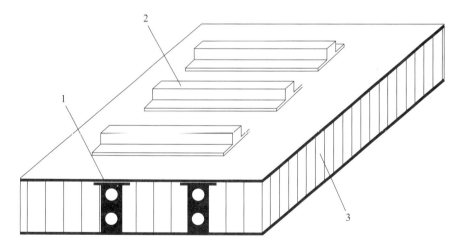

Fig. 4.23 Application of ammonia-containing axially grooved heat pipe in the isothermy of equipment mounting plate. 1—Embedded heat pipe; 2—External bonded heat pipe; 3—Honeycomb panel

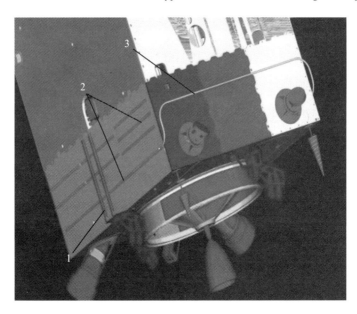

Fig. 4.24 Application of ammonia-containing axially grooved heat pipes in thermal coupling connection between different radiating surfaces of satellite. 1—External bonded heat pipe; 2—Embedded heat pipe; 3—Coupling heat pipe for south and north plates

2) Application cases of loop heat pipe

Case 1: Application of loop heat pipe in payload heat dissipation and highaccuracy temperature control.

On November 3, 2019, China's first sub-meter resolution stereo mapping satellite GF-7 with laser altimeter as one of its key payloads was successfully launched. The semiconductor laser in the laser altimeter is a device with high power density and is very sensitive to the change of operating temperature. The weak current caused by temperature change will lead to a great change of output energy and the change of device parameters (such as lasing wavelength and noise performance). Therefore, the laser temperature needs to be controlled at 20 °C ± 1.5 °C. On the other hand, the heating components of the laser altimeter body (mainly including laser, back light path electronics and front electronics) are all concentrated in the back light path area, so that the peak heat flux per cubic meter of the back light path components reaches 525.6 W/m^3. In order to ensure the temperature level of the laser and other key components and reduce the influence of the temperature of the light path components on the optical-mechanical structure, a capillary pumped fluid loop system based on loop heat pipe is designed as shown in Fig. 4.25. The working fluid in loop heat pipe is ammonia.

The working principle of the system is as follows. The heat generated by the front electronics, the back light path electronics and the laser is transferred to the capillary pump and heat collectors 1–3 through the grooved heat pipes. The loop heat pipe is

4.2 Heat Transfer Technology

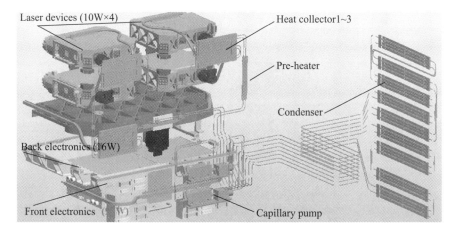

Fig. 4.25 Schematic diagram of loop heat pipe layout for payload heat dissipation and high-accuracy temperature control

driven by the heat dissipated by the front electronics and the compensation heater. The vapor condenses into a liquid state or a vapor-liquid two-phase state after the heat is dissipated by the radiating surface. The working fluid is heated and saturated by a preheater and then sequentially flows through the heat collectors 1–3 to absorb the heat of the laser and the back light path electronics. The vapor–liquid two-phase working fluid is condensed into supercooled liquid after dissipating heat on the heat radiating surface again and finally returns to the capillary pump. The operating temperature level of loop heat pipe is controlled by a temperature-controlled heater installed on the outer surface of the liquid accumulator.

The on-orbit data shows that two-phase temperature of the loop heat pipe is controlled within 19 °C ± 1 °C so that the laser temperature is controlled within 19.8–20.4 °C, as shown in Fig. 4.26.

Case 2: Application of loop heat pipe in reentry spacecraft.

In the mission of Chang'e 5 lunar high-speed reentry spacecraft, the laser IMU and the external wall of reentry spacecraft are coupled with two ammonia-containing loop heat pipes, as shown in Fig. 4.27, in order to solve the contradiction between the high-power heat dissipation and low-power heat insulation of Inertial Measurement Unit (IMU) at different stages and the high-temperature IMU insulation during high-speed returning, The inner wall of the reentry spacecraft is completely covered with MLI, so the loop heat pipe becomes the only heat transfer channel connecting internal equipment and external radiating surface. Through reversible switching between on-orbit temperature control mode and cutoff mode, the loop heat pipe could adapt to the change of on-orbit IMU heat dissipation from 70 W in operation to 0 W in non-operation. The temperature control range of IMU is from −10°C to 45 °C.

On October 24, 2014, the Chang'e 5 lunar high-speed reentry spacecraft was launched from Xichang Satellite Launch Center successfully. On November 1, it

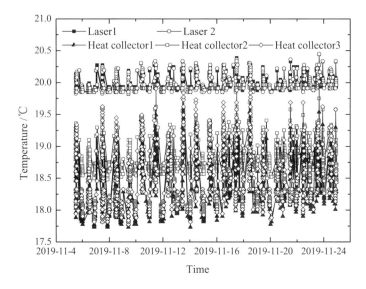

Fig. 4.26 Typical temperature of the laser altimeter on orbit

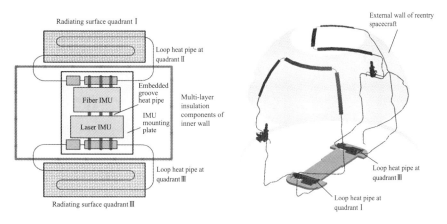

Fig. 4.27 Schematic diagram of layout of heat dissipation surface and loop heat pipe in reentry spacecraft

returned to Siziwangqi in Inner Mongolia after experiencing the Earth–moon transfer stage, lunar near-side turning stage, moon–Earth transfer stage, returning stage and landing stage. According to the on-orbit data, when two IMUs were shut down, their temperature was 3.7–6.0 °C; during operation, the temperature of optical fiber IMU was stabilized at 3.7–8.6 °C, and the temperature of laser IMU was stabilized at 10.8–14.5 °C. During the entire flight, the operating characteristics of loop heat pipes are consistent with what are expected. The loop heat pipes and the on-orbit flight temperature data of the IMUs are shown in Figs. 4.28, 4.29 and 4.30.

4.2 Heat Transfer Technology

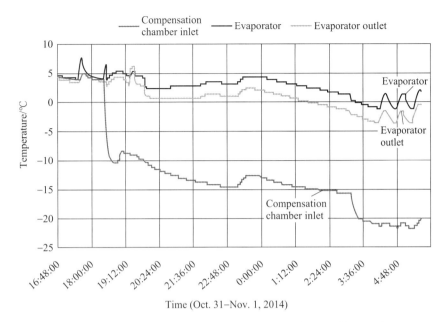

Fig. 4.28 Typical temperature of loop heat pipe at quadrant I during on-orbit operation

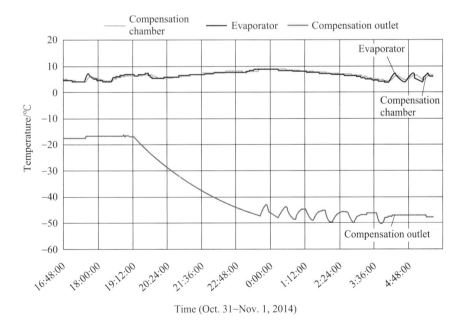

Fig. 4.29 Typical temperature of loop heat pipe at quadrant III during on-orbit cutoff

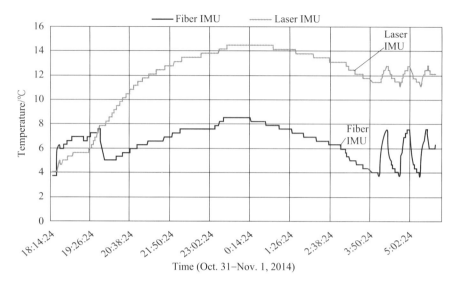

Fig. 4.30 Typical temperature change of IMU during on-orbit calibration

3) Application cases of VCHP

VCHP is a kind of heat pipe that could adapt the size of condensation section to the changing heat load and keep the temperature of evaporation section basically constant. This type of heat pipe has been widely used in temperature control as the main body of variable-radiation radiators and as the heat transfer element of the components, devices and equipments with larger heat load variation and higher temperature control requirements. Consequently, VCHP has a unique status and function in the field of spaceborne thermal control, especially in deep space detection and the thermal control design in which thermal power is difficult to determine accurately.

In the mission of Chang'e 3, a VCHP-based variable-conductance radiator technology is adopted to adaptively control the heat dissipation and leakage of the lander, as shown in Fig. 4.31. In the VCHP, the heat-transferring fluid is ammonia and the control gas is helium. During lunar day, VCHP could transfer the equipment heat to radiator efficiently to realize heat dissipation and temperature control. When lunar night comes, the gradual decreasing of equipment temperature inside the cabin would gradually reduce the effective heat transfer length of the condensation section that couples VCHP with the radiating panel. When the effective heat transfer length is completely cut off, the heat dissipation channel between equipment and radiating surface will be blocked, leaving only the convection of internal gas and the conduction by the heat pipe shell to realize heat leakage control at lunar night. The layout of VCHPs in the $+Y$ cabin of the lander and on the $+Y$ radiator is shown in Fig. 4.32. Furthermore, some ammonia-containing constant-conductance heat pipes are added to the radiator and are thermally coupled with the VCHPs so as to improve the

4.2 Heat Transfer Technology

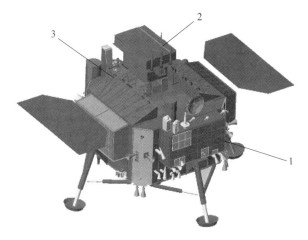

Fig. 4.31 Layout of VCHPs in lander. 1—Lander; 2—Lunar rover; 3—VCHP

Fig. 4.32 Layout of VCHPs at $+Y$ cabin and $+Y$ radiator. 1— $+Y$ cabin; 2— $+Y$ radiator; 3—VCHP; 4—Heat pipe; 5—Equipment

temperature uniformity and heat dissipation capacity of the radiator during lunar day.

In the Chang'e 3 lander, temperature measuring points are set up in the gas storage chamber, the middle part of condensation section and evaporation section of one of the VCHPs to monitor on-orbit VCHP temperature. The data of four typical stages are selected for analysis. The temperature distributions of the VCHPs at the four stages (i.e., Earth–moon transfer stage, circumlunar stage, lander–rover separation stage and typical lunar day) are shown in Figs. 4.33, 4.34, 4.35 and 4.36.

It is shown in Fig. 4.33 that VCHP could operate steadily after start-up, since environmental heat flux and equipment heat dissipation at the Earth–moon transfer stage are more stable. Temperature of the middle part of condensation section is close to that of the evaporation section, but is much different from the temperature

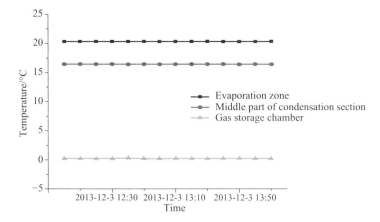

Fig. 4.33 Temperature distribution of VCHP at Earth–moon transfer stage

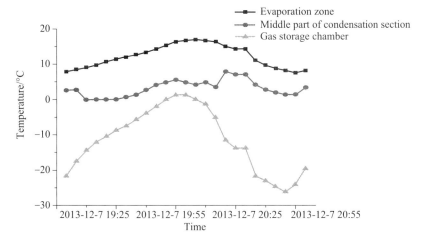

Fig. 4.34 Temperature distribution of VCHP at circumlunar stage

of gas storage chamber. This indicates that the middle part of condensation section has not been blocked by the control gas yet. Ammonia is used as the working fluid in the middle part of condensation section to dissipate heat effectively. As shown in Fig. 4.34, the temperature of the middle part of condensation section in the lunar shadow is very close to that of evaporation section, and the temperature of gas storage chamber varies greatly with heat flux changes. However, the temperature of the middle part of condensation section does not change with the temperature of gas storage chamber, indicating that the middle part of condensation section has not been blocked by the control gas yet. Ammonia is used as the working fluid in the middle part of condensation section to dissipate heat effectively. Earth–moon transfer stage and circumlunar stage are under microgravity, and the working condition of VCHP is consistent with its design condition at the two stages. It is shown in Fig. 4.35 that

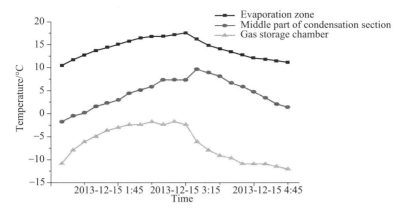

Fig. 4.35 Temperature distribution of VCHP at lander–rover separation stage

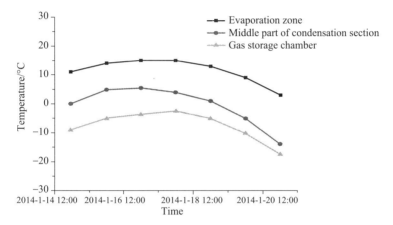

Fig. 4.36 Temperature distribution of VCHP on typical lunar day

when the temperature of VCHP evaporation section is lower during the separation of lander and rover on lunar day, the temperature of the middle part of condensation section is close to that of the gas storage chamber, indicating that this area is occupied by control gas. But with the increase of temperature of the evaporation section, the temperature of the middle part of condensation section gradually gets closer to that of the evaporation section, which means that the control gas in the middle of condensation section is gradually replaced by ammonia and the heat dissipation capability is obtained. As illustrated in Fig. 4.36, the temperature difference between the middle part of condensation section and the evaporation section is the smallest around lunar noon, manifesting that the adaptive control of heat dissipation capacity is maximized at this stage. This temperature difference increases significantly as the morning and night draw near, i.e., the effective radiation area decreases automatically due to the influence of control gas. This means that the adaptive control of heat dissipation of the VCHP has been realized. During the lunar day that the lander went through, adaptive temperature control and heat dissipation allocation of the variable-conductance radiator were achieved.

By May 2017, the lander had successfully spent 40 lunar days and lunar nights, which indicates that the variable-conductance radiator has achieved desirable adaptive control.

4) Application cases of flexible heat pipe

Case 1: Application of flexible heat pipe in flexible installation.

To improve the heat dissipation of one gyroscope line box on some satellite and to adapt to flexible installation, China developed a flexible heat pipe that used ammonia as working fluid, as shown in Fig. 4.37. Stainless steel bellow is used as the flexible section, and stainless steel meshwork is adopted as the capillary wick. The heat transfer capacity of this flexible heat pipe at 30 °C is about 30 W, and the total heat resistance is 0.17 K/W.

Case 2: Application of flexible heat pipe in the heat dissipation of swinging camera.

In 2006, China developed a flexible heat pipe for the visible light camera on Tiangong-1 to meet the requirement of camera focus movement, as shown in Fig. 4.38. The operating temperature range of the flexible heat pipe is from −20 °C to 60 °C, and its heat transfer capacity at 20 °C is no less than 20 W. With the swinging frequency lower than 40 times/min, the flexible heat pipe underwent 25,000 swinging cycles at ±5° pendulum angle and 20,000 swinging cycles at ±0.1° pendulum angle and at 5° off the center.

5) Application case of PCM (phase change material) heat pipe

The PCM heat pipe formed by coupling n-dodecane with ammonia-containing axially grooved heat pipe was adopted first on the Chang'e 1 in 2007 to control the heat dissipation and temperature fluctuation of CCD 3D camera. The on-orbit remote data shows that heat dissipation and temperature fluctuation suppression has been realized by using PCM heat pipes. The PCM heat pipe is shown in Fig. 4.39.

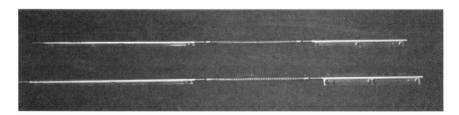

Fig. 4.37 Ammonia-containing flexible heat pipe

4.2 Heat Transfer Technology 101

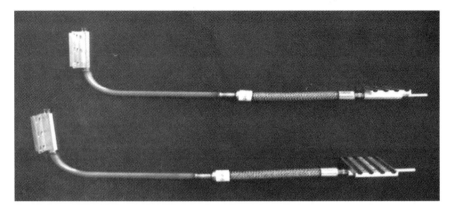

Fig. 4.38 Application of ammonia-containing flexible heat pipe in visible light camera on Tiangong-1

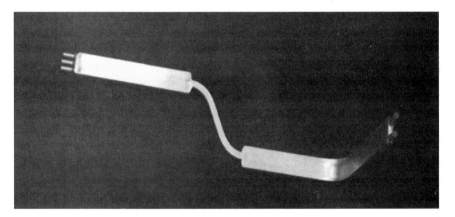

Fig. 4.39 PCM heat pipe for Chang'e 1

4.2.4 Thermal Interface Fillers

4.2.4.1 Functions and Working Principles

Thermal interface fillers are widely applied in electronics and components to intensify the heat exchange between solid interfaces. The working principle of their application is to improve the contact between two solid walls, increase effective contact area and effectively reduce thermal contact resistance on interface by using thermal interface fillers, thus intensifying the heat exchange between interfaces.

Thermal contact resistance is caused by incomplete contact which is produced by microcosmic unevenness of two contact surfaces or worse macroscopic flatness index. When two surfaces contact with each other under certain pressure, the valid contact area is a function of the surface condition and physical property, surface flatness, roughness and contact pressure of the contact material. The thermal contact resistance is formed because the actual contact area of two contact surfaces is only a tiny part of the apparent contact area of the two surfaces, as shown in Fig. 4.40.

The value of thermal contact resistance is related to many factors, such as the material, connection mode, surface condition, contact pressure and temperature of the contact surfaces. The thermal contact resistance is an uncertainty in spacecraft-specific thermal analysis.

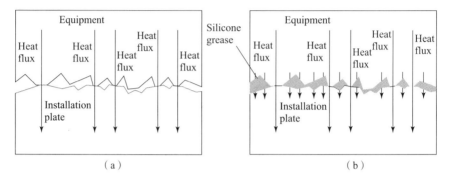

Fig. 4.40 Typical microcosmic view of solid surface contact. (a) Dry contact between equipment and mounting plate; (b) application of heat-conducting silicone grease between equipment and mounting plate

4.2.4.2 Categories and Characteristics

Thermal interface fillers mainly include thermal greases, thermal adhesives and thermal pads. Among them, the thermal grease is usually applied to the spacecraft, because it will not solidify after being coated and is convenient for equipment redisassembly. In contrast, thermal adhesive will be solidified after being coated, and is inconvenient for equipment redisassembly. The silicone rubbers solidified at normal temperature are usually used on the spacecraft. It is convenient to use thermal pads to disassemble the equipment again, so silicone rubber pads and indium foil are usually used in the spacecraft. These fillers, except for indium foil, are non-metallic.

As an enhancement method of heat transfer on interfaces, thermal interface fillers have the following basic characteristics:

(1) Thermal interface fillers are easy to deform after being pressurized to ensure good filling effect on the solid–solid installation interfaces.
(2) Thermal conductivity or contact heat transfer coefficient is the key index that is much different among different categories of the thermal interface fillers.
(3) Different thermal interface fillers have different applicable temperature ranges. For example, the upper temperature limit of thermal grease is generally not more than 80 °C, while that of silicone rubber pad could be higher than 200 °C.

(4) Mass loss percent is a feature of non-metallic conductive filler, and different thermal interface fillers have different mass loss percent.

4.2.4.3 Selection Principles and Application Cases

As the enhancement means of interface heat transfer, thermal interface fillers, such as thermal silicone grease, silicone rubber, silicone rubber pad and indium foil, have played an important role in the spacecraft engineering practice. Among them, thermal grease and silicone rubber pad are widely used in the spacecraft assembling.

1. **Selection principles**

The selection of thermal interface fillers shall observe the following basic principles:

(1) The filler shall meet the requirements of equipment thermal design. Whether thermal interface filler is used to intensify the heat transfer on the equipment mounting surface or not or which kind of thermal filler is adopted depends on equipment temperature requirement, heat dissipation index, installation location, surface status, installation condition and other factors. There is no unique criterion for it, and the final status shall be determined by thermal analysis. In other words, the installation state and categories of thermal interface fillers could be determined according to the design criterion, i.e., the analyzed temperature level plus design margin falls into the range of temperature index.
(2) For the equipment based on thermal convection dissipation and the high infrared emissivity equipment with a small ratio of heat dissipation to surface area, the dry contact without thermal interface fillers can be considered as the installation state.
(3) In the case that the ratio of heat dissipation to surface area is large or the heat flux on the installation interface is higher than $1\ 000\ W/m^2$, thermal interface fillers shall be used on the installation interface to intensify heat transfer.
(4) Because thermal grease, silicone rubber, silicone rubber pad and indium foil have different contact heat transfer coefficients, they shall be chosen according to the requirement of thermal design.
(5) Since optical equipment, high-power microwave switch, OSR and other devices are sensitive to the pollution caused by the condensable volatile matter of thermal interface fillers, they shall be screened according to the characteristics of the application locations.
(6) Indium foil has a low melting point and is relatively fragile and conductive. Particular preventive measures shall be taken during application.
(7) For the equipment that needs to be disassembled and assembled for many times, it is recommended to select thermal grease, silicone rubber pad, and indium foil, etc.

2. **Application case**

Case 1: Thermal grease.

On a satellite, the heat dissipation of the power controller is 110 W and the heat flux on the installation surface is 0.12 W/cm^2. RKTL-DRZ-1 thermal grease is coated over the installation surface to intensify the heat exchange between the equipment and the installation surface on structural panel, as shown in Fig. 4.41. According to the temperature telemetry data during on-orbit flight, the operating temperature range of the power controller in a year is 20–32 °C, which meets the requirement of not exceeding 45 °C. The on-orbit operation of the power controller is in good condition.

Case 2: Silicone rubber pad.

To meet the on-orbit maintenance and replacement requirement of a certain spacecraft, silicone rubber pads are used to intensify the heat exchange between the charging–discharging regulator and the cold plate, as shown in Fig. 4.42. The heat dissipation of this equipment is 125 W, and the heat flux on the installation surface is 0.1 W/cm^2. According to the data of the heat balance test, the operating temperature of the charging–discharging regulator is no more than 23 °C, 2.5 °C different from the temperature of the cold plate, indicating that the equipment temperature meets the requirement of not exceeding 40 °C.

Fig. 4.41 Thermal grease coated on power controller

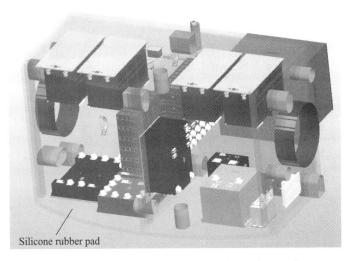

Fig. 4.42 Installation of silicone rubber pads on charging–discharging regulator

Case 3: Silicone rubber.

During the thermal control design of payload bay of Shijian-13 satellite, eight U-shaped ammonia-containing axially grooved heat pipes in four groups are adopted to achieve thermal coupling of $\pm Y$ decks. Heat transfer capacity of one single heat pipe is 80 W, and heat flux is 0.3 W/cm^2. To intensify heat exchange between the U-shaped ammonia axially grooved heat pipe and $\pm Y$ decks, GD414 silicone rubber is applied on the installation interface, as shown in Fig. 4.43. According to the telemetry data of on-orbit flight temperature, the temperature difference between the south plate and the north plate in the payload bay is 5–10 °C.

4.2.5 Thermal Control Coating

4.2.5.1 Functions and Working Principles

In daily life, people always use the surface colors to regulate the temperature of an object. For example, it is cooler to wear white clothes in hot summer than dark clothes; while in winter, dark clothes could absorb more heat from sunlight. When parking in summer, a piece of aluminized light barrier placed behind the windshield could reflect most of sunlight, thus reducing the temperature inside the compartment. The surface of refrigerated truck is generally coated with white or silver gray paint. The surface of solar water heater is generally in black with high absorptivity. The selection of these colors is actually to realize temperature control by using the thermo-optical performance of material surfaces.

Fig. 4.43 Installation of ammonia-containing axially grooved heat pipe using silicone rubber

In the field of spaceflight, these materials are called thermal control coatings. The thermo-optical characteristic of the surface of thermal control coating is known as heat radiation performance. The heat exchange between spacecraft and external space is basically carried out in the form of radiation due to the ultra-high vacuum (UHV) of cosmic space. Therefore, thermal control coating is the key to the temperature control of spacecraft surface.

For a spacecraft in vacuum environment, radiation heat transfer is the main form of heat exchange between its inside and outside. However, the surface thermophysical properties of onboard structural components or equipment play a decisive role in radiation heat exchange. Consequently, special regulations on the surface materials or thermophysical properties of relevant structural components or equipment are formulated in the thermal design of spacecraft. These surface materials are called thermal control coatings.

For an unmanned spacecraft, almost all visible external surfaces are thermal control coatings. Actually, all products absorb and dissipate energy by radiation. The function of thermal control design is making the temperature of equipment and structures balanced as needed by choosing the coatings reasonably and organizing the route, process and capacity of heat exchange between the internal and external surfaces of the spacecraft.

Regarding an isothermal object on the orbit of the Earth, if only direct solar radiation is taken into consideration, and the solar albedo and infrared radiation of other planets (including the Earth, the moon, etc.) on this object as well as the influence of inner heat source of this object are ignored, the heat flux of sunlight absorbed by this object will be equal to the heat radiated by it:

$$\alpha_s A_s E_s = \sigma \varepsilon A_\varepsilon T^4 \tag{4.1}$$

The surface temperature T of the object is

$$T = [(\alpha_s A_s E_s)/(\sigma \varepsilon A_\varepsilon)]^{1/4} \tag{4.2}$$

where α_s refers to the solar absorption ratio of the surface of the object; E_s refers to the heat flux that the sun projects on the object surface, in W/m^2; A_s refers to the surface area of the object vertically exposed to sunlight, in m^2; A_ε refers to the radiation surface area of the object, in m^2; σ is a Stefan–Boltzmann constant which is 5.67×10^{-8} W/(m^2 K^4); ε refers to the infrared emissivity on the object surface (the infrared emissivity mentioned in this book refers to hemispherical emissivity unless otherwise specified); T refers to the thermodynamics temperature, in K.

The influence of environmental factors on coating performance shall be taken into consideration when selecting the thermal control coating. According to different application locations, thermal control coatings fall into two categories: internal coating and the external coating. Since there is no sunlight inside the spacecraft and the heat is exchanged in the form of infrared radiation, only infrared emissivity should be considered. The external coating will be exposed to the thermal radiation of the sun, the Earth and even other stars at the same time. On the one hand, it could be irradiated by the sun directly or indirectly; on the other hand, it may be exposed to infrared radiation of the Earth or other stars. In the end, thermal balance could be realized under the combined action of direct sunlight, the reflection of sunlight from the Earth, the infrared radiation of the Earth as well as the infrared radiation of the spacecraft itself. Solar absorption ratio and infrared emissivity of the coatings play a decisive role in temperature balancing. In general, the coating characteristics are described by the ratio of solar absorption ratio to infrared emissivity, namely α_s/ε.

4.2.5.2 Categories and Characteristics

Thermal control coatings are usually classified according to material technology or thermo-optical property (usually characterized by α_s/ε, which is also known as heat radiation performance). The two most important performance parameters of the thermal control coating are solar absorption ratio α_s and infrared emissivity ε. In other words, all materials have their own solar absorption ratio and infrared emissivity, so their functions shall not be limited to thermal control only. The key is how to make use of the characteristics of these materials. Therefore, it is more systemic and objective to classify thermal control coatings according to material technology.

1. **Thermal control coating for metal matrix material**

Using the surface status of metal matrix material to regulate its heat radiation performance is the simplest way to use thermal control coating. For the same kind of metal, its thermal radiation performance may vary greatly under different surface conditions. In practical application, a technique with stable thermal radiation performance shall be chosen. There are usually two treatment techniques: polishing and sandblast.

1) Polished metal surface

The metal surface that is polished mechanically, chemically or electrically would reflect sunlight strongly. Its solar absorption ratio is 0.10–0.20, but its infrared emissivity is lower. Therefore, its α_s/ε is very high. Because the polished metal surface is very stable and sensitive to pollution, it shall be protected carefully before application.

2) Sandblasted metal surface

The sandblasted surface is rough, so it has higher solar absorption ratio and infrared emissivity. For example, the solar absorption ratio of sandblasted stainless steel surface is 0.58, and its infrared emissivity is 0.38, which is 2–3 times higher than that of the polished surface. Since sandblasted metal surface is rough and wearproof, its stability is better than the polished metal surface. However, during the process of sandblast, the metal parts, especially thin metal parts, will deform, which limits the application of this technique.

2. **Electrochemical thermal control coating**

1) Anodic oxidation coating

For most of the metals, when they are exposed to air, a layer of oxidation film would form on their surface. This layer of oxidation film is transparent to visible light, but it could absorb infrared ray better. Therefore, the oxidation film increases the infrared emissivity on the metal surface without changing the solar absorption ratio of the metal substrate. As the oxidation film becomes thicker, its infrared absorption tends to increase. Therefore, this characteristic could be used to adjust the α_s/ε of metal surface.

Natural oxidation film is often loose and unstable. A special anodizing technique needs to be adopted to get more compact and stable oxidation films.

Anodizing is a process of electrolytic oxidation in which a layer of oxidation film could develop on the surface of metal or alloy. This layer of oxidation film is protective or decorative, even with other functional features.

In addition to the common anodizing technique, there is also a "coloring" anodizing technique, i.e., changing the oxidation film into different colors by adopting the coloring technique and preparing it into the coating with high solar absorption ratio and different emissivities.

The natural color anodizing of aluminum alloy, the black anodizing of aluminum alloy, the microarc oxidation of titanium alloy and other anodizing techniques are commonly used in the aerospace field.

2) Electroplating coating

Electrodeposited coating is a common preparation technique of metal coating, namely plating a layer of metal film on the material with a specialized technique. According to the characteristics of different coating metals, the surface physical properties of matrix materials could be changed to get different solar absorption ratios and infrared emissivities. Electrogilding, black nickel electroplating and other electroplating techniques are commonly applied in the spaceflight field.

3. **Metal coatings formed by evaporation and sputtering**

Various metal coatings made of aluminum, silver, gold, copper, chromium, cadmium, platinum, rhodium and other metals could be produced through the processes such as vacuum evaporation–deposition (evaporation), magnetron sputtering and so forth. It is easy for some metals to react with oxygen, such as aluminum, silver and copper, so these electroplating processes are usually completed in high vacuum.

Base materials could be various metals or non-metals. Heat radiation properties of the formed metal coatings are similar to those of the polished metal surfaces. The coatings shall also be protected from pollution during application.

4. **Paint-type thermal coating**

Paint-type thermal coating is the most widely used thermal control coating and is composed of base stock and pigment. By using different pigments and mixing ratios, a series of thermal control coatings with different heat radiation performances could be produced. Generally, there are organic varnish coating and inorganic varnish coating.

1) Organic varnish coating

Organic varnish coating is widely used in various metal or non-metal base materials. The construction technique of organic varnish, either spraying or brushing, is simple. The cured coating has good adhesion and cleaning performance. With the addition of different pigments, a series of organic varnishes in different colors such as white paint, gray paint, black paint, celadon paint and organic metal paint could be produced. The variation range of α_s/ε is 0.2–3.3.

2) Inorganic varnish

Inorganic varnish is composed of inorganic base stock and inorganic pigment. Similar to the organic varnish, a series of inorganic varnishes in various colors or with various heat radiation performances could be produced by using different mixing ratios.

5. **Secondary surface mirror-type thermal control coating**

Different from the common surface coating, the heat radiation performance of secondary surface mirror-type thermal control coating depends on the characteristics of two surfaces, one of which is a transparent surface that has strong absorptivity toward visible light and infrared ray, and the other of which is a metal base that has strong reflection toward visible light. When visible light passes through transparent surface and reaches the non-transparent metal base, the solar absorption ratio α_s depends on the metal base and the infrared emissivity ε depends on the transparent surface that has strong absorptivity toward infrared ray. The bright aluminum anodized coating as mentioned aforesaid is a type of secondary surface mirror-type thermal control coating actually. The surface oxidation film is just the transparent finish, and the unoxidized polished aluminum base is just the metal base.

The secondary surface mirrors commonly used in the aerospace field are classified into glass type, plastic film type, paint type and evaporation–deposition type.

1) Glass-type secondary surface mirror

Glass-type secondary surface mirror is a secondary surface mirror-type thermal control coating that uses glass material as transparent finish. In general, the glass material is quartz glass with high luminousness, but it is very brittle and easy to break. To increase the tenacity of glass, the quartz glass mixed with "cerium" has been developed to greatly improve the maneuverability. Increasing the glass thickness would enlarge the infrared emissivity of the secondary surface mirror. For example, when quartz glass is 2 μm thick, the infrared emissivity of the coating is about 0.6; when it is 50 μm thick, the infrared emissivity of the coating could be 0.8.

The metal base is usually a layer of thin metal film that is formed on one side of the glass through evaporation or sputtering. Aluminum and silver are commonly used and their solar absorption ratio is less than 0.1 in general.

Glass-type secondary surface mirror has excellent resistance to ultraviolet radiation and space particle radiation, namely it has good space stability. Therefore, it is widely used in the aerospace field, especially as the heat radiation coating exposed to space.

2) Plastic film-type secondary surface mirror

Plastic film-type secondary surface mirror is a secondary surface mirror-type thermal control coating that uses plastic film as transparent finish. Different finishes shall be chosen according to different needs, usually including F46 film, Kapton film and so forth. Generally, the transparency of plastic film is not as good as that of quartz glass. Therefore, the solar absorption ratio of plastic film-type secondary surface mirror is higher than that of glass-type secondary surface mirror, and its infrared emissivity varies with the finish material.

Generally speaking, the space environment stability of plastic film-type secondary surface mirror is worse than that of the glass type. When it is applied in the aerospace field, the influence of heat radiation degeneration of the material at the end of its service life on the overall thermal control design shall be taken into consideration.

3) Paint-type secondary surface mirror

A paint-type secondary surface mirror could be produced by directly spraying a layer of transparent polymer on the metal base. As a secondary procedure like pasting is unnecessary, this type of secondary surface mirror specially fits irregular curves with complicated shapes. However, ordinary polymer has worse radiation resistance, which needs to be considered during application.

4) Evaporation–deposition-type secondary surface mirror

For the surfaces with complex shapes, evaporation–deposition-type secondary surface mirror could be directly prepared on the metal base by using the evaporation–deposition medium film technique.

Both aluminum–silicon dioxide and aluminum–aluminum oxide are evaporation-deposition-type secondary surface mirrors that are easy to obtain and have better adhesiveness and UV resistance.

6. **Thermal control tape**

Thermal control tape is a kind of carrier for different coatings instead of a coating with special performance.

Generally, the thermal control tape is composed of the base film (such as aluminum foil, polyester film, Kapton film, F46 film and so forth) and the coating (such as the aluminum, silver and gold coatings formed through vacuum evaporation–deposition, the sprayed paint film and so forth). To be implemented conveniently, PSA (such as rubber, acrylic acid resin adhesive, silicone resin adhesive, etc.) is usually composited on the back of the thermal control tape. Some thermal control tapes also directly use tinsel as a coating and have PSA composited on the back. In this case, heat radiation is the performance of the tinsel.

Owing to multiple choices of coating performance, simple and flexible implementation and convenient application, a series of thermal control tape products have been developed, mainly including gilded thermal control tape, one-side aluminized polyester film thermal control tape, one-side germanium-plated Kapton thermal control tape, Kapton aluminized thermal control tape, aluminum-based tape, copper-based tape, etc.

4.2.5.3 Selection Principles and Application Cases

1. **Selection principles**

1) Thermal and temperature control

During the comparison and selection of thermal control coating, whether the heat radiation performance of thermal control coating could meet the requirements of thermal control design shall be taken into consideration. The higher the coating's solar absorption ratio is, the more solar energy it would absorb. The higher the infrared emissivity is, the more heat radiation there will be. The ratio of solar absorption ratio to infrared emissivity could affect surface temperature under sunlight directly. Increasing this ratio could raise the surface temperature.

2) Radiation resistance performance

For the spacecraft operating on GEO at an altitude of more than 10,000 km from the Earth, the space ionization effect cannot be ignored. The effects of UV radiation and space charged particles on the coating shall be considered emphatically and the coating with the corresponding protection capacity shall be chosen.

Regarding the selection of thermal control coating, the influence of application environment characteristics on the coating shall be considered as well. Different environments will lead to the changes in the heat radiation performance and physical performance of the coating, such as the changes of heat radiation parameters, coating density or adhesive force. Under long-term UV radiation and the radiation of space charged particles, the coating may become brittle and cracked and even fall off, and its solar absorption ratio may increase. Practice has proved that the space adaptability of glass-type secondary surface mirror is the best, followed by plastic film-type secondary surface mirror, anodic oxidation, inorganic paint, etc. The vacuum evaporation–deposition-type secondary surface mirror coating also has good radiation resistance, but the performance of organic paint and paint-type secondary surface mirror is ordinary. During selection, the application environmental characteristics of the coating shall be taken into consideration.

3) Anti-atomic oxygen

For the spacecraft operating on an orbit lower than 1,000 km, especially for the spacecraft with operational orbit lower than 600 km (generally referred to as LEO spacecraft), the erosion and oxidation of atomic oxygen on the surface of thermal control coating shall be taken into consideration. The materials with anti-atomic-oxygen protective coating shall be chosen.

4) Anti-static performance

In order to avoid EMI, discharging, dielectric breakdown and other bad influences caused by the accumulation of electrostatic charges on the product surface, the thermal control coating is sometimes required to be anti-static. For example, the secondary surface mirrors with ITO conductive coating, including glass-type secondary surface mirror, F46 secondary surface mirror and Kapton film, could be adopted.

ITO coating is a widely used transparent conductive coating, which is usually deposited on glass-type/thin-film-type secondary surface mirror to increase conductivity without affecting optical performance significantly.

What's more, the heat radiation coating on antenna reflector is usually anti-static ACR-1 white paint or germanium-plated Kapton film, which can play an anti-static role while transmitting waves.

5) Coating degeneration

When choosing the external coating for spacecraft, it is necessary to consider its ability to adapt to space environment and the degeneration of its thermal parameters in space environment, apart from its thermo-optical parameters. The solar absorption ratio of the thermal control coating is different in the initial, final and even middle stages.

6) Outgassing rate

Organic coatings would leak gas in vacuum. If the outgassing product recondenses on the precision optical surface, it will bring many problems such as the degeneration of optical surface performance and the failure of electric contact component. In order to avoid the pollution caused by material outgassing, the outgassing rate of spacecraft materials is strictly regulated as follows: The total weight loss in vacuum is less than 1% and the condensable volatile is less than 0.1%.

2. **Application cases**

Case 1: Application of high-emissivity thermal control coating.

The universe is in a high vacuum state and heat is transferred mainly by conduction and radiation. The heat dissipations of various equipment inside the spacecraft cabin are different, and even very different. Intensifying the heat radiation exchange between equipment could effectively balance their temperature. Hence, the inner surface of cabin deck and the surface of equipment shells shall be high-emissivity surfaces with ε no less than 0.85.

The deck is an aluminum honeycomb panel whose surface is aluminum alloy sheet. If coatings are not used, the infrared emissivity ε will be about 0.05, which is too low to meet the requirement of intensifying the radiation heat transfer. Hence, organic white paint shall be sprayed over the surface of aluminum alloy. The equipment inside the cabin are also covered with high-emissivity thermal control coating such as black paint or black anodized coating on aluminum alloy surface. A typical application case of high-emissivity thermal control coating is shown in Fig. 4.44.

Case 2: Application of low-emissivity thermal control coating.

Except for intensifying the radiation heat exchange between equipment, some equipment needs to reduce the heat exchange with the outside. For example, the nickel–cadmium battery, unlike common electronics, cannot be affected too much by other equipment since it shall operate in a relatively stable environment. So the heat exchange between nickel–cadmium battery and other equipment shall be reduced, for example, by placing the nickel–cadmium battery in an independent compartment. The inner surface of the compartment could be made of polished aluminum alloy or pasted with a layer of one-side aluminized polyester film thermal control belt to keep the infrared emissivity ε less than or equal to 0.1.

Case 3: Application of heat radiation coating.

The heat dissipation coating is characterized by low solar absorption ratio and high infrared emissivity, i.e., low ratio of solar absorption ratio to infrared emissivity. When it is used in the places where the sun shines, it can reduce the absorption of

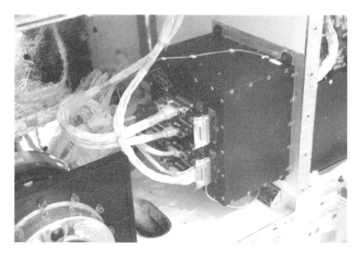

Fig. 4.44 Spray organic white paint over inner surface of deck, and black paint over outer surface of equipment

sunlight energy and increase its outward radiation capacity. The common heat radiation coatings mainly include SR107-ZK white paint, glass-type secondary surface mirror, F46 film secondary surface mirror, etc. A typical application case of heat dissipation coating is shown in Fig. 4.45.

Case 4: Application of special thermal control coating.

The design of thermal control coating shall not only consider thermal control functions, but also consider other special functions of the products. For example, the parabolic receiving surface of a data transmission antenna is usually coated with white paint for heat radiation. However, since the antenna has high requirements for anti-static property, its coating shall have special anti-static function in addition to meeting the requirements for the ratio of solar absorption ratio to infrared emissivity.

Fig. 4.45 Typical application case of heat radiation coating. (a) External surface of deck is a cerium glass silvered secondary surface mirror; (b) External surface of rotary joint is SR107-ZK white paint

Hence, the coating shall be anti-static ACR-1 white paint instead of common white paint, as shown in Fig. 4.46.

Some antennas need thermal control coating with high emissivity and also require high wave permeability, so their coating shall be made of the material with wave-transparent effect, such as Kapton germanium-plated film. While conducting thermal analysis for this thermal control status, not only solar absorption ratio and infrared emissivity but also solar transmittance shall be considered.

Fig. 4.46 Anti-static ACR-1 white paint on surface of antenna

Case 5: Combined use of thermal control coatings.

Thermal control coating in one area may not be a single coating. For example, rigid glass-type secondary surface mirror (such as quartz glass silvered secondary surface mirror, quartz glass aluminized secondary surface mirror) has low rigidity and high brittleness, so its installation base must be flat. When the flexible thin-film-type secondary surface mirror (such as F46 film silvered/aluminized secondary surface mirror) is used as heat dissipation coating, it is usually applied on a curve or irregular surface as an auxiliary material for glass-type secondary surface mirror, or used for partial repair of heat radiation surface in the final assembly stage of the spacecraft.

Paint-type thermal coatings in different colors have different solar absorption ratios. A comprehensive solar absorption ratio could be obtained to improve the coating flexibility by combining the coatings in different colors.

4.2.6 Fluid Loop

4.2.6.1 Functions and Working Principles

Fluid loop is an important technical means for spacecraft thermal control. It is used to collect and transfer the waste heat inside spacecraft and finally dissipate it to the outer space through radiator to realize the comprehensive heat management of the spacecraft.

1. **Mechanically pumped single-phase loop (SPL)**

Mechanically pumped single-phase fluid loop mainly includes mechanical pump and its controller, bypass regulation temperature control valve, accumulator, filter, pipeline, cold plate, etc. The mechanical pump drives single-phase working fluid to circulate in the loop, and the cold working fluid exchanges heat with the heating equipment through cold plate. Then, the temperature of the working fluid increases, and the sensible heat is taken away by the working fluid. After releasing and dissipating the accumulated heat through radiator or intermediate heat sink/cold plate, the hot working fluid reverts to cold working fluid and flows back into the mechanical pump, forming a heat exchange circulation. In the loop, bypass regulation temperature control valve is used for regulating the flow of hot working fluid that flows through the radiator or the middle heat sink to realize systemic temperature control. With the change of heat load and external heat environment, the temperature of working fluid changes as well, which leads to contraction and expansion of the working fluid. Accumulator is used to compensate or recover the working liquid to the fluid loop in accordance with the above conditions, and also compensate slight leakage of the working fluid. Besides, the back-pressure in accumulator could keep certain static pressure inside the fluid loop and ensure that the NPSH (net positive suction head) of the working fluid meets the engineering requirements before flowing into the mechanical pump. The principle of a typical mechanically pumped single-phase fluid loop is shown in Fig. 4.47.

2. **Mechanically pumped two-phase fluid loop**

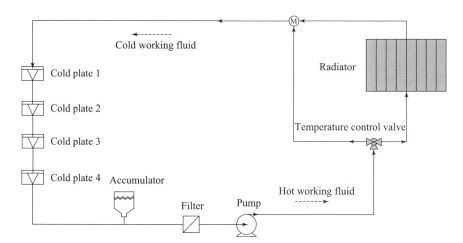

Fig. 4.47 Schematic diagram of a typical mechanically pumped SPL

Mechanically pumped two-phase fluid loop mainly includes mechanical pump and its controller, accumulator, evaporator, filter, pipeline, condenser, etc. The mechanical pump drives the single-phase working fluid to flow to an evaporator that couples with a heating device. In the evaporator, the working fluid absorbs heat, evaporates and turns into vapor–liquid two-phase status. By exchanging heat with a radiator or other heat sinks through the condenser, the two-phase working fluid could release and dissipate the heat collected from the heating device, revert to sub-cooled single-phase working fluid, and flow back to the mechanical pump so as to form a flowing and heat exchange circulation. In the loop, the accumulator is used to compensate or recycle the working liquid to the loop so as to adapt to the working liquid temperature changes, vapor–liquid distribution changes and slight leakage caused by the changes of heat load and external heat environment. In normal condition, the inside of both accumulator and evaporator is always kept in vapor–liquid status. According to Clausius–Clapeyron equation, the temperature and pressure of vapor–liquid two-phase working fluid are in one-to-one correspondence relation. Hence, for the identified working fluid, the temperature difference between accumulator and evaporator depends on the pressure difference between them. As a result, the other main function of the accumulator is to realize the evaporator temperature control by controlling its own temperature (heating and cooling regulation), thus controlling the temperature of the heating device. The schematic diagram of a typical mechanically pumped two-phase fluid loop is shown in Fig. 4.48.

3. **Gravity-driven two-phase fluid loop**

Generally, gravity-driven two-phase fluid loop is a closed loop that is composed of evaporator, vapor line, condenser, liquid line, accumulator, control valve, etc. The system is driven by gravity field, such as the gravity field of the moon or Mars. The working fluid inside the loop could absorb heat from heat sources (heating equipment,

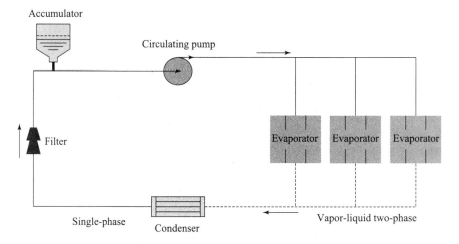

Fig. 4.48 Schematic diagram of a typical mechanically pumped two-phase fluid loop

RHU) and evaporate through evaporator. The working fluid changes from liquid into vapor or vapor–liquid, resulting in the density difference of working fluid inside the pipeline and further forming a natural circulation flow in the gravity field. The vapor releases heat in the condenser and is cooled into liquid. Then, it flows back to the evaporator through liquid line to absorb heat and evaporate again, thus realizing the heat collection, transfer and release. The schematic diagram of gravity-driven two-phase fluid loop is shown in Fig. 4.49.

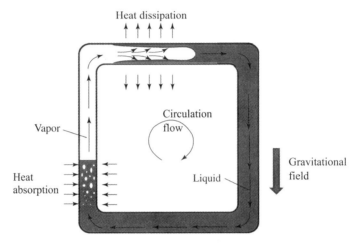

Fig. 4.49 Schematic diagram of gravity-driven two-phase fluid loop

4.2.6.2 Categories and Characteristics

Macroscopically, spaceborne fluid loops are classified into single-phase fluid loops and two-phase fluid loops. The latter include mechanically pumped two-phase fluid loop and gravity-driven two-phase fluid loop.

Typical characteristic indexes of the fluid loop mainly include appearance, size, mass, operating temperature zone, heat transfer capacity, temperature control index, heat flux, working fluid leakage rate, power consumption, service life and "six properties" (such as safety, reliability, maintainability). The specific indexes vary with different products and application purposes. The categories and characteristics of spaceborne fluid loops are shown in Table 4.2, in which the working fluid is for reference.

4.2 Heat Transfer Technology

Table 4.2 Categories and characteristics of spaceborne fluid loop

Name of fluid loop	Working fluid	Recommended operating temperature zone/°C	Special notes
Mechanically pumped SPL	Water	4–50	The working fluid has non-toxicity, high specific heat, high thermal conductivity and small density, and applies to the thermal control of the sealed cabin with astronauts. Please pay special attention to the chemical compatibility between working fluid (water) and loop material, the bacteria breeding, and the freezing under low temperature condition or fault condition
	Ammonia	−60 to 40	The working fluid has toxicity, small density, wide temperature zone and high pressure, and applies to manned spacecraft (except for the sealed cabin) or unmanned spacecraft. Please pay special attention to the chemical compatibility between ammonia and loop material, as well as the safety
	Ethylene glycol aqueous solution	−20 to 60	The working fluid has little toxicity, small density and high specific heat and applies to the loop inside a manned spacecraft or to an unmanned spacecraft. Please pay special attention to the chemical compatibility between ethylene glycol aqueous solution and loop material, as well as the safety after leakage

(continued)

Table 4.2 (continued)

Name of fluid loop	Working fluid	Recommended operating temperature zone/°C	Special notes
	Freon CFC-11	−100 to 100	The working fluid has non-toxicity, wide temperature zone and high pressure, and applies to the loop outside a manned spacecraft or to an unmanned spacecraft. Please pay special attention to safety issues
	Perfluorotriethylamine	−100 to 50	With a wide temperature zone, the working fluid is toxic-free and inexplosive and applies to the loop outside a manned spacecraft or to an unmanned spacecraft. Please pay special attention to its adaptation to space radiation environment
	Perfluorocycloether	−100 to 80	With a wide temperature zone, the working fluid is toxic-free and inexplosive and applies to the loop outside a manned spacecraft or to an unmanned spacecraft. Please pay special attention to its adaptation to space radiation environment

(continued)

Table 4.2 (continued)

Name of fluid loop	Working fluid	Recommended operating temperature zone/°C	Special notes
Mechanically pumped two-phase fluid loop	Ammonia	−60 to 80	The working fluid has toxicity, small density, wide temperature zone and high pressure and applies to the spacecraft with high load heat flux and high temperature consistency between different equipment. Please pay special attention to the chemical compatibility between ammonia and loop material, the safety and the dryness control of working fluid at the outlet of the evaporator
	Carbon dioxide	−40 to 25	The working fluid has non-toxicity, narrow temperature zone and high pressure and applies to the spacecraft with high load heat flux and high temperature consistency between different equipment. Please pay special attention to the safety and the dryness control of working fluid at the outlet of the evaporator

(continued)

Table 4.2 (continued)

Name of fluid loop	Working fluid	Recommended operating temperature zone/°C	Special notes
Gravity-driven two-phase fluid loop	Ammonia	−60 to 80	The working fluid has toxicity, small density, wide temperature zone and high pressure. It applies to the landers/rovers under proper gravitational fields of the moon, Mars, etc. Please pay special attention to the chemical compatibility between ammonia and loop material, the safety, the heat transfer capacity under different gravitational fields and its ground equivalent verification

4.2.6.3 Selection Principles and Application Cases

1. **Selection principles**

The fluid loop technology is mainly applied in manned spacecraft, deep space exploration, high-power satellite and other fields. Its selection principles are listed as follows:

(1) Mechanically pumped single-phase fluid loop and two-phase fluid loop apply to the spacecraft requiring high-power heat collection, transfer and dissipation, such as manned spacecraft and high-power communication satellite.
(2) Mechanically pumped single-phase fluid loop and two-phase fluid loop are typical active thermal control technologies, especially applying to the spacecraft featuring complicated/unknown space environment changes and multiple load operation conditions, such as deep space exploration spacecraft.
(3) Gravity-driven two-phase fluid loop is applied to the landers/rovers under proper gravitational fields of the moon, Mars, etc.
(4) The selection of working fluid in fluid loop depends on the requirements for operating temperature zone and heat transfer capacity, and is limited by many factors, such as the adaptability to space environment and the maturity of matched hardware products, which shall be considered comprehensively.
(5) The compatibility between the working fluid in fluid loop and the system materials, especially the chemical compatibility, is the key factor that affects the service life and reliability of fluid loop, and shall be confirmed during selection and design. Test verification shall be arranged if necessary.
(6) Mechanical pump is the key part for mechanically pumped single-phase fluid loop and two-phase fluid loop. For its selection, the service life, reliability and the key factors that could cause mechanical pump failure shall be considered. And attention must be paid to the system design and application process. Generally, the mechanical pump with backups is adopted in the design.
(7) CC or accumulator is the key to ensure that the fluid loop can adapt to space environment changes, working load changes and slow fluid leakage in its life cycle. Their designs shall match to the system requirement.
(8) When mechanically pumped fluid loop is applied in the manned spacecraft, special attention shall also be paid to modular design and maintenance design.
(9) During the application of mechanically pumped two-phase fluid loop, the dryness of the working fluid at the evaporator outlet cannot be too high. Otherwise, it is easy to cause temperature fluctuation and system instability. It is better to control the working fluid dryness under 0.3, and the specific data shall be determined according to the ground test.
(10) The heat transfer of two-phase fluid loop on the ground is different from that in space microgravity or other gravitational fields. For example, the changes of two-phase flow patterns caused by gravitational fields could further affect evaporation and condensation heat transfer—especially the condensation heat

transfer, i.e., the condensation heat transfer coefficient decreases under microgravity, which will affect the design of the condenser and the verification of ground equivalent test.

2. **Application cases**

Case 1: Shenzhou manned spacecraft series.

Shenzhou manned spacecraft series is currently China's main tool for astronauts to shuttle between space and the Earth. Up to date, 11 Shenzhou spacecraft have been successfully launched. The thermal control technology of fluid loop is a key technology in the thermal control subsystem of manned spacecraft and has been successfully applied in previous flights. The thermal control fluid loop system of spacecraft, which is a single-phase double-loop system, is shown in Fig. 4.50 The inner loop runs through the orbital module, return module and the propulsion module, and the working fluid is ethylene glycol aqueous solution. Either of the orbital module and return module is equipped with a condensing dryer, and four cold plates are set up downstream the condensing dryer of the return cabin. The inner loop pump assembly is mounted in the propulsion cabin, and the inner loop and the outer loop are connected through an intermediate heat exchanger installed in the propulsion cabin. The inner loop absorbs the heat dissipated by people and equipments in the cabin and transfers it to the outer loop through the intermediate heat exchanger. Then, the heat could be dissipated to the space by radiator in the outer loop, thus achieving the temperature controlling of the spacecraft. Temperature of the outer loop could be controlled by adjusting the main bypass flow of the radiator with the temperature control valve.

Case 2: Two-phase fluid loop is used for the spacecraft for scientific detection.

On May 16, 2011, the high-precision particle probe "AMS-02" (Alpha Magnetic Spectrometer 2) was successfully launched on the American spacecraft "Endeavour" and operated at ISS. The mechanically pumped two-phase fluid loop system applied

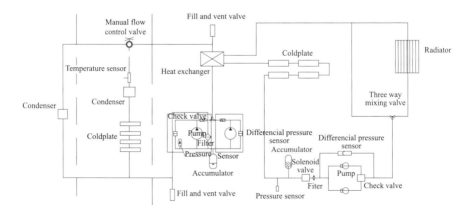

Fig. 4.50 Schematic diagram of thermal control fluid loop in rendezvous and docking spacecraft

in AMS-02 particle probe is the first set of key technologies in the world that operates on orbit and represents an important development tendency of spaceborne thermal control technology in future.

AMS-02 is a large international cooperative scientific experiment project led by Samuel Ting, a famous Chinese American scientist, with the aim of seeking for antimatter and dark matter. The partial probe is the key scientific detection load of AMS-02, including 192 silicon trackers, each of which dissipates 0.75 W heat, totaling 144 W. To guarantee detection accuracy, the silicon trackers shall dissipate heat and realize the high stability and uniformity control.

The mechanically pumped two-phase fluid loop, of which the working fluid is CO_2, is applied in AMS-02 to realize the heat dissipation and temperature control of silicon trackers. The system has four mechanical pumps (with redundancy design). The main reasons for choosing CO_2 two-phase fluid loop system are listed as follows. CO_2 is a HP working fluid whose saturation temperature changes little within a certain range of pressure difference, which is good for realizing the temperature uniformity of 192 components. Meanwhile, the two-phase system could miniaturize the evaporator (which is a thin pipeline actually), which is beneficial for realizing the high integration of partial probe. AMS-02 partial probe requires that the operating temperature range of two-phase fluid loop is from −15 °C to 25 °C and the storage temperature range is 20–40 °C. The temperature stability of 192 probe components in each orbital period shall be better than 3 °C, and their temperature uniformity shall be better than 1 °C/9 m (the length of the evaporator pipeline coupled with probe components). The maximum temperature difference between probes on each layer shall be less than 10 °C. The mechanically pumped two-phase fluid loop mainly includes mechanical pump, accumulator, heat regenerator, evaporator, condenser, preheater, pipe and so on.

Case 3: Apply two-phase fluid loop in Chang'e 3 lunar probe.

The method of "RHU + lunar gravity-driven two-phase fluid loop" is adopted to supply heat to Chang'e 3 lunar probe to help it survive during the lunar night. The RHU could provide heat and is installed on the outside of the lunar probe. The evaporator of lunar gravity-driven two phase fluid loop is coupled with RHU to absorb heat from RHU, and the condenser is coupled with the inside of the probe to release heat to its inside, thus realizing the heat supply, as shown in Fig. 4.51.

The two-phase fluid loop integrates heat accumulation, heat transportation and heat dissipation, uses ammonia as its working fluid and operates under the drive of lunar gravity field. During lunar day, the control valve of two-phase fluid loop is turned off, so the loop does not operate, thus realizing the thermal isolation between RHU and the inside of the probe. During the transition from lunar day to lunar night, the control valve of two-phase fluid loop is opened, so the loop could be started by lunar gravity and transfer heat to the inside of the probe to ensure heat compensation during the entire lunar night. During the transition from lunar night to lunar day, the loop could be turned off by control valve, thus ensuring the thermal disconnection between RHU and the inside of the probe during the lunar day, as shown in Fig. 4.52.

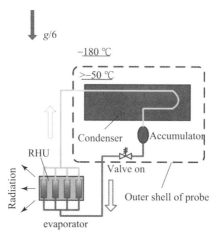

Fig. 4.51 Schematic diagram of the operation of heat supply system at lunar night

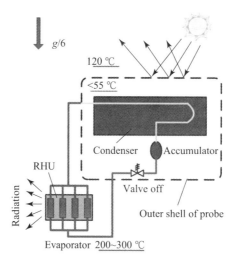

Fig. 4.52 Schematic diagram of the shutdown of heat supply system at on lunar day

During the research of lunar gravity-driven two-phase fluid loop, the theoretical and experimental researches with regard to the gravity field's influence on two-phase flow and heat exchange with the working fluid of ammonia, to ground equivalent test method, to operation stability, to performance degradation laws and to suppression techniques, are carried out to realize the heat transfer capacity of two-phase fluid loop and guarantee the heat transfer temperature difference. The theoretical analysis shows that for the two-phase fluid loop system adopted on the probe, the flow resistance in the lunar gravity field of 1/6 g is a little higher than that in the Earth gravity field of 1 g, but not more than 10%. Based on the aforesaid, the ground equivalent test method of heat transfer capacity has been formulated. And the test results show that the heat transfer capacity of two-phase fluid loop is qualified for engineering requirements. The analysis results based on two-phase fluid loop model show that the system could operate steadily for a long term within the working temperature range, which has been verified in the test. The non-condensable gas generated by ammonia decomposition at high temperature is the major reason for system performance degradation, whose symptom is the increasing of heat transfer temperature difference. The research results show that when the evaporator temperature does not exceed 235 °C, the increasing of heat transfer temperature difference is small and the heat transfer performance of two-phase fluid loop at the end of its service life could meet the engineering requirements. The aforesaid basic research work strongly supports the engineering development of lunar gravity-driven two-phase fluid loop.

The main features of heat supply system of Chang'e 3 lunar probe are listed as follows. ① It has a small weight of only 15 kg, including 10 kg for lander and 5 kg for rover. ② Since its operation is driven by lunar gravity rather than power, it adapts to the special cases when there is no power supply at lunar night. ③ There are no long-term running parts, and the inherent reliability is high.

The Chang'e 3 lunar probe was launched on December 2, 2013. It has successfully passed 40 diurnal cycles and 6 lunar eclipses by May 2017. Its thermal control system works normally and all of its functions and performance meet the requirements. The lunar gravity-driven two-phase fluid loop could be turned on and off normally, and operate steadily. The RHU temperature and heat supply are stable. The heat supply system of lander for surviving during the lunar night provides an essential guarantee for continuous lunar probe tasks.

4.2.7 Convection Ventilation Device

4.2.7.1 Functions and Working Principles

The convection ventilation device is applied to the spacecraft with a sealed cabin structure. These devices are widely applied on manned spacecraft, as well as some deep space probes and recoverable satellites. The forced gas convection circulation formed by the fan achieves the function of heat transfer between different equipments in the sealed cabin and between heat exchangers and heat radiation surfaces. Moreover, for manned spacecraft, some convection ventilation devices shall work with condensation dryer to achieve the effect of humidity control in manned sealed cabin. Consequently, the service objects of convection ventilation devices are equipments and astronauts. A typical spacecraft flow field based on convection ventilation technology is shown in Figs. 4.53 and 4.54.

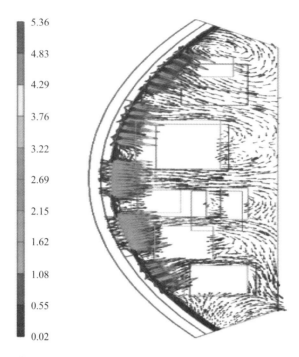

Fig. 4.53 Convection ventilation and heat exchange between equipment and cabin wall

4.2.7.2 Categories and Characteristics

The convection ventilation devices generally include gas treatment equipments (including fan, heat exchanger, filter), gas transportation line, gas distribution device, etc. According to the setup of gas treatment equipment, there are two types of ventilations: centralized ventilation and distributed ventilation. In the centralized ventilation system, all gas treatment equipments are set up in a central sealed area; thus, gas could be sent to the cabin through the ventilation pipe. The distributed ventilation system is a system where gas treatment equipments and cold/heat sources without gas transportation line and gas distribution device are set up flexibly and dispersedly inside the cabin as required, rather than in individual confined spaces.

4.2.7.3 Selection Principles and Application Cases

1. **Selection principles**
(1) The centralized convection ventilation device has the advantages such as convenient and accurate temperature adjustment control, strong adaptability and high efficiency. It adapts to the sealed cabins in long-term stay on orbit, where the

4.2 Heat Transfer Technology

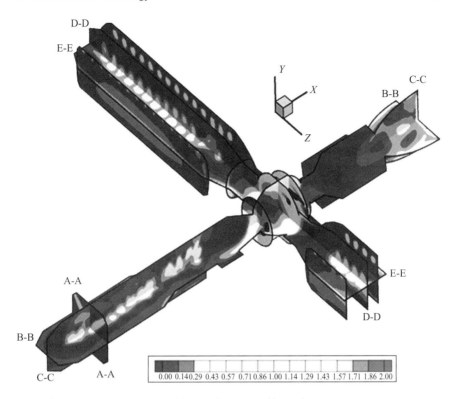

Fig. 4.54 Convection ventilation and heat exchange at cabin section

device area and the human activity area are separated and the requirement for accuracy of temperature control is high.

(2) The distributed convection ventilation device is an open type of forced-ventilation heat exchange, which has the advantages such as simple system, few equipments and light weight. It is suitable for the sealed cabins in short-term stay on orbit, with the integration design of device area and human activity area, and lower requirements for temperature control precision.

2. **Application cases**

Case: Distributed convection ventilation device.

The distributed convection ventilation device is generally applied to the reentry satellite sealing cabins, and the space laboratories and cargo ships in small and medium sizes, and is also adopted in the spacecraft in China.

The thermal control system of a reentry satellite capsule in China adopts the thermal control technology that is based on convection ventilation and active electrical heating. Payloads are set up on the control board of the capsule, and some wall surfaces are heat dissipation surfaces so that a convective circulation could be formed by the driving effect of fans to transfer the heat generated by equipments on the control board to the cabin walls and then radiate it to the space. Besides, the equipment temperature inside the capsule could be guaranteed by using electric heating at low temperature. The layout of capsule fans is shown in Fig. 4.55.

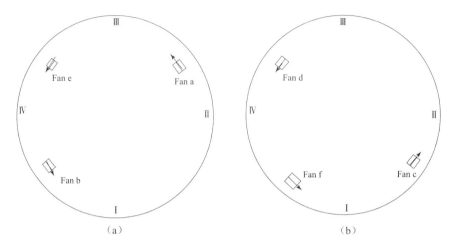

Fig. 4.55 Layout of fans in sealed cabin. (a) Layout of two-layer fans; (b) Layout of three-layer fans

4.2.8 Radiator

4.2.8.1 Functions and Working Principles

For a spacecraft operating in vacuum environment, most of its waste heat is dissipated to the outer space by radiator. The capacity of radiating heat dissipation depends on radiating surface area, infrared emissivity, solar absorptance and temperature level. Most of the radiators are covered with the coatings with high infrared emissivity and low solar absorptance in order to dissipate heat to the outside farthest as well as reduce the heat load from the sun. The capacity of radiating heat dissipation is a strong function of temperature and is proportional to the biquadrate of temperature, and will increase rapidly as the temperature increases.

4.2.8.2 Categories and Characteristics

Radiators are generally categorized into fixed radiators and deployable radiators depending on whether or not they are deployable.

The fixed radiator usually makes use of the structural board on the surface of spacecraft, which plays the role of both structural support and thermal control. In special cases, the fixed radiator could be designed independently so it only has thermal control function. The deployable radiator is a radiator that is folded at launching stage and deployed after entering the orbit. It is mainly used for the heat dissipation of high-power spacecraft and is an important means of the compensation and expansion of heat dissipation capacity for the spacecraft that uses the fixed radiator to dissipate heat. From the perspective of heat transfer modes, the deployable radiators include loop heat pipe-based radiator, pumped based SPL radiator and pumped based two-phase fluid loop radiator.

Depending on whether the heat dissipation capacity of radiators is adjustable or not, the radiators could be categorized into passive radiators and active radiators.

Most of the common fixed radiators applied in spacecraft, such as those embedded or attached to ammonia-containing axially grooved heat pipes, are passive radiators whose heat dissipation capacity cannot be adjusted actively. In some cases, heat dissipation could be adjusted for specific fixed radiators, such as those that adopts variable-conductance heat pipe, loop heat pipe, mechanically pumped two-phase/single-phase loop to transfer heat, or those that adopts louver or intelligent coating to change the characteristics of radiation surface. Among them, the fixed radiators that adopt VCHP, loop heat pipe, mechanically pumped two-phase fluid loop for heat transfer could regulate the heat dissipation by adjusting the length/area of the thermal coupling between the two-phase condensation section of heat transfer system and the radiator. Meanwhile, the fixed radiators that adopt mechanically pumped SPL for heat transfer could realize the heat dissipation adjustment by adjusting the working fluid flow through temperature control valve and bypass system. The fixed radiators that adopt louver for heat transfer could realize the active adjustment of heat dissipation by adjusting the radiating surfaces that are shielded by the louver. The intelligent coating based fixed radiators could dissipate heat by changing the infrared emissivity and solar absorptance. Generally, deployable radiators belong to active radiators, and the loop heat pipes and fluid loops thermally coupled with them are capable of heat dissipation adjustment.

The representative indexes of radiators mainly include overall heat dissipation capacity, heat dissipation capacity per unit weight, operating temperature range, total weight, heat dissipation area, heat dissipation capacity range, etc. Regarding the performance assessment of different radiators, their heat dissipation capacities per unit weight shall be assessed and compared on the same temperature benchmark. Of course, the control range of heat dissipation capacity is also a key index in the cases where heat dissipation capacity needs to be adjusted.

4.2.8.3 Selection Principles and Application Cases

1. **Selection principles**

(1) According to the comprehensive consideration of spacecraft orbit and attitude, equipment layout, temperature level and surface coating features and the reasonable design and thermal analysis assessment, if spacecraft surface structure (which generally refers to mounting plate) is used as radiator to dissipate heat

to space, the fixed radiator could be chosen. This is a common radiating dissipation method adopted by most spacecraft. Furthermore, a fixed radiator could be designed separately for the spacecraft or its external equipment if necessary.

(2) When the equipment heat dissipation and environmental heat flux change in a large range, the adjustable fixed radiator based on VCHP, loop heat pipe, fluid loop, louver or intelligent coating could be adopted in order to reduce the demand for electrical heating compensation power.

(3) When the heat dissipation of spacecraft is so large that the adoption of fixed radiator alone cannot meet the demand for heat dissipation, the deployable radiator shall be adopted as the means to support the heat loss through radiation, or even as the main way to reject heat. The deployable radiator could be loop heat pipe based radiator, pumped SPL-based radiator or pumped two-phase fluid loop radiator. Generally speaking, loop heat pipe-based radiator is more suitable for high-power point heat source. If the heat source is dispersed, heat shall be collected by groove heat pipe. Heat transfer link is complicated, and heat transfer temperature difference of the system is obvious. It is recommended to use pumped SPL radiator and pumped two-phase fluid loop-based radiator for distributed heat sources. From the perspective of system weight, loop heat pipe-based radiator and pumped two-phase fluid loop-based radiator are lighter than the pumped SPL radiator.

2. **Application cases**

Case 1: Fixed radiator-based on spacecraft surface structure.

As shown in Fig. 4.56, the most common and simplest radiator is a fixed radiator that uses the existing aluminum honeycomb panel on spacecraft surface. This aluminum honeycomb panel is a part of the structure and also plays the role of radiating the heat. Generally, it has the embedded heat pipe or external heat pipe, both of which could be adopted if necessary in order to balance the temperature of radiators and

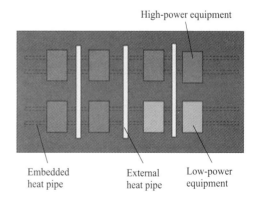

Fig. 4.56 Fixed radiator

4.2 Heat Transfer Technology

improve the heat radiation capacity. The inner surface of the fixed radiator could be used as the mounting surface of electronic equipments, while the outer surface could be used as a radiating surface. Heat could be transferred from the inner surface of radiator to the outer surface through aluminum honeycomb core and heat pipe, thus realizing heat dissipation in the end.

Besides, the external load of some spacecraft needs independent fixed radiator for heat dissipation. And heat pipe is adopted for thermal coupling between load and radiator.

Case 2: Louver-based fixed radiator.

When the outer surface (radiating surface) of radiator is used with the louver, the radiator has the capability of heat dissipation regulation. This type of radiator is applied in the spaces where environmental heat flux and equipment heat dissipation change a lot. As introduced in the aforesaid text, the louver-based fixed radiator could realize the active regulation of heat dissipation by adjusting the area of heat radiation surface that is shielded by the louver. In engineering, this regulation capacity is represented by the equivalent emissivity variation of heat radiation surface. Generally, the blade surface has a low-emissivity thermal control coating. For example, the gilded and polished surface has an infrared emissivity of 0.02–0.03; and the heat radiation surface with high-emissivity coating (such as OSR) has an infrared emissivity of 0.79. A large range of equivalent emissivity regulation could be realized by the combination of high and low emissivities and the opening and closing of blades. Equivalent emissivity is defined as the ratio of the net heat exchange amount that is radiated from louver surface to the black-body radiation amount at the same temperature when there is no environmental heat flux. Besides, equivalent emissivity is also an important heat radiation parameter of louver, which is defined as the share of solar energy incident on the surface of the louver and absorbed by its unit area.

The louver is generally composed of blades, driver, shaft and bearings, framework, etc., as shown in Fig. 4.57 [4]. Among them, blades are major components for adjusting the equivalent emissivity in the louver. Continuous emissivity adjustment can be realized through the change of the opening angle of the blades. The change of louver equivalent emissivity is 0.6–0.8 in general. The driver is a device for rotating the blades, whose rotation angle is restricted by temperature control requirements.

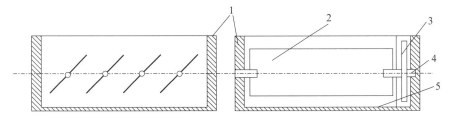

Fig. 4.57 Schematic diagram of louver. 1—Framework; 2—Blade; 3—Driver; 4—Bearing; 5—Baseplate

There are many ways to drive the louver, such as bimetallic drive, bellows drive, motor drive, memory alloys drive and so forth.

In China, louvers have been applied in Shijian 1, Shijian 2 and Shenzhou spacecraft. The orbital module of Shenzhou spacecraft has two states: on-orbit autonomous flight and parking-orbit flight. The heat loads in orbital module in these two states are much different. To meet thermal control requirements in these states, two sets of electric louvers are installed symmetrically on the radiation surface of the outer wall of orbital module, each having the area and weight of 1.5 m^2 and 10 kg, respectively. Electric louver is composed of framework, blades, stepping motor, harmonic reducer, angular displacement sensor, pull rod, rod-supporting device, connecting rod and louver controller. The infrared emissivity ε of the F46 aluminum-laminated film that is pasted and drilled on blade surface is 0.05. On the radiating surface of cabin wall corresponding to electric louvers, KS-Z white paint is sprayed and its infrared emissivity ε is 0.92. The opening angle of the louver is 0.5°–89.5° and the range of equivalent emissivity variation is 0.16–0.80. During autonomous flight with low heat load, the blades are closed and cover the radiating surface; while during parking-orbit flight with high heat load, the blades are opened and the radiating surface on the skin is exposed so as to meet the heat dissipation requirements.

The heat leakage of the louver is less than 15 W/m^2 and the power dissipation is less than 5 W/m^2.

Case 3: Pumped single-phase loop (SPL)-based deployable radiator.

Compared with the conventional fixed radiator, the deployable radiator breaks through the restriction of star surface area and has the double-surface heat dissipation capacity. During launching, the deployable radiator is tightly locked against spacecraft surface. After orbit insertion, it is deployed by using the deployable mechanism. Figure 4.58 is a schematic diagram of pumped SPL-based deployable radiator that is designed on DFH-5 platform, and the heat dissipation of each radiator is over 1,700 W. The radiation panel is composed of a honeycomb panel and fluid pipes inside it. The inner fluid pipes are connected with the fluid loop outside of the radiator through metal hose. During operation, a circulating pump drives the single-phase working fluid into the fluid loop to transfer heat. The glass-type silvered secondary surface mirror is pasted on the surface of radiator to realize heat dissipation.

Case 4: Loop heat pipe-based deployable radiator.

SJ-17, a scientific experiment satellite, was launched in 2016. A deployable radiator-based on two LHPs was used to transfer the heat of devices to the radiator. The heat transfer capability of each LHP is over 400 W. The radiator was deployed after the satellite had been on orbit. As shown in Fig. 4.59, the embedded heat pipes are responsible for accumulating the heat of devices, and the LHPs can transfer the heat to the radiator.

4.2 Heat Transfer Technology

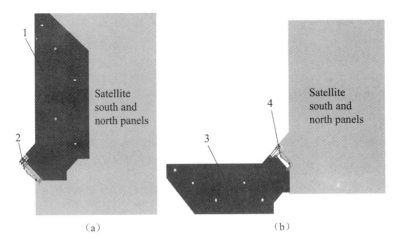

Fig. 4.58 Schematic diagram of deployable radiator on DFH-5 platform. (a) Closed status; (b) Deployed status. 1, 3—Radiation panel; 2, 4—Deployment mechanism

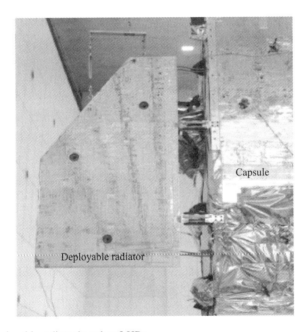

Fig. 4.59 Deployable radiator based on LHPs

4.2.9 Consumable Heat Dissipating Device

4.2.9.1 Functions and Working Principles

Heat radiation is the most important way for spacecraft to dissipate heat to the outer space. Special spacecraft even adopt consumable heat dissipating devices to dissipate heat, as a supplement to heat radiation. Sublimator and evaporator are two typical consumable heat dissipating devices that can cool down the equipments by evaporating or subliming a consumable substance to space and making use of its latent heat of phase change.

A typical phase change of sublimator is liquid–solid–vapor conversion. A typical phase change of evaporator is liquid–vapor conversion. Considering many factors such as latent heat of phase change, operating temperature zone, triple point pressure and safety, water is adopted as the main consumable substance during application in vacuum. Hence, water sublimator and water evaporator are common consumable heat dissipating devices. When they are applied in sub-orbit (there shall be some atmospheric pressure at some point), ammonia, Freon and other high-pressure working fluids could be used as well.

1. **Sublimator assembly**

Sublimator assembly is a consumable heat dissipating device with the integrated functions of heat collection, heat transfer and heat dissipation, mainly including working fluid storage tank, charging–discharging valve, pressure reducing valve, self-locking valve, sublimation heat exchanger, pipeline, working fluid, pressure sensor, etc. The composition of sublimator assembly is shown in Fig. 4.60.

The functions of each component in water sublimator assembly are listed below. Working fluid storage tank is used for the storage and supply of sublimation working fluid. Charging–discharging valve is used for charging/discharging the gaseous/liquid working medium in the storage tank. Pressure sensor is used for monitoring the pressure of working fluid before it flows into the sublimation cold plate of the sublimation heat exchanger. Self-locking valve is used for controlling the on-off of the sublimator. Pressure-reducing valve is used for reducing the pressure of working liquid to the working pressure of sublimation cold plate of the sublimation heat exchanger. Sublimation heat exchanger is used to realize the evaporation/sublimation heat dissipation of the working fluid. Pipeline is the channel through which the storage tank provides the working fluid to sublimation heat exchanger. Among them, the key functional part of water sublimator assembly is water sublimation heat exchanger.

The water sublimator could only operate normally in vacuum, and the outer surface of perforated plate of water sublimator needs to be exposed to vacuum. For the sublimator that uses water as working fluid, the supplied water would flow into the water chamber of water sublimation heat exchanger under certain pressure during operation. The heat exchanger absorbs heat from the cooling loop and transfers it to water in the water supply chamber. The water entering the water supply chamber or permeating the perforated plate would rapidly evaporate and cool in the vacuum

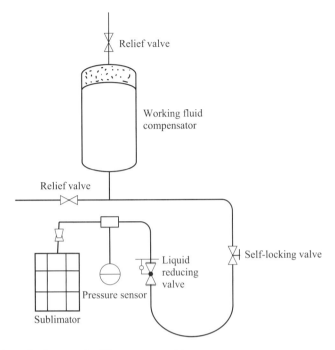

Fig. 4.60 Schematic diagram of sublimator composition

environment. When the pressure of saturated vapor inside the water chamber or perforated plate reaches (or is lower than) the triple point pressure, the water close to the vacuum side will freeze and sublimate into vapor and then dissipate in vacuum. Thus, a stable sublimation process is formed to take away the waste heat of the equipments. The schematic diagram of water sublimator is shown in Fig. 4.61.

There are four operation modes for water sublimator according to different heat load conditions:

(1) Sublimation mode: An ice layer is formed on the inner surface (one the water supply side) of the perforated plate and sublimates on the interface between ice and perforated plate. When heat load is lower or external environmental

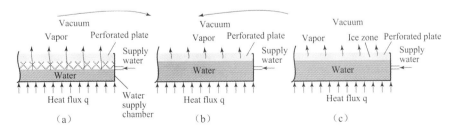

Fig. 4.61 Schematic diagram of operation of water sublimator

pressure is low enough, the pressure drop caused by the vapor generated by sublimation flowing through the perforated plate is small. Therefore, the vapor pressure on sublimation surface is lower than the triple point pressure of water, and ice keeps sublimating inside the perforated plate. The thickness of the ice layer depends on heat load and the pressure drop of the vapor which is produced by sublimation and flows through the perforated plate. As ice sublimates, the supplied water will condense on the water–ice interface at the same rate. Driven by water supply pressure, the ice layer will keep slipping toward the perforated plate to ensure that the physical location of sublimation interface is relatively stable. Sublimation is the most desired operation mode of water sublimator.

(2) Evaporation mode: When heat load and environmental pressure are larger, the pressure drop of the vapor, which is generated by sublimation and flows through the perforated plate, may be higher than triple point pressure, so that no ice will form and water can directly flow into the perforated plate. Water in the holes of perforated plate will produce certain surface tension which could hold water inside the perforated plate. In this case, the sublimator could dissipate heat by water evaporation. However, the surface tension on the water–hole interface is inversely proportional to the hole size. If the hole diameter of the perforated plate is too large or the water supply pressure is high, "breakdown" will happen, namely water will flow directly through the perforated plate into vacuum environment without phase change, thus causing the sublimator to lose its ability to dissipate heat.

(3) Mixed mode: Since the shapes and sizes of holes of the perforated plate are different and randomly distributed, some holes will have no ice and the phase change will happen in the way of evaporation when the heat load increases; while the other holes will have ice and the phase change will happen through sublimation. This is called mixed mode. However, researches show that this mode could only happen when the perforated plate materials cannot be wetted by water, but majority of the perforated plate materials can be wetted by water.

(4) Periodic mode: The concept of "Periodic mode" was put forward firstly by J. Alan. When pressure drop of the vapor that flows by the perforated plate is higher than the triple point pressure, ice will not form in water chamber; and water will flow into the capillary holes of the perforated plate until the pressure drop after passing the rest of the holes is lower than the triple point pressure. At this moment, water will condense into ice. Then, ice interface will transfer heat and medium in the perforated plate in the sublimation mode. When the sublimation of ice layer makes the sublimation interface retreat until the vapor pressure is higher than the triple point pressure of water, ice will disappear. Then, the water in the holes flows toward the lower-pressure direction and causes the evaporation effect instantly. When the vapor pressure re-reaches the triple point pressure, the interface temperature will drop to the freezing point and the sublimation state will occur again to transfer the heat and medium. So a new cycle will start. Periodic mode is also the combination of sublimation and evaporation in alternation. Generally, the ice layer in the perforated plate will not slip under the driving force of water supply pressure, and only the ice

interface will retreat. However, when the pressure for water supply is higher, the slipping of ice layer may happen.

2. **Evaporator assembly**

Like the sublimator assembly, evaporator assembly is also a consumable heat dissipating device with the integrated functions of heat collection, heat transfer and heat dissipation, mainly including working fluid compensator, charging–discharging valve, pump, pressure-reducing valve, back-pressure-regulating valve, evaporator, pipeline, working fluid, etc. The composition of sublimator assembly is shown in Fig. 4.62.

The functions of each component in evaporator assembly are listed as follows. Working fluid compensator is used for the storage and supply of the working fluid in circulation and evaporation. Charging–discharging valve is used for charging/discharging the gaseous/liquid working medium. Pressure sensor is used for monitoring the pressure of working fluid before it flows into the evaporator. Back-pressure-regulating valve is used for controlling the operational back-pressure of evaporator. The function of pressure reducing valve is the same as that of sublimator, that is, to reduce the pressure of working fluid to the working pressure of evaporator. Evaporator is used for realizing the evaporation and heat dissipation of the working fluid. Pipeline is a channel for the entire circulation system to provide the working fluid to evaporator. Its key functional part is evaporator.

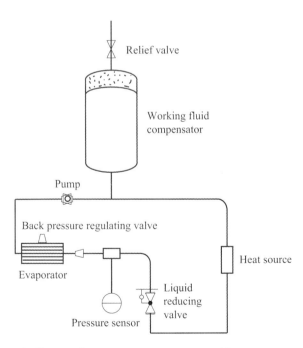

Fig. 4.62 Schematic diagram of evaporator component composition

Evaporator could take away heat by evaporating the working fluid to the outer environment through a special perforated plate structure. When the pressure of outer environment of the perforated structure is low enough, the liquid working medium will boil on the surface of the perforated structure and absorb heat. The produced vapor will dissipate out of the chamber. As long as the vapor dissipation rate is fast enough to avoid the increase of the chamber pressure, the phase change can continue and thus the working fluid itself can be cooled down. Evaporator could better adapt to the environment—not only the vacuum environment, but also a specific atmospheric environment. The schematic diagram of evaporator operation is shown in Fig. 4.63.

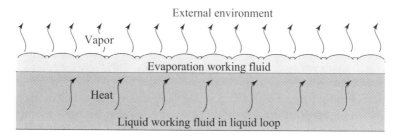

Fig. 4.63 Schematic diagram of evaporator operation

4.2.9.2 Categories and Characteristics

Consumable heat dissipating devices include sublimator and evaporator. Since their operation is based on the consumption of the working fluid, they are mainly used for the dissipation of high heat generation in short-term missions such as lunar exploration and space–Earth returning spacecraft, instead of long-term missions.

The advantages of sublimator are as follows: small volume, high efficacy, low cost and reliable operation under the conditions of weightlessness and heat load changes. It has been successfully applied in thermal control for many times. The typical indexes of sublimator and evaporator mainly include heat dissipation power, operation temperature, working fluid utilization, operation stability, continuous working capacity, weight, etc.

4.2.9.3 Selection Principles and Application Cases

1. **Selection principles**

(1) Sublimators are applied in vacuum or rare atmospheric environment. The pressure of rare atmospheric environment is clearly lower than the triple point pressure of working fluid. Whether sublimators are adaptive or not shall be confirmed by analysis or tests.

(2) Evaporators could be applied in both vacuum and rare atmospheric environment. The saturated vapor pressure of the working fluid within the operation temperature range shall be obviously higher than the atmospheric environment pressure. Whether evaporators are adaptive or not shall be confirmed by analysis or tests.
(3) The working fluid for evaporator and sublimator shall have high latent heat of vaporization so as to reduce the mass of the carried working fluid, such as water, ammonia.
(4) In addition to the heat dissipation, temperature level and working time of the heat source, the influence of acceleration, gravitational field and other factors should also be concerned during the selection of evaporator and sublimator.
(5) Attention shall be paid to the influence of vapor emission on spacecraft attitude.

2. **Application cases**

In 1960s, the USA took the lead in the research of water sublimator, which was actually applied in the subsequent Apollo project for the first time. The Space Shuttle, Apollo Lunar Module, GEMINI exploration, extravehicular spacesuit and so forth have utilized the water sublimators. And the Russia's Orlan spacesuit also adopts water sublimator as a main thermal control means. In the early 1990s, the thermal control system of the European "Hermes" project also adopted water sublimators as a thermal control solution. At present, the research institutions for water sublimators are concentrated in the USA, such as the Johnson Space Center of NASA, the Glenn Research Center, the American Research Consortium and the Paragon Space Institute. The water sublimator is also applied in China's extravehicular spacesuit as an auxiliary thermal control means. Meanwhile, Chang'e 5 lunar probe adopts water sublimator system as an important supplement to heat radiation.

The technical researches of water sublimator in China started in 1990s and succeeded in the spacesuit thermal control during China's manned spacecraft mission. Chang'e 5 is the first spacecraft to sample lunar soil and return from the moon. The thermal control, heat collection and heat rejection systems of its lander and ascender mainly include mechanically pumped fluid loop, water sublimator, heat pipe, radiator, etc. Water sublimator mainly includes water tank, pressure reducing valve, self locking valve, pressure sensor, water sublimation heat exchanger and so forth. The entire system adopts two sets of water sublimation heat exchangers that could work independently or in combination so as to realize the adjustment of heat dissipation capacity. Water sublimator is mainly used for realizing the auxiliary heat dissipation of the probe at lunar noon and for guaranteeing low temperature for the ascender before launching. To adapt to high-temperature environment on the moon and try to tap the heat dissipation potential of radiator as much as possible, water sublimator has been designed with high-temperature start-up/operation capability at 40 °C. The heat dissipation capacity of each set of sublimator is about 400 W. The schematic diagram of Chang'e 5 water sublimator system is shown in Fig. 4.64, and the water sublimation heat exchanger is shown in Fig. 4.65.

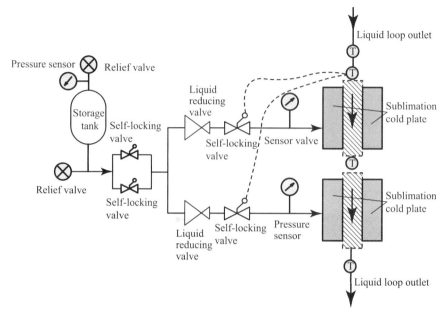

Fig. 4.64 Schematic diagram of Chang'e 5 water sublimator system

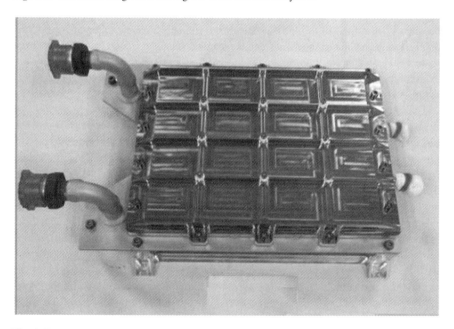

Fig. 4.65 Water sublimation heat exchanger

4.2.10 Phase Change Material (PCM) Device

4.2.10.1 Functions and Working Principles

Phase change material, also known as latent heat storage material, refers to a material that can absorb or release heat when a substance undergoes phase change, while the temperature of the substance itself does not change or changes little. When the temperature is higher than the phase change point, this type of material would absorb heat and cause phase change (energy storage progress). On the contrary, when the temperature drops and is lower than the phase change point, inverse phase change (energy release progress) will happen. Because of the aforesaid features, PCM could be used in the spacecraft whose inner heat source or external environment changes periodically, so as to keep the temperature of equipment relatively stable. PCM can also be used to realize the energy storage and utilization in some special missions. PCM device is a device that uses the characteristics of PCM to store or release heat during the phase change process, thus controlling the temperature of instruments and equipments. It has the functions of energy storage and temperature control. A typical temperature change during the phase change of PCM is shown in Fig. 4.66.

PCM is the simplest way for short-term equipment to realize thermal control at the launching and reentry stages of the spacecraft. Although this type of equipment is only used for once, its heat dissipation is high. This heat shall be dissipated to avoid the operation failure due to overheating. PCM could provide thermal protection for this type of equipment. The heat produced by electronic devices could be absorbed as latent heat of fusion by PCM without increasing the equipment temperature significantly. This system is completely passive and very reliable.

PCM is generally applied in the thermal control of periodically working electronic equipment. The heat produced during equipment start-up could be stored in PCM by phase change. When the equipment is turned off, the fusion heat is dissipated by radiator, heat pipe, conductive tape or other ways, so PCM could be cooled down

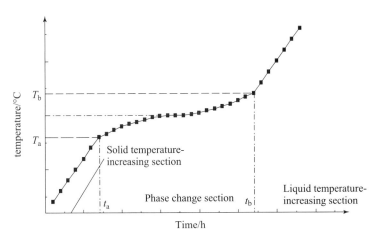

Fig. 4.66 Schematic diagram of typical temperature change during the phase change of PCM

and ready for the next use. The alternating fusion and condensation of PCM could control the operating temperature of the equipment within a narrow range.

4.2.10.2 Categories and Characteristics

There are four types of PCM phase change: solid–solid, solid–liquid, liquid–vapor and solid–vapor phase changes. The latent heat of phase changes increases gradually in the aforesaid order. The last two phase changes are rarely used for heat storage, as they produce a large amount of vapor and cause a great change in the volume of phase change substance. At present, the most commonly used is solid–liquid PCM.

PCMs could be categorized into inorganic PCM, organic PCM and mixed PCM. Inorganic PCMs include crystal salt hydrate, fuse salt, metal alloy and other inorganic matters. Organic PCMs include paraffin, fatty acids and other organics. Mixed PCMs mainly include the mixing materials of organic and inorganic cofusion PCMs.

According to different phase change temperatures, PCMs could fall into two types: low-temperature PCM (<200 °C) and high-temperature PCM (≥200 °C). High-temperature PCMs are mainly high-temperature melting salt, salt mixture, metal, alloy, etc. The common high-temperature melting salts include fluoride salt, chloride, nitrate and carbonate, which have a wide phase change temperature coverage and high latent heat of fusion, but strong corrosion to cause chemical reaction on the vessel surface. Low-temperature PCMs are mainly organic PCMs, in which paraffin is the most widely used.

Paraffin is mainly the mixture of straight-chain paraffins, and its common molecular formula is C_nH_{2n+2}. The fusion point of short-chain paraffin is low, for example, the fusion point of ethane (C_6H_{14}) is -95.4 °C. As the carbon chain increases, the fusion point of paraffin will increase. It increases rapidly in the beginning, then slowly (e.g., the fusion point of $C_{30}H_{62}$ is 65.4 °C and that of $C_{40}H_{82}$ is 81.5 °C) and finally toward a certain value. The paraffin fusion heat increases with the chain length. As a kind of PCM, paraffin has many advantages, such as phase change point close to room temperature, large latent heat of phase change, no supercooling phenomenon, no corrosion, repeatable fusion and crystallization, stable chemical and physical properties. Therefore, paraffin PCM is widely applied in the field of spaceborne thermal control. However, paraffin PCM has the disadvantages like small conductivity, low density, etc. To increase the thermal conductivity of phase change storage device, the framework of thermal conductivity could be designed to improve thermal conductance performance by using metal, graphite and so on. Metal materials such as aluminum, in the forms of foam, fin and honeycomb, have the best comprehensive performance.

4.2.10.3 Selection Principles and Application Cases

1. **Selection principles**

(1) The key to the design of PCM device is the selection of PCM and the heat transfer performance of phase change devices. PCM device is mainly used for temperature control of internal equipment of spacecraft. Hence, the selection of PCMs should first consider whether the phase change point is close to the operating temperature of equipment. The operating temperature of the equipment is affected by many factors such as operation orbit and attitude of spacecraft, the position and size of the radiating surface, and the operation mode of equipment.

(2) The thermal conductivity of PCM is small. Therefore, on the premise of ensuring the structure of phase change device is compact and its weight is small, how to design and select the phase change device to store heat when the equipment is operating or the environmental heat flux is large, and how to completely release heat when equipment is not operating or the environmental heat flux is small is of vital importance.

2. **Application cases**

Case 1: Apply PCM device in restriction of environmental heat flux changes.

By using the phase change heat transfer plate on the Yaogan-9, the temperature control of rubidium clock when space environmental heat flux changes tempestuously is solved. The thermal control design of rubidium clock is shown in Fig. 4.67. The phase change device (i.e., phase change heat transfer plate) is connected with the honeycomb panel, and the contact surfaces are filled with enhancement conductive filler. The other side faces toward the cryogenic space, and the surface is a coating with low solar absorption and high infrared emissivity, so heat of equipment can

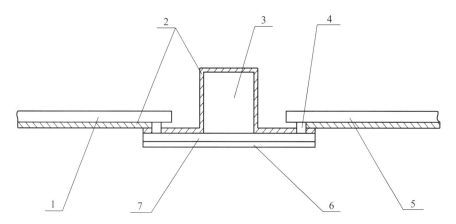

Fig. 4.67 Schematic diagram of temperature control system of rubidium clock of Yaogan-9. 1, 5—−Z plate; 2—Multi-layer heat shield; 3—Rubidium clock; 4—Crew and heat shield; 6—Phase change heat transfer plate; 7—Installation panel of rubidium clock

be dissipated. PCM uses n-Tridecane, and on-orbit temperature fluctuation of the rubidium clock is better than 2 °C/orbit. The actual product of phase change device is shown in Fig. 4.68.

Fig. 4.68 Phase change heat transfer plate of Yaogan-9

Case 2: Apply PCM device in storage and application of heat.

In China's Mars exploration, a high-temperature PCM device and a cryogenic one are designed in the thermal control system, which are embedded in the installation panel of equipment. The high-temperature PCM device is used to limit the temperature rising of short-term working equipment with large heat dissipation; the low-temperature PCM device is used to store solar energy collected by the solar thermal collector on Mars in the daytime and release the stored heat for heat isolation of equipment on Mars at night. High-temperature PCM device adopts n-Octadecane as PCM, while the low-temperature PCM device adopts n-Undecane as PCM.

Due to the high weight requirements of Mars exploration, the conventional welding processed PCM accounts for less than 30% of the total weight, which requires a new structure. In order to improve the filling ratio of PCM, PCM device of Mars rover was integrally shaped by adopting 3D lattice structure AM technology. Considering welding performance and material intensity comprehensively, AlSi10Mg powder is adopted as basic material for 3D printing, so PCM filling ratio can be as high as 60%. 3D printing shell of Mars PCM device is shown in Fig. 4.69, and 3D printing heat-conductive honeycomb structure of inside of the device is shown in Fig. 4.70.

4.2 Heat Transfer Technology

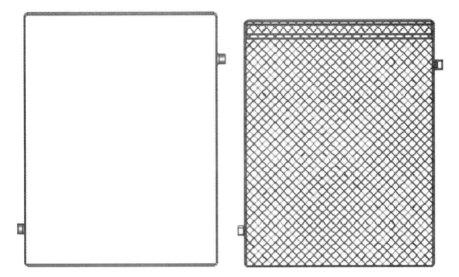

Fig. 4.69 3D printing shell of Mars PCM device

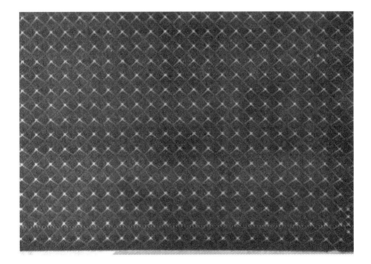

Fig. 4.70 3D printing inside honeycomb structure of Mars PCM device

4.2.11 Thermal Switch

4.2.11.1 Functions and Working Theory

The thermal switch is a thermal control device, which can set up and cut off thermal connection between the two parts, and automatically connect or disconnect the heat transfer route as required. Thermal switch technology plays an important role in spacecraft thermal design that needs frequent maneuver or drastic changes in the external environment changes. It can passively control temperature of electronic equipment or devices instead of using thermostatic controller and heater, thus reducing demand for power, heater control circuit and control software, which is of great significance to optimize the thermal control of spacecraft.

Generally, there are three modes for the thermal switch to set up and cut off thermal connection between two connected parts. The first mode is to make moving parts of thermal switch stretch or displace by driving force or deformation produced by devices/materials with effect of temperature, magnetic and static, thus changing contact status between the moving parts and the fixed parts, and realizing connection/disconnection of heat transfer link. For thermal switches such as paraffin-driven thermal switch, memory alloy thermal switch, microdilatancy thermal switch, static-driven thermal switch, they are defined as "interface contact-type thermal switch" which is usually known as "mechanical thermal switch" in other literatures. The second mode is to realize connection/disconnection of heat transfer link by filling or removing gas/liquid that is for thermal connection inside the inner clearance of thermal switch, such as air-gap/liquid-gap thermal switch. This kind of thermal switch can be defined as "medium filling-type thermal switch." The third mode is to realize connection/disconnection of heat transfer link by largely changing conductivity of connection material inside the thermal switch with special physical effect, such as superconducting thermal switch. This kind of thermal switch can be defined as "conductivity-adjustable thermal switch." In addition, VCHP and loop heat pipes also have the characteristics of thermal switch, but they are fallen into heat pipe according to the classification.

In the following, we take paraffin-driven thermal switch and superconducting thermal switch as examples to introduce working principles of different types of thermal switches.

1. **Working principles of paraffin-driven thermal switch**

The middle bellows of paraffin-driven base-type thermal switch is filled with paraffin. When the heat load on the hot end of thermal switch is small or zero, the temperature of paraffin PCM in the bellows cavity is low and remains in the solid state. The middle bellows cavity and the pressure-bearing cold plate on it will be isolated by a small gap. For heat transfer route of the thermal switch from the hot end to the cold end, there are only radiant "leak heat" of heat insulation material of the external support structure and the gap, which has small heat conductivity, thus making heat leakage from the hot end to the cold end very small, i.e., the thermal switch is "disconnected." When heat load on hot end of the thermal switch is high, paraffin inside the bellows will melt to compel the bellows to expand, and to close the gap between pressure-bearing cold plate and bellows cavity. With the increase of the thermal conductance, the main heat transfer relationship between the hot end and cold end of the thermal switch becomes heat conduction, which indicates that the thermal switch is "connected." The "connection" thermal conductance of the thermal switch can be adjusted automatically. In other words, the higher the temperature, the greater the proportion of paraffin melting and the higher the interface pressure. Because the interface contact thermal conductance is related to the pressure on it, the thermal conductance will increase with the increase of heat load.

In addition, paraffin-driven thermal switch also includes diaphragm-type sheet thermal switch, fastener-type thermal switch and washer-type thermal switch. It working principle is basically the same as that of base-type thermal switch.

2. **Working principles of superconducting thermal switch**

Electrons in superconductor do not participate in heat transfer. Hence, the heat conduction of superconductors only depends on a few phonons inside the superconductor. By adding magnetic field on the superconductor, superconducting performance of the superconductor can be damaged, and electrons can participate in heat transfer, and the thermal conductivity of material can be increased by 1,000 times. Based on the above mechanism, the superconducting thermal switch is developed. The superconducting thermal switch should work at very low temperature, which is the key part of ultra-low-temperature equipment to realize the heat connection and disconnection. The commonly used superconducting materials are Pb, Sn, In, Al, Zn, etc., the latter two materials are usually applied when temperature is lower than 1 K since their superconducting switch temperature is very low.

4.2.11.2 Categories and Characteristics

As mentioned above, according to the physical mechanism of thermal switch to set up and cut off thermal connection, this book divides thermal switch into interface contact-type thermal switch, medium filling-type thermal switch and conductivity-adjustable thermal switch; while thermal switch is generally divided into normal-temperature thermal switch and low-temperature thermal switch according to the working temperature. In the design of spacecraft, the thermal switch is usually selected according to the working temperature first, and performance indexes such as switch ratio and response time of thermal switch as well as necessary resources or restrictions during application are considered.

No matter how to classify thermal switches, the basic characteristics of thermal switches mainly include working temperature scope, switch ratio, make and break of thermal conductance. Among them, the switch ratio refers to the ratio of make/break of thermal conductance, which is an index to show thermal conductance adjustment capacity of thermal switch. When it is necessary to adjust the heat in a large range, a higher switch ratio is required. The thermal conductance is the index of heat transfer capacity per unit temperature difference when the thermal switch is connected, and the unit is W/°C. In general, when heat dissipation of equipment is high, a higher thermal conductivity is required. The thermal break is the index of heat leakage characteristics under unit temperature difference when the thermal switch is disconnected, and the unit is W/°C. Generally, when the spacecraft is in a deep cryogenic environment, in order to reduce the heat leakage of the system, the lower the thermal break, the better.

4.2.11.3 Selection Principles and Application Cases

1. **Selection principles**

Generally, the thermal conductivity indicator of the thermal switch is small, which is only suitable for the special cases of small heat dissipation of the equipment. In addition, thermal switches are often used in deep space exploration spacecraft so as to adapt to large changes in ambient temperature. The selection principles are as follows:

(1) Different types of thermal switches have different applicable scopes, switch ratio and different assistance resources, which should be selected as needed.
(2) Paraffin-driven thermal switch is usually applicable to room temperature environment and without additional resources, especially for deep space exploration spacecraft with rare resources.
(3) Air-gap thermal switch, microdilatancy thermal switch and superconducting thermal switch are low-temperature thermal switches, which are generally used in profound hypothermia load.

2. **Application cases**

Thermal switch technology emerged with the development of space probes, especially deep space probes. After more than 40 years of development, many kinds of thermal switches such as microdilatancy thermal switch, air-gap thermal switch and paraffin-driven thermal switch emerged. The operating temperature of the thermal switch is mostly between 0.1 K and 300 K.

Case: Paraffin-driven base-type thermal switch.

Paraffin-driven base-type thermal switch is the most successful one among all those thermal switches.

Paraffin-driven base-type thermal switch is studied in China and is carried on XY-1 new scientific experiment satellite. N-heptadecane is adopted as PCM, and phase change point is 22 °C. The actual measured weight is 159.56 g, and the volume is ϕ56.03 mm × 27.27 mm, as shown in Fig. 4.71.

Figure 4.72 indicates the temperature change of the thermal switch during the on-orbit test. Thermal conductivity of the thermal switch is bigger than 0.3 W/K when the temperature exceeds 30 °C, while it is less than 0.025 W/K when the temperature is less than 18 °C. "Switch ratio" can be greater than 12, and a total of 50 on-orbit switch function verifications have been carried out.

4.3 Thermal Insulation Technology

Fig. 4.71 Product of paraffin-driven base-type thermal switch of China

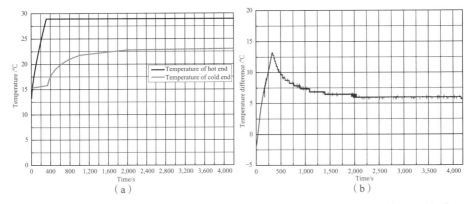

Fig. 4.72 Temperature change of thermal switch on hot end and cold end in on-orbit test. (a) The temperature of hot end and cold end changes with time; (b) The temperature difference of hot end and cold end changes with time

4.3 Thermal Insulation Technology

4.3.1 Introduction

Thermal insulation design is usually carried out in the following cases:

(1) The temperature indexes of different equipment vary greatly. When coupling thermal design of each equipment costs too much, the thermal insulation can be carried out between the equipment with harsher temperature requirements and

its surrounding equipment. For example, generally, temperature of the battery pack should be kept in a narrower range, and its allowable working temperature is far lower than other equipment. Here, insulation between battery pack and its surrounding equipment is a better plan.

(2) When equipment with small heat dissipation is set on radiation surface or structure with large temperature fluctuation, thermal insulation shall be taken. In general, radiation surface is in lower temperature. If the equipment with small heat dissipation is installed on the radiation surface directly, its temperature will also be reduced. In order to improve the temperature condition of the equipment at low temperature, thermal insulation can be taken between the equipment and the radiation surface.

(3) In order to reduce thermal impact of high/low-temperature equipment (outside the spacecraft) on the interior of the spacecraft, thermal insulation design can be carried out. For example, the temperature range of the spacecraft antenna is very wide, and its high/low-temperature level is far higher/lower than the inside one, so the thermal insulation between antenna and the spacecraft must be carried out. For another example, due to the high operating temperature of spacecraft attitude and orbit control motors, thermal insulation should also be taken between the motors and the spacecraft.

(4) Generally, thermal insulation shall be taken when the equipment needs thermostatic control. For example, the atomic clock of BDS requires that the temperature within one orbital period be kept within ± 1 °C. Therefore, thermal control shall be designed for the atomic clock independently, and thermal insulation shall be taken between it and its surrounding equipment to reduce the power, weight and heat dissipation area of its thermal control system needed.

(5) When the housing of spacecraft is not all radiation surfaces, the non-radiation surfaces shall take thermal insulation. The temperature inside and outside the spacecraft is very different. In addition to the radiation surfaces, the non-radiation surface usually takes thermal insulation to reduce the radiation exchange between this surface and the space.

The basic modes of heat transfer are conduction, convection and radiation. Hence, thermal insulation design is mainly used to inhibit conduction, convection and radiation. Conduction insulation, convection insulation in air environment and radiation insulation are usually adopted as thermal controls of spacecraft. Radiation insulation is often used in engineering, sometimes combining the above two or all aforesaid thermal insulation measures.

In addition, regarding loop heat pipe, VCHP, fluid loop, radiator, thermal switch, sublimator and evaporator based on the above three heat transfer technologies as introduced in Sect. 4.2, in addition to realizing main functions like heat transfer or heat dissipation, they also have active or passive adjustment ability, which can block heat transfer route to realize thermal insulation under extreme conditions. For example, the loop heat pipe can stop operating by turning on the heater on the reservoir to realize the thermal insulation between the equipment and the radiation surface. Please refer to Sect. 4.2 for more details.

4.3.2 Radiation Insulation

4.3.2.1 Function and Working Principle

1. **Basic principle**

The purpose of radiation insulation is to limit the radiation exchange between object and its surrounding environment. It is assumed that the object is isothermal and heat exchange surface conforms to the assumption of diffuse graybody and average effective radiation [5], the net radiation heat of the object to the surrounding environment can be expressed by the following formula:

$$q = \varepsilon E_b - \alpha G \tag{4.3}$$

where q refers to net radiation heat of the object to the surrounding; E_b refers to blackbody radiation force of which temperature is equivalent to the object temperature; G refers to projection radiation on surface of the object; and α refers to object surface absorption ratio for projection radiation.

If G is infrared radiation, then $\alpha = \varepsilon$; if G is solar radiation, α is object surface absorption ratio for solar radiation.

As shown in Formula (4.3):

(1) The most important part for thermal insulation design is to reduce ε and α as far as possible. In other words, it makes special design for thermology and phonology parameters of object surface in order to reduce the radiation toward the surrounding environment and decrease the absorption of radiation from the surrounding environment (including infrared radiation and solar radiation).
(2) When εE_b is equivalent to αG, q is also small. Aiming at this special case, it does not mean that it is unnecessary to take radiation insulation, in other words, the effect of the radiation insulation is not obvious. If αG changes the object temperature obviously, radiation insulation shall be taken as follows.

It assumes that heat dissipation of object is zero and it only has radiation with the surrounding environment, q in Formula (4.3) is zero and the absolute temperature T of the object can be given by the following formula:

$$T = \sqrt[4]{\frac{\alpha}{\varepsilon \sigma} G} \tag{4.4}$$

If object temperature T is out of range, ε and α shall be adjusted. Some measures such as reducing ε or α, belong to radiation insulation. MLI (multi-layer insulator) is the common radiation insulation measure.

2. **Principles of MLI**

MLI (multi-layer Insulator) is the most common used radiation insulation measure for spacecraft. In general, it is formed by the alternatively overlapped reflecting screen with low emissivity and the isolated layer with low conductivity, as shown in Fig. 4.73. Its thermal insulation principles are shown in Fig. 4.74. Since radiation insulation performances of MLI in vacuum (below 1×10^{-3} Pa) is very good and the space itself is a highly vacuum environment, MLI is a commonly used material on spacecraft for shielding the impact of high/low-temperature environment on equipment.

In vacuum (below 1×10^{-3} Pa), the calculation formula of the radiation heat flux of N-layers of non-touching reflecting screens set up between two infinite parallel surfaces is given in Formula (4.5):

$$q = \frac{\sigma\left(T_h^4 - T_c^4\right)}{(N+1)\left(\frac{2}{\varepsilon} - 1\right)} \tag{4.5}$$

where T_h and T_c refer to thermodynamics temperature on two surfaces, which serves as in T_{hot} and T_{cold} in Fig. 4.74.

As shown in the aforesaid formula, when there are enough layers of reflecting screens and surface emissivity is small enough, radiation exchange through MLI can be controlled at a very low level.

3. **Equivalent treatment of MLI**

In fact, the heat exchange through MLI includes the radiation between the reflecting screens, and the touching conduction between the isolated layer and the reflecting screen. When there is gas, the conduction and convection of the gas will increase the heat exchange through MLI. Hence, it is hard to calculate heat exchange through

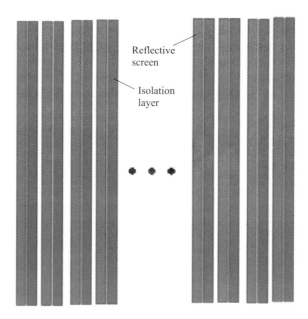

Fig. 4.73 Schematic diagram of MLI

4.3 Thermal Insulation Technology

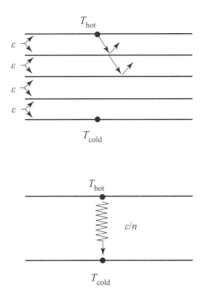

Fig. 4.74 Schematic diagram of principles of MLI

MLI directly by Formula (4.5). In order to meet the engineering design requirements, a variety of equivalent parameters are proposed to characterize the thermal insulation performances of MLI [5, 6], as follows.

(1) Equivalent thermal conductivity model.

In the equivalent thermal conductivity model, MLI is regarded as a continuous medium with thickness δ and thermal conductivity λ_{eq}. Fourier's law is used to describe the heat transfer between the cold and hot surfaces of MLI:

$$q = \lambda_{eq} \frac{T_h - T_c}{\delta} \tag{4.6}$$

(2) Equivalent heat transfer coefficient model.

Equivalent heat transfer coefficient model takes the following formula to describe heat transfer between cold and hot surfaces with MLI:

$$q = h_{eq}(T_h - T_e) \tag{4.7}$$

where h_{eq} refers to equivalent heat transfer coefficient of MLI.

(3) Equivalent emittance model.

Equivalent emittance model takes the following formula to describe heat transfer between cold and hot surfaces with MLI:

$$q = \varepsilon_{eq}\sigma\left(T_h^4 - T_e^4\right) \tag{4.8}$$

where ε_{eq} refers to equivalent emittance of MLI.

(4) Effective emissivity model.

The effective emissivity model makes the MLI equivalent to the coating layer on the hot surface, and its emissivity is ε_{eff}. By using this coating layer, the hot surface can exchange heat with the outside directly. The so-called equivalent means that the actual temperature of the hot surface in the MLI state is the same as the assumed coating condition.

(5) Assessment of several models.

The aforesaid three models regard the surface covered with MLI and the outer surface of MLI as two infinite parallel surfaces, and heat transfer through MLI is, respectively, regarded as one-dimensional conduction, radiation and contact heat transfer between these two infinite parallel surfaces.

It is shown from Formulas (4.6)–(4.8) that the equivalent thermal conductivity and equivalent heat transfer coefficient shall not be regarded as irrelevant constants to temperature. Relation between them and temperature shall be calibrated by tests in pursuit of accuracy, and it should be iterated as a parameter affected by temperature in thermal analysis. It is achieved done in engineering presently, and in most cases, the equivalent thermal conductivity and equivalent heat transfer coefficient are regarded as constants in analysis. However, it cannot make sure the application conditions are consistent with the testing conditions, which would cause analysis error, not to mention that the higher the temperature difference is, the bigger the error will be. Regarding application of equivalent thermal conductivity model and equivalent heat transfer coefficient model in engineering, both have the disadvantage that convenience and accuracy are difficult to be taken into consideration. Furthermore, one additional disadvantage of equivalent thermal conductivity model is that the parameter of MLI thickness is introduced. In fact, the actual MLI thickness has influence on test results of equivalent thermal conductivity. Therefore, even if theoretical MLI thickness in application condition is as same as that in testing conditions, the actual thickness in both conditions cannot be precisely the same. These factors are not reflected in equivalent thermal conductivity model.

In contrast, the equivalent emittance can be approximated as a constant within a certain temperature range. Then again, as mentioned above, the actual MLI is not an ideal insulating shield with only radiation, so the equivalent emittance can only come from experimental measurement. Generally, if MLI and test pieces have the same structure and differences of installations are ignored, the estimated value can be adopted whatever temperature on cold surfaces and hot surfaces is the same as estimated. This is also the reason why the equivalent emittance model is widely used. Although the effective emissivity model is often used in engineering, it is mainly introduced for the convenience of calculation.

4.3 Thermal Insulation Technology

(6) The relationship between the equivalent emittance model and the effective emissivity model.

This section gives a simple example to show the relationship between the equivalent emittance model and the effective emissivity model.

As shown in Figs. 4.75 and 4.76, it is assumed that the temperature of the object covered by MLI is T_h, the temperature of the outer surface of MLI is T_c, the hemispherical emissivity of outer surface of MLI is ε, and there is no heat flux on the outer surface of MLI (i.e., $q_e = 0$).

According to the equivalent emittance model, heat flux q that flowing through MLI is:

$$q = \varepsilon_{eq}\sigma\left(T_h^4 - T_c^4\right) \quad (4.9)$$

According to the effective emissivity model, heat flux q that flowing through MLI is:

$$q = \varepsilon_{eff}\sigma T_h^4 \quad (4.10)$$

The energy balance relation of the outer surface of MLI is listed as follows:

$$q = \varepsilon\sigma T_c^4 \quad (4.11)$$

The following result can be deduced from the aforesaid three formulas:

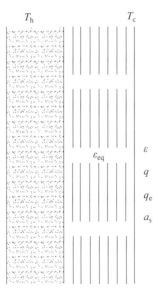

Fig. 4.75 Schematic diagram of equivalent emittance model for MLI radiation

Fig. 4.76 Schematic diagram of effective emissivity model for MLI radiation

$$\varepsilon_{eq} = \frac{\varepsilon_{eff}}{1 - \frac{\varepsilon_{eff}}{\varepsilon}} = \frac{\varepsilon \cdot \varepsilon_{eff}}{\varepsilon - \varepsilon_{eff}} \quad (4.12)$$

or

$$\varepsilon_{eff} = \frac{\varepsilon_{eq}}{1 + \frac{\varepsilon_{eq}}{\varepsilon}} = \frac{\varepsilon_{eq}}{\varepsilon + \varepsilon_{eq}} \varepsilon \quad (4.13)$$

When MLI on the outer surface of spacecraft adopts the effective emissivity model, effective solar absorption α_{eff} shall be introduced to show influence of environmental heat flux like sunlight. The relation of α_{eff} and MLI effective emissivity, hemispherical emissivity ε of outer surface of MLI and solar absorption α_s are given in the following text.

Still taking the aforesaid case as an example, assuming that the solar heat flux on the outer surface of MLI is q_s and the infrared heat flux is zero, the energy balance relationship derived from the effective emissivity model is as follows:

$$q + \alpha_{eff} q_s = \varepsilon_{eff} \sigma T_h^4 \quad (4.14)$$

The energy balance relation of outer surface of MLI is:

$$q + \alpha_s, q_s = \varepsilon \sigma T_c^4 \quad (4.15)$$

By substituting T_h^4 and T_c^4 that is calculated by Formulas (4.14) and (4.15) into Formula (4.9), we have:

4.3 Thermal Insulation Technology

$$q_s\left(1 - \frac{\varepsilon_{eq}}{\varepsilon_{eff}} + \frac{\varepsilon_{eq}}{\varepsilon}\right) = \left(\frac{\alpha_{eff}}{\varepsilon_{eff}} - \frac{\alpha_s}{\varepsilon}\right)q_s \quad (4.16)$$

By substituting Formula (4.12) into Formula (4.16), the left side of Formula (4.16) is equal to zero. Therefore,

$$\frac{\alpha_{eff}}{\varepsilon_{eff}} - \frac{\alpha_s}{\varepsilon} = 0 \quad (4.17)$$

This shows that

$$\frac{\alpha_{eff}}{\varepsilon_{eff}} = \frac{\alpha_s}{\varepsilon} \quad (4.18)$$

or

$$\alpha_{eff} = \frac{\alpha_s}{\varepsilon}\varepsilon_{eff} \quad (4.19)$$

By substituting Formula (4.13) into the above formula, we can get:

$$\alpha_{eff} = \frac{\varepsilon_{eq}}{\varepsilon_{eq} + \varepsilon}\alpha_s \quad (4.20)$$

The above analysis shows that the equivalent emittance model of MLI is closer to the actual situation. Except that the equivalent emittance between the outer surface of MLI and the covered object is a virtual parameter, the thermophysical properties of all outer surfaces are actual parameters. Therefore, the radiation exchange between MLI and the surrounding environment is not changed on matter whether heat flux on the outer surface is infrared radiation or solar radiation.

ε_{eff} and α_{eff} in the effective emissivity model are virtual parameters. When it is based on assumption of diffuse reflection-diffuse gray surface and the heat flux reaching the surface is infrared radiation, the radiation exchange between MLI and the surrounding environment is close to the actual situation. If there is sunlight in the heat flux reaching the surface, the radiation exchange between MLI and the surrounding environment will be quite different from the actual one, which is because the effective solar absorption is far lower than the actual solar absorption of outer surface of MLI in most cases. Hence, if it is treated as an effective emissivity model, most of the solar heat flux will be reflected as sunlight after it reaching the surface of MLI, which is quite different from the actual situation. According to the equivalent emittance model, a part of solar heat flux reaching on the surface will be absorbed and further radiated toward the surrounding environment as infrared heat flux. The proportions of solar heat flux that leaves the outer surface of MLI in these two models in the total heat flux are different, thereby the influence on the surrounding radiation is different. In severe cases, the effective emissivity model will lead to a large difference between the calculated temperature of surrounding environment and the actual situation, even if the error between the temperatures of the objects covered with MLI is not much different from the actual situation.

So far, the effective emissivity model of MLI has not been strictly proved yet. Theoretically, it is only reasonable in special circumstances [7]; in more common cases, the problem is obviously more complicated. Therefore, the effective emissivity model shall be used carefully.

4.3.2.2 Categories and Characteristics

According to different temperature tolerances, MLI is usually divided into low-temperature, middle-temperature and high-temperature MLI.

1. **Category of MLI**

1) Low-temperature MLI

(1) Aluminized polyester film/Dacron separator MLI.

In general, the reflecting screen of low-temperature MLI of the spacecraft adopts leveling and smooth double-sided aluminized polyester film, and the isolated layer adopts Dacron netting. Generally, one reflecting screen and one isolated layer can form one unit, as shown in Fig. 4.77.

The performance parameters of double-sided aluminized polyester film that is commonly used in spacecraft of China are shown in Table 4.3. The commonly used Dacron netting which areal density is about 9.16 g/m² is knitted by 20d/1F Dacron yarns, and the long-term application temperature is no more than 120 °C.

Formula (4.5) shows that, theoretically, the thermal insulation effect of MLI is directly proportional to the reciprocal of the number of layers plus 1. However, when the number of layers increase to a certain value, the thermal insulation performance will not be infinitely improved since as the number of layers increases, the thermal loss

Fig. 4.77 Aluminized polyester film/Dacron separator MLI

4.3 Thermal Insulation Technology

Table 4.3 Performance parameters of commonly used double-sided aluminizing polyester film

No.	Thickness/μm	Unit area weight/ (g·m^{-2})	Coat thickness /μm	Hemispherical emissivity	Solar absorption	Long-term use temperature/°C	Short-term use temperature/°C	Surface resistivity/ (Ω·sq^{-1})
1	6	8.3	0.09 ± 0.01	$0.04^{+0.02}_{-0.01}$	0.09 ± 0.02	−196 to 120	≤150	3
2	12	16.6						
3	18	25						
4	20	27.7						

through the radiation heat transfer is very small, comparing with that of the interlayer thermal conductance and other thermal losses. Considering the above factors, 25 units are usually taken to realize the best thermal insulation effect. During the engineering design, considering different spacecraft have different sensitivities to the heat leakage of MLI, generally, MLI on the outer surface of spacecraft usually uses 10–30 units.

(2) Hollow glass microsphere isolated layer MLI.

In order to reduce the weight of MLI, Technical Institute of Physics and Chemistry CAS in China adopts light and high-intensity hollow glass microsphere with low conduction as the raw material and paste it on the double-sided aluminized polyester film by using a special technique, thereby preparing light MLI without the isolated layer [8]. Since the previous Dacron netting is replaced by the hollow glass microsphere, this MLI can ensure better thermal insulation performance as well as reduce its weight at the same time. Comparing with the conventional "6 μm double-sided aluminized polyester film + Dacron netting" structure, the areal density of each unit decreases from 20 g/m^2 to about 11 g/m^2, but its equivalent emittance is still no more than 0.02, which is within the acceptable range in engineering, as shown in Table 4.4.

(3) Embossed reflecting screen MLI.

The embossed reflecting screen MLI adopts the double-sided aluminized polyester film with repousse as the isolated layer. The characteristic of this type of MLI is the integration of the reflecting screen and the isolated layer, which can reduce the weight of MLI. Only a few points will contact between layers due to the repousse, so heat conductance between layers can be reduced. The double-sided aluminized makes the polyester film work as the wall with low conductivity. Tests conducted by JPL show that comparing with the conventional leveling aluminized polyester film/Dacron separator MLI, the equivalent emittance of this type of embossed reflecting screen MLI without the isolated layer increases by 19% [9].

The hot press molding repousse aluminized polyester film is shown in Fig. 4.78. Compared with the hollow glass microsphere aluminized polyester film, the areal density of embossed reflecting screen with the same thickness is smaller, but it is as same as that of weight reduction effect is more obvious the leveling aluminized polyester film with the same thickness. Hence, the weight is reduced obviously after the wall is removed.

The thermal insulation test results of both conventional and embossed reflecting screen MLI are listed in Tables 4.5 and 4.6, respectively. The hemispherical emissivity of the outer surface of MLI is 0.67 [10].

The aforesaid data shows that the thermal insulation effect of both embossed reflecting screen MLI and hollow glass microsphere MLI are worse than that of the conventional MLI, but their thermal insulation performances are still within the acceptable range in engineering, so they can be adopted in the thermal insulation design of spacecraft. In pursuit of weight reduction and thermal insulation performance, it is necessary to weigh whether to use the light MLI. It should be noted that

4.3 Thermal Insulation Technology

Table 4.4 Performance test results of 15 layers of conventional MLI and hollow glass microsphere MLI

Power/W	Conventional MLI			Hollow glass microsphere MLI		
	Temperature on high-temperature surface T_H/°C	Temperature on low-temperature surface T_C/°C	Equivalent emittance ε_{eq}	Temperature on high-temperature surface T_H/°C	Temperature on low-temperature surface T_C/°C	Equivalent emittance ε_{eq}
0.25	21.6	−68.7	0.0127	17.6	−38.1	0.0183
0.3	35.4	−62.7	0.0128	30.8	−28.3	0.0183
0.4	52.7	−54.9	0.0131	46.4	−16.8	0.0192
0.5	71.4	−46.3	0.0131	60.7	−8.8	0.0200

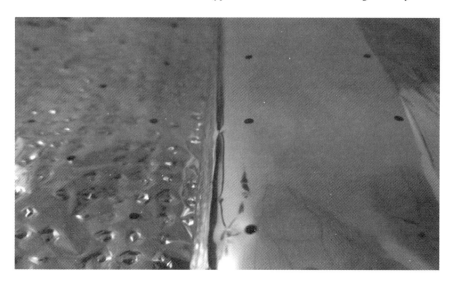

Fig. 4.78 Repousse double-sided aluminizing polyester film (left: repousse film; right: smooth leveling film)

Table 4.5 Thermal insulation test results of 10 units conventional MLI

Test piece temperature /°C	Effective emissivity ε_{eff}	Equivalent emittance ε_{eq}	ε_{eq} calculated according to Formula (4.12)
−10.3	0.0130	0.0133	0.0133
2.2	0.0126	0.0128	0.0128
20.7	0.0114	0.0116	0.0116
40.1	0.0112	0.0114	0.0114

Table 4.6 Thermal insulation test results of 10 units embossed reflecting screen MLI

Test piece temperature /°C	Effective emissivity ε_{eff}	Equivalent emittance ε_{eq}	ε_{eq} calculated according to Formula (4.12)
−11.1	0.0178	0.0181	0.0183
0.8	0.0176	0.0179	0.0181
19.8	0.0169	0.0172	0.0173
39.9	0.0161	0.0163	0.0165

since walls are removed, MLI is sensitive to the compressive load or the tightness of covering. Therefore, those MLIs are suitable for thermal insulation on large-area panel since it is easy to maintain MLI in a naturally loose status.

4.3 Thermal Insulation Technology

(4) Selection of low-temperature MLI surface film.

During the manufacturing of the low-temperature MLI, the inner reflecting screen generally uses a 6-μm-thick double-sided aluminized polyester film to reduce its weight as far as possible. In order to make sure MLI is rigid enough to adapt to the following installation and other operations, thick films are usually used on the outer surfaces of both sides.

For the MLI inside the spacecraft, the 20-μm-thick double-sided aluminized polyester film is usually chosen as the surface film. If requirements for insulation protection on the covered objects are proposed, one side that faces to the covered object prefers to adopt the one-side aluminized film and the non-aluminized shall face to the covered object; or it could adopt insulation film with both sides non-aluminized.

When MLI is applied on the outside of spacecraft, the side facing to the covered object still adopts the 20-μm-thick double-sided aluminized polyester film; while for the space-facing film (which is usually known as the outer surface film), the space adaptability of it shall be considered. Generally, factors such as temperature, radiation (charged particle radiation, ultraviolet radiation, etc.), atomic oxygen and so forth shall be considered. From the perspective of spacecraft overall design, it is generally required that the outer surface film can prevent static accumulation, limit stray light. When the MLI is under sunlight, the outer surface film shall be opaque; otherwise, the temperature of the reflecting screen on the secondary outer surface will be high. In order to reduce solar thermal influence on the covered object as small as possible, the film with low α_s/ε shall be adopted. Aluminized film is not suitable to be used as an outer surface film since α_s/ε of its surface is much larger than 1. Because the temperature of the outer surface film is very high under the sunlight, the reliability of the film (including aluminized coating) at high temperature would be affected, even damaged or out of control even if the thermal influence on the covered object could be accepted. Note that although α_s/ε of the one-side aluminized polyester film is very low, it is not suitable for being used as an outer surface film because of its poor UV radiation resistance ability [11].

As for the low-temperature MLI outside the spacecraft, the outer surface film commonly used in China's spacecraft is shown in Table 4.7. For the thermal design of spacecraft in China, it is worthwhile to draw lessons from the fact that when F46 film is chosen as the outer surface film of low-temperature MLI on surface of spacecraft, it would be adhered to a solid supporting material, such as Kapton, since F46 film would lose its mechanical intensity under the long-term function of the charge particles and thermal circulation [9].

2) Middle-Temperature MLI

The term "Middle-Temperature" of the middle-temperature MLI usually refers to a temperature less than 300 °C in a long term. At such a high temperature, the polyester film reflecting screen is no longer applicable. Generally, the reflecting screen adopts double-sided aluminized Kapton film with allowable long-term application temperature no more than 300 °C and the allowable short-term one no more than 400 °C.

Table 4.7 Performance parameters of outer surface films for common low-temperature MLI

Material performance	ITO-type one-side aluminized Kapton film	ITO-type one-side silvering F46 film	Black Kapton film [12]	Anti-atom oxygen composite film (gray)	Anti-atom oxygen composite film (white)
Description	One side facing to the outside shall be plated with diaphanous ITO, and the other side facing to the inside shall be aluminizing	One side facing to the outside shall be plated with diaphanous ITO, and the other side facing to the inside shall be aluminizing	Naked Kapton film. Generally, both sides are not plated	Knitted glass fiber cloth, dipped in Teflon with 4% nanosilica	
Common thickness /μm	25	75	27	50	
Solar absorption on film surface	≤0.44	≤0.15	≥0.90	0.51	0.21
Hemispherical emissivity of film surface	≥0.60	≥0.62	≥0.78	0.83	0.81
Film resistivity /(kΩ·sq^{-1})	≤250	≤250	$1 \times 10^3 – 1 \times 10^8$	–	–
Volume resistivity /(Ω·m)	–	–	$10 – 1 \times 10^5$	–	–
Unit area weight/(g·m^{-2})	35	195	38	120	
Environmental compatibility	Short-term atomic oxygen, long-term proton, electron and UV radiation	Short-term atomic oxygen, long-term proton, electron and UV radiation	Short-term atomic oxygen, long-term proton, electron and UV radiation	Short-term atomic oxygen, long-term proton, electron and UV radiation	

Performance parameters of double-sided aluminized Kapton film that is commonly used on spacecraft of China are listed in Table 4.8.

There are few options for the isolated layer of the middle-temperature MLI. In general, spacecraft in China chooses high-temperature MLI isolated layer: glass fiber cloth or aluminum silicate cloth. Although temperature resistance of the isolated layer is not a problem and the allowable temperature of it is far higher than that of Kapton reflecting screen, its weight is higher.

By referring to the idea of weight reduction of the low-temperature MLI, the reflecting screen with double-sided aluminized Kapton film can be squeezed with

4.3 Thermal Insulation Technology

Table 4.8 Performance parameters of common double-sided aluminized Kapton film

No.	Thickness /μm	Unit area weight /(g·m^{-2})	Coat thickness /μm	Hemispherical emissivity	Solar absorption	Long-term use temperature /°C	Short-term use temperature /°C	Surface resistivity /(Ω·sq^{-1})
1	12	16.8	0.09 ± 0.01	$0.04^{+0.02}_{-0.01}$	0.09 ± 0.02	−196 to 300	≤400	3
2	20	29.4						
3	25	38.4						
4	50	83.3						

repousse by using hot-press molding technique, thereby eliminating the isolated layer. In fact, it can only consider to use the leveling reflecting screen and to make sure the required thermal insulation performance by adding layers of reflecting screens [13].

Generally, the middle-temperature MLI is not used alone, but it is usually used in combination with the high-temperature MLI. The entire MLI would be seamed or riveted tightly to make sure its mechanical property during manufacturing, so that the compressive load of MLI is higher. Hence, it is not easy to eliminate the isolated layer of the middle-temperature MLI. In recent years, with the development of Kapton fiber technique, the cost has been reduced. Kapton fiber can be knitted as a mesh structure to work as the wall of the middle-temperature MLI, so as to realize the temperature resistance matching between the reflecting screen and the wall, and greatly reducing its weight.

For the middle-temperature MLI that is applied outside spacecraft separately, its outer surface film usually chooses ITO type one-side aluminized Kapton film, Black Kapton film or β cloth as listed in Table 4.7.

3) High-Temperature MLI

When the temperature of long-term application is over 300 °C, it is necessary to consider using the high-temperature MLI. The reflecting screen of high-temperature MLI is usually a variety of metal foils with low emissivity, such as aluminum foil, nickel foil, stainless steel foil and tungsten foil. Considering the restriction of weight and the requirement of intensity, the thickness of tinsel is usually 10–15 μm. Temperature resistance performances of the typical tinsel are listed in Table 4.9.

Table 4.9 Highest allowable application temperature of typical tinsel

Material	Aluminum foil	Nickel foil	Stainless steel foil	Tungsten foil
Maximum allowable application temperature /°C	550	900	1,400	3,000

The isolated layer of the high-temperature MLI of spacecraft in China is usually glass fiber cloth or aluminum silicate cloth. The allowable application temperature of glass fiber cloth could be as high as 900 °C, and that of aluminum silicate cloth could be as high as 1,400 °C. When the temperature is required to be higher, mesh structures made of molybdenum wire, stainless steel wire and tungsten wire can be used as the wall. For example, the allowable application temperature of tungsten mesh could be as high as 3,000 °C.

The high-temperature MLI is usually used for high-temperature radiation protection or plume thermal protection when the spacecraft engine is working. In the case that there may be motor plume on the outer surface of MLI, tinsel of the outer surface needs to be coated with high-temperature anti-oxidation coating by using special technique in case that the residual oxidizing agent or other substance in the plume would corrode tinsel at high temperature. The commonly used coating on the high-temperature MLI of the spacecraft of China is gray chemical transferring coating made of stainless steel foil, and its basic information is shown in Table 4.10.

4.3 Thermal Insulation Technology

Table 4.10 Basic information of chemical transferring coating made of stainless steel foil (gray)

Base material basic information	1Cr18Ni9Ti stainless steel foil	8Cr18Ni10Ti stainless steel foil
Thickness of common basic material /μm	50, 200	
Length and width of basic material /mm	Maximum width: 500; length: 270	
Solar absorption	0.79 ± 0.02	0.84 ± 0.02
Hemispherical emissivity	0.78 ± 0.02	
Temperature resistance performance	1. It shall be able to keep thermal insulation for 2 h in atmospheric pressure environment at 800–900 °C. The coating appearance shall be uniform and without crack, bubble or falling off; it shall have no spot that would affect thermal radiation propertied and mechanical intensity and its solar absorption and hemispherical emissivity changes shall be less than 0.02 2. In atmospheric pressure environment, after is has been through cold and hot alternation, being soaked in liquid nitrogen for 30 s for 150 times continuously and baked at 900 °C for 90 s, the coating appearance shall be uniform and without crack, bubble or falling off; it shall have no spot that would affect thermal radiation propertied and mechanical intensity and its solar absorption and hemispherical emissivity changes shall be less than 0.02	

2. **Functions, performance influence factors of MLI**

1) Influence of gas pressure

According to thermal insulation principles of MLI, the best thermal insulation performance of MLI could be achieved in vacuum (below 1×10^{-3} Pa). With the increase of air pressure, the heat leakage ratio through gas conduction inside MLI increases. When air pressure increases from 1×10^{-3} Pa to atmospheric pressure, gas conduction becomes the main heat transfer method in MLI. Under atmospheric pressure, the equivalent thermal conductivity of MLI is kept still and is approximately equal to the thermal conductivity of gas.

The variation of equivalent thermal conductivity of MLI as vacuum changes is tested by the testing device as shown in Fig. 4.79 [14]. 1D adiabatic slab test method is used in the test: use thin-film-type heater as a heat flux simulative heater and lay it on the surface of the slab; cover the heat flux simulative heater with the tested MLI and paste cold and hot boundary measuring points on both sides of the tested MLI; install tracking operational heater and adiabatic boundary MLI on the other side of the slab. When the current is charged on the slab heater, temperatures of both slab and MLI will increase at the same time and when MLI is steady, the heat is considered to be transferred only through MLI if the temperature of the tracking heat control point and that of the hot boundary temperature are in good consistency. It is better to take

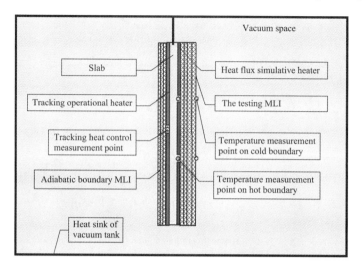

Fig. 4.79 Schematic diagram of test device

equivalent conductivity model since the increasing pressure enlarges the proportion of the gas thermal conduction.

5, 10, 15 and 20 layers, including surface film, reflecting screen, isolated layer and stylolite, are used to test MLI. 18- and 6-μm-thick double-sided aluminized polyester film are, respectively, used as the surface film and reflecting screen, and Dacron netting that have been processed with heat-setting (unit area weight: 10 g/m^2, 4 holes/cm) is used as the wall. Outgassing of MLI for testing is uniform (aperture is 3 mm), and the opening rate of it is 0.5%. The variation of measurement values of MLI equivalent conductivity that changes with the vacuum degree is shown in Fig. 4.80. The test results of the equivalent thermal conductivity are shown in Table 4.11.

As can be seen from Fig. 4.80 and Table 4.11:

(1) Equivalent conductivity changes most obviously when MLI is within pressure scope of 0.1–100 Pa, which increases with the increase of pressure.
(2) In the medium vacuum environment, thermal insulation performance of MLI drops obviously, comparing with that in the high vacuum environment. When vacuum degree changes from 1 to 100 Pa, the equivalent conductivity of 15-layer and that of 20-layer MLI will increase by 8.4 times.
(3) The smaller the MLI layers, the greater the effect of vacuum degree on its equivalent conductivity is. For MLI with more than 15 layers, different layers have little effect on the thermal conductivity.
(4) Vacuum degree of MLI shall be at least higher than 10^{-2} Pa to ignore the influence of gas conduction.
(5) When the pressure is better than 10^{-3} Pa, the change of equivalent conductivity is approaching to be zero. When the pressure is within 0.1–100 Pa, the equivalent conductivity will increase rapidly. When the pressure is higher than 100 Pa, the vacuum degree changes' influence on the thermal conductivity will decrease.

4.3 Thermal Insulation Technology

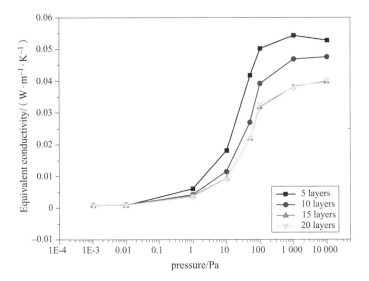

Fig. 4.80 Equivalent conductivity changes with vacuum degree

Table 4.11 Test results of equivalent conductivity   W/(m·K)

MLI, layer number	0.001 Pa	0.01 Pa	1 Pa	10 Pa	50 Pa	100 Pa	1000 Pa	10,000 Pa
5	9.75×10^{-4}	1.01×10^{-3}	6.16×10^{-3}	1.82×10^{-2}	4.19×10^{-2}	5.02×10^{-2}	5.43×10^{-2}	5.28×10^{-2}
10	8.33×10^{-4}	9.78×10^{-4}	4.42×10^{-3}	1.15×10^{-2}	2.70×10^{-2}	3.92×10^{-2}	4.70×10^{-2}	4.75×10^{-2}
15	8.50×10^{-4}	9.26×10^{-4}	3.78×10^{-3}	9.43×10^{-3}	2.25×10^{-2}	3.17×10^{-2}	3.80×10^{-2}	3.99×10^{-2}
20	1.03×10^{-3}	1.01×10^{-3}	3.68×10^{-3}	9.06×10^{-3}	2.21×10^{-2}	3.10×10^{-2}	3.79×10^{-2}	4.01×10^{-2}

2) Influence of temperature

Temperature change will affect surface emissivity of the reflecting screen, thermal conductivity of reflecting screen and the isolated layer, and interlayer contact thermal resistance, etc., which will further affect the equivalent thermal conductivity of MLI. In Formula (4.21), the relation of low-temperature MLI equivalent emittance changing with temperature is given [9]:

$$\varepsilon_{eq} = \left(0.000,136 \cdot \frac{1}{4\sigma T_m^2} + 0.000,121 \cdot T_m^{0.667}\right) \cdot f_N \cdot f_A \cdot f_P \qquad (4.21)$$

where T_m refers to layer average temperature of MLI and it is defined as

$$4T_m^3 = (T_h^2 + T_c^2)(T_h + T_c) \tag{4.22}$$

f_N is a coefficient that is related to MLI layer numbers; f_A is a coefficient that is related to MLI area; and f_P is a coefficient that is related to MLI layer numbers.

When T_m is in 133–413 K, Formula (4.21) is correct. For a specific MLI, f_N, f_A and f_P are constants. f_T is defined as follows:

$$f_T = 0.000,136 \cdot \frac{1}{4\sigma T_m^2} + 0.000,121 \cdot T_m^{0.667} \tag{4.23}$$

The f_T changing with T_m is shown in Fig. 4.81. It is shown in this figure that as layer average temperature of MLI increases, f_T tends to decline. When T_m increases from 130 K to 410 K, f_T declines from 0.039 to 0.010.

It is assumed that T_h or T_c of MLI does not change, the average temperature of the layers of MLI will increase with the increase of T_c or T_h and the thermal insulation performance will improve accordingly as well. From Formula (4.22), the following could be deducted:

$$\frac{d(4T_m^3)}{dT_h} - \frac{d(4T_m^3)}{dT_c} = 2(T_h^2 - T_c^2) \gg 0 \tag{4.24}$$

It shows that the influence of the temperature change of the MLI hot surface on the equivalent emittance is much bigger than that of the cold surface.

The following is a brief discussion of the changing relationship between equivalent thermal conductivity of the low-temperature MLI and the average temperature of

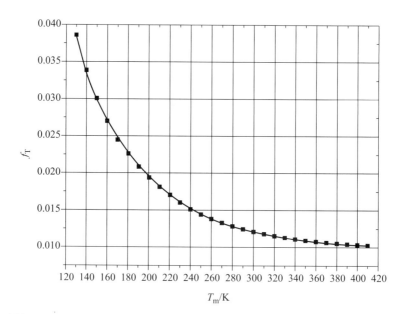

Fig. 4.81 f_T changing with T_m

4.3 Thermal Insulation Technology

layers. The following could be deducted according to Formulas (4.7) and (4.8):

$$h_{eq} = 4\varepsilon_{eq}\sigma T_m^3 \tag{4.25}$$

By substituting Formula (4.22) into Formula (4.25), we get:

$$h_{eq} = \left(0.000,136 \cdot T_m + 0.000,484\sigma \cdot T_m^{3.667}\right) \cdot f_N \cdot f_A \cdot f_P \tag{4.26}$$

Obviously, it is shown in Formula (4.26) that the equivalent thermal conductivity of low-temperature MLI increases with the increase of temperature. f_T' is defined as follows:

$$f_T' = 0.000,136 \cdot T_m + 0.000,484\sigma \cdot T_m^{3.667} \tag{4.27}$$

The f_T' changing with the average temperature of layers of MLI is shown in Fig. 4.82. It is shown in this figure that when T_m increases from 130 K to 410 K, f_T' increases from 0.019 to 0.16. Rangeability of f_T' changing with the average temperature of layers is much bigger than f_T. Figures 4.83 and 4.84 put forward MLI equivalent emittance and equivalent thermal conductivity, which changes with the average temperature of layers of MLI with 10 layers, as measured by Société Nationale Industrielle Aerospatiale, Francais (France National Aerospatial Industry Company), respectively. The laws also meet the above requirements [15].

Because the maximum working temperature of the middle-temperature MLI is only 100 °C higher than that of the low-temperature MLI, and the temperature is not

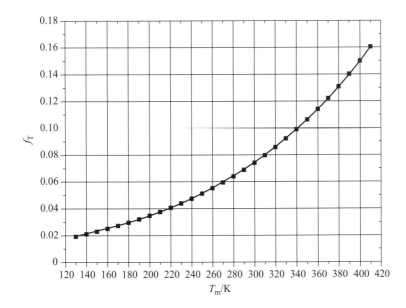

Fig. 4.82 f_T' changing with layer average temperature of MLI

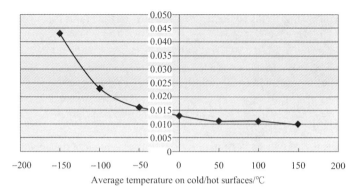

Fig. 4.83 Equivalent emittance changing with layer average temperature

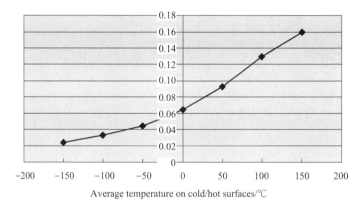

Fig. 4.84 Equivalent thermal conductivity changing with layer average temperature

high, the variation law of its equivalent emittance and equivalent thermal conductivity with the average temperature of layers conforms to that of the low-temperature MLI, that is, the equivalent emittance decreases with the increase of average layer temperature, while the equivalent thermal conductivity increases with the increase of the average layer temperature.

The highest working temperature of high-temperature MLI is much higher than that of the middle-temperature MLI and low-temperature MLI. The emissivity of reflecting screen, the thermal conductivity of the reflecting screen and the isolated layer, and the interlayer contact thermal resistance at high temperature will change more obviously than that at room temperature. All these comprehensive changes result in the variation law of the equivalent emittance of the high-temperature MLI different from that of the middle-temperature and low-temperature MLI. The variation of equivalent emittance measurements of the high-temperature MLI with the average temperature of layers is shown in Fig. 4.85 [16]. Test results of the equivalent emittance and equivalent thermal conductivity are listed in Table 4.12. The reflecting

4.3 Thermal Insulation Technology

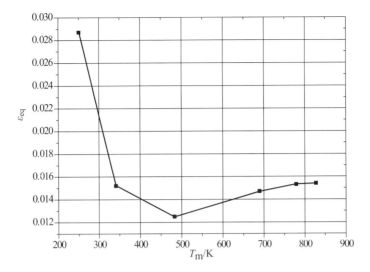

Fig. 4.85 High-temperature MLI equivalent emittance changing with layer average temperature

Table 4.12 Measurement values of equivalent emittance and equivalent thermal conductivity of high-temperature MLI

Test condition	Hot surface temperature /K	Cold surface temperature /K	Layer average temperature /K	Equivalent emittance	Equivalent thermal conductivity /(W·m^{-2}·K^{-1})
Condition 1	294.2	200.7	250.359,763	0.0287	0.129
Condition 2	431.7	231.2	341.263,731	0.0152	0.137
Condition 3	626.0	303.7	482.772,146	0.0125	0.318
Condition 4	891.3	437.7	689.361,235	0.0147	1.095
Condition 5	1006.7	495.0	778.852,717	0.0153	1.637
Condition 6	1069.8	526.0	827.661,201	0.0154	1.987

screen of high-temperature MLI is 15-μm-thick nickel foil and the isolated layer is 100-μm-thick U-Silica, with a total of 5 units. The 1D adiabatic slab test method is used in the test, as shown in Fig. 4.79.

As shown in Fig. 4.85, the equivalent emittance of the high-temperature MLI firstly decreases with the increase of the layer average temperature. It reaches to the minimum value when layer average temperature is about 480 K (the corresponding hot surface is about 620 K and cold surface temperature is about 300 K); then the layer average temperature increases with the increase of the temperature. It can be seen from Table 4.12 that the equivalent thermal conductivity of high-temperature MLI increases with the increase of the layer average temperature.

3) Influence of installation modes

The common installation modes of low-temperature MLI are nylon mesh binding, Velcro lapping and pin-tableting component fixation, as shown in Figs. 4.86, 4.87 and 4.88.

Nylon mesh binding is generally used to fix the MLI with components in complicated shapes, including MLI fixations on the surfaces of sphere, cylinder, cone and their combo. Since most of the covered objects are hook faces and tightness of the nylon mesh is hard to control, which often leads to the decrease of thermal insulation.

The fixation of Velcro lapping and pin-tableting component are usually used to fix MLI that needs to be assembled and disassembled for many times, especially for the large-area fixation. Although these two fixation modes are easy to keep MLI in a naturally loose status, the heat leakage through the pin or the Velcro is considerable. Care should be taken when it is sensitive to heat leakage of MLI. See Table 4.13 heat leakage measurement results through the pin or the Velcro [17].

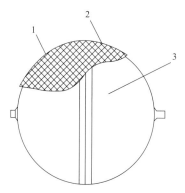

Fig. 4.86 Schematic diagram of binding nylon mesh. 1—Nylon mesh; 2—MLI; 3—Equipment

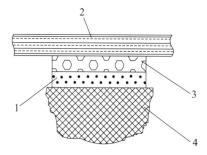

Fig. 4.87 Schematic diagram of Velcro lapping mode. 1—"Hook" of Velcro; 2—MLI; 3—"Ring" of Velcro; 4—Spacecraft structure

4.3 Thermal Insulation Technology

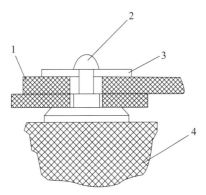

Fig. 4.88 Schematic diagram of pin-tableting component fixation mode. 1—MLI; 2—Pin; 3—Tableting; 4—Deck

Table 4.13 Heat leakage through pin or Velcro

MLI fixation mode	MLI hot surface temperature /°C	Heat leakage
Velcro	0.7	0.27 W/m
	37.2	0.39 W/m
Pin-tableting component	1.1	0.021 W/piece
	37.4	0.028 W/piece

4) Influence of vent hole

Research shows that when the size of MLI is large, it is difficult to draw air through the edge of MLI. In order to ensure that the gas between MLI layers could be exhausted ASAP to realize high vacuum and obtain necessary thermal insulation performance after completing the ground vacuum thermal test and the launching, holes are usually made on the reflecting screen and films to reduce resistance for air exhaust. However, making holes on the film will reduce reflection and thermal resistance, and affect the thermal insulation performance of MLI to a certain extent [18]. Generally, aspiration and thermal insulation effects are to be considered and compared comprehensively.

Although small holes are good for deflate and suction, heat flux through MLI with small holes is larger than that through the MLI with big holes at the same aperture ratio (percentage of opening area in the total area). Then, when the ratio of aperture and the heat radiation wavelength is less than 0.4 (e.g., when the temperature is lower than 300 K, the aperture of the reflecting screen with 6–12-μm-thick aluminum foil or aluminizing polyester film is 2–50 μm), heat radiation actually does not penetrate through the small holes because of diffraction. However, the air exhaust conditions are greatly improved. In engineering, the diameter of the film vent hold is millimeter-level.

Relevant researches show that the aperture ratio plays an important role in the heat flux through MLI. When the opening diameter is constant, the relationship between the measurement values of MLI equivalent emittance and the change of aperture ratio is shown in Fig. 4.89. The MLI test pieces are 45 layers with a thickness of 10 mm. The outer surfaces are all single-sided aluminizing polyester films (film surface faces the heat sink); the diameter of film vent hole is 1.5 mm. The hot surface temperatures of the MLI test pieces are shown in Fig. 4.90 (the heating power for the test pieces is the same) [19].

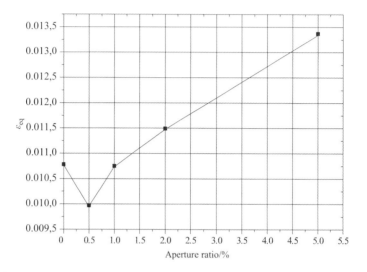

Fig. 4.89 MLI equivalent emittance changing with aperture ratio

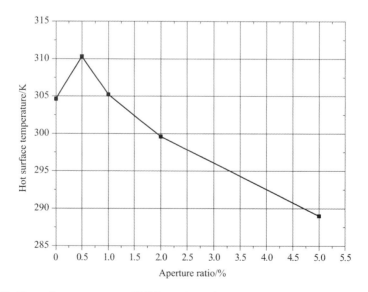

Fig. 4.90 Hot surface temperature of MLI test pieces changing with aperture ratio

As can be seen from Figs. 4.89 and 4.90, under the condition of the same vent hole diameter, the test piece has an optimal opening rate and the MLI has the best thermal insulation performance, showing the lowest equivalent emittance or the highest hot surface temperature. Generally, when the aperture ratio is 0–1%, it can ensure the MLI has the best thermal insulation performance; when the aperture ratio is greater than 1%, thermal insulation performance of MLI decreases with the increase of aperture ratio, i.e., the equivalent emittance increases with the increase of aperture ratio increases.

It can be seen from the above that the air pressure has a great influence on the MLI performance. Therefore, vent holes are set up to speed up the exhausting gas inside MLI, thus establishing a vacuum environment for the operation of MLI ASAP. On the other hand, it improves MLI's adaptability to decompression in the launching phase.

During the launch phase, air pressure outside the satellite drops very fast. Status of CZ-3 and CZ-5 launch vehicles of which fairing column pressure changes with time at the launching phase is shown in Fig. 4.91. It can be seen from this figure that the fairing pressure drops from air pressure to a minimum one every 100 s, and the maximum decompression ratio is about 6.8 kPa/s. When the flow resistance of air inside the MLI on the surface of the spacecraft is large, the internal decompression speed will be lower than the decompression rate in the fairing, thereby causing a certain pressure difference between the inside of MLI and the fairing, especially considering that a considerable amount of gas inside the spacecraft will also be exhausted outward through MLI. Hence, the design of exhaust duct shall be fully considered in the MLI design; otherwise, the MLI will swell when the air pressure at the launch active stage drops rapidly, even be damaged or separate from the spacecraft in severe cases.

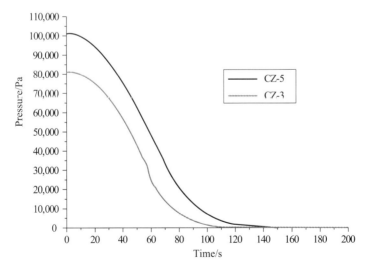

Fig. 4.91 Change of fairing column pressure of CZ-3/CZ-5 at launching phase

The results of JPL show that when a vent hole area of 0.11 cm^2 is provided for each liter of stagnant gas, the requirement for the pressure drop ratio of 8.6 kPa/s can be met. When the pressure attenuation test is conducted at the rate of 15 kPa/s, the MLI designed according to the above codes will not swell.

CAST (China Academy of Space Technology) used the test system as shown in Fig. 4.92 to study the adaptability of MLI to decompression at the launching active phase [20]. A sealing container with a size of 1400 mm × 550 mm × 550 mm shall be put inside the vacuum tank as shown in Fig. 4.92. Except for an opening on one side of the sealed container, which covers MLI, the joint seam of six sides shall be sealed with taps. The aperture ratio of each film of the MLI is 0.5%, which is equivalent to providing 0.09 cm^2 of vent hole area for each liter of stagnant gas. The test shows that when nylon clasp is used to fix MLI and the decompression rate is over 6.8 kPa/s, MLI is well fixed without falling off and without other damages like partial fracture, etc.; under the same decompression rate, although MLI fixed with pin-tableting component does not fall off, MLI on the pin-tableting holes are fractured. Therefore, by adopting a proper fixation method, the vent hole area of 0.09 cm^2 is provided per liter of stagnant gas, which can adapt to the decompression environment at the launching stage of the CZ series rockets.

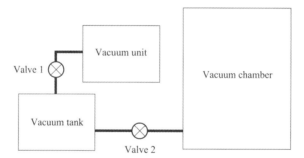

Fig. 4.92 Schematic diagram of MLI's adaptability research test system at the launching phase

4.3.2.3 Selection Principles and Application Cases

1. **Selection principles**

(1) Low-temperature, middle-temperature and high-temperature MLI or the combination of these three shall be chosen according to application occasions.
(2) Surface films for the low-temperature and middle-temperature MLI shall be chosen according to inside and outside features of spacecraft. For MLI surface film on the outer surface of spacecraft, film shrinkage influence at extreme low-temperature condition shall be considered if necessary.
(3) The number of layers of MLI shall be confirmed comprehensively according to the thermal design, weight restriction and implementation technique, etc.

4.3 Thermal Insulation Technology

(4) During application of MLI, air pressure, temperature, vent hole, installation mode influence shall be concerned if necessary.

2. **Application cases**

Case 1: Apply MLI inside satellite.

Status of DFH-4 platform cylinder that is covered with MLI is shown in Fig. 4.93. Because it is installed inside satellite, outer surface film is 20-μm-thick double-sided aluminized Kapton film. It is hard to realize skin coverage of MLI on the sphere hook face. Hence, fixing MLI on the surface of MLI with nylon mesh can ensure MLI is docile and does not cock partially.

Case 2: Apply low-temperature MLI outside satellite.

The status of the low-temperature MLI on a part of the outer surface of DFH-4 platform satellite is shown in Fig. 4.94. The surface film of satellite MLI is ITO-type one-sided aluminized Kapton film. The MLI film on the back of antenna reflector is Black Kapton film, which aims to eliminate stray light interference on the solar sensor nearby when the antenna is deployed.

Case 3: Apply MLI in middle/high-temperature MLI.

The schematic diagram of engine thermal insulation shield is shown in Fig. 4.95. The thermal insulation shield uses high-temperature MLI on the outermost layers of the engine and gradually transits to the medium-temperature and the low-temperature MLI according to the decrease of temperature, which can not only ensure the thermal insulation effect, but also reduce the weight of the whole thermal insulation shield. Status of high-temperature thermal insulation shield film near by the engine of some

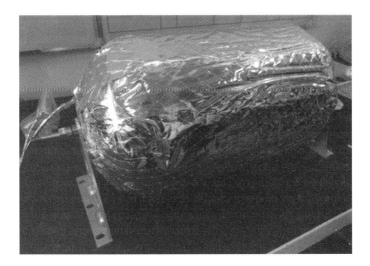

Fig. 4.93 MLI coverage figure of DFH-4 platform cylinder

Fig. 4.94 Status of low-temperature MLI on surface of DFH-4 platform satellite

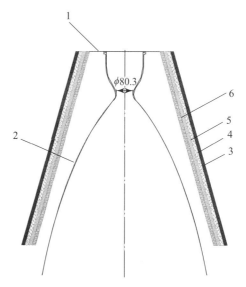

Fig. 4.95 Schematic diagram of engine thermal insulation shield. 1—Connect with structural thermal insulation; 2—Engine effuser; 3—Support of thermal insulation shield; 4—Low-temperature MLI; 5—Middle-temperature MLI; 6—High-temperature MLI

spacecraft is shown in Fig. 4.96. This film is made of stainless steel material and coated with high-temperature oxidation-resistant coating.

Fig. 4.96 Status of high-temperature thermal insulation shield film of some spacecraft

4.3.3 Thermal Insulation of Heat Conductance

4.3.3.1 Functions and Working Theory

In some cases, heat conductance through the contact surface shall be restricted to reduce heat transfer and make sure that temperature of one side could meet requirements, which belongs to the design of heat conduction and insulation. The formula of heat transfer between two contact surfaces is given in Formula (4.28):

$$Q = h \cdot A \cdot \Delta T \qquad (4.28)$$

where A stands for the contact area; h stands for the heat transfer coefficient; and ΔT stands for the temperature difference of two contact surfaces.

Formula (4.28) shows that reducing contact area or thermal conductivity can reduce the heat transfer between the contact surfaces. The design of heat conduction and insulation is to take measures to reduce contact area or thermal conductivity to reduce heat conductance or to increase the contact thermal resistance between two contact surfaces. The simplest and most commonly used measure is to use low-conductivity gasket at the joint of both sides, as shown in Fig. 4.97.

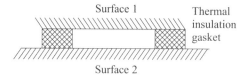

Fig. 4.97 Schematic diagram of thermal insulation gasket on two contact surfaces

After adding the thermal insulation gasket, the surfaces that were in direct contact only have contact at the connection point, thus greatly reducing the contact area. Because the thermal insulation gasket is made of materials with low conductivity, and there is contact thermal resistance between the thermal insulation gasket and surface 1 and surface 2. As a result, the thermal resistance on the joint is also much higher than that without the thermal insulation gasket. The common application of thermal insulation gasket on the installation ear-flake of equipment (to be clear, gasket for screw installation is ignored) is shown in Fig. 4.98 (a). It can be seen from this figure that heat transfer on the joint includes heat transfer through the thermal insulation gasket and that through the screw. Sometimes, even if the screw conductivity is low, the heat transfer cannot be ignored. Hence, the method shownin Fig. 4.98 (b) is adopted, and thermal insulation gasket is also used between the screw and the equipment.

For the thermal insulation method shown in Fig. 4.98 (b), a thermal resistance network can be set on the joint, as shown in Fig. 4.99. In this figure, T_1 and T_2 refer to temperatures of equipment and structure, respectively. Meanings of R_{11}, R_{12}, R_{13}, R_{21}, R_{22}, R_{23} and R_{24} are listed in Table 4.14. Obviously, the total thermal resistance on the joint can be increased by increasing thermal insulation gasket thickness, reducing its sectional area and selecting low-conductivity material. When the

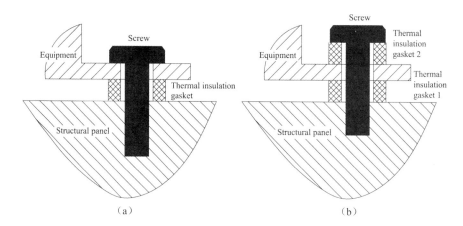

Fig. 4.98 Application method of common thermal insulation gasket on installation ear-flake of equipment

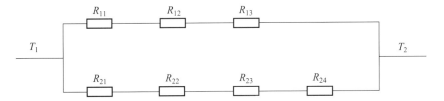

Fig. 4.99 Schematic diagram of thermal resistance network on the joint

mechanical property is allowed, the screw shall be made of materials with low thermal conductivity (such as replacing titanium alloy screw with Kapton screw) to increase the total thermal resistance. When the contact thermal resistance of several contact surfaces makes up a large proportion, it is a common measure to increase contact thermal resistance by replacing a thick insulation gasket with multiple thin ones when the mechanical property and stability are allowed.

4.3.3.2 Categories and Characteristics

Common materials for the thermal insulation gasket include Kapton, fiber-reinforced polymer (or epoxy fiber glass), TC4 titanium alloy, 304L stainless steel, etc. Properties of common thermal insulation materials are listed in Table 4.15.

4.3.3.3 Selection Principles and Application Cases

1. **Selection principles**

The selection principles of thermal insulation materials are listed as follows:
(1) The temperature resistance of the material shall be matched with the temperature conditions of the heat insulation object, and the heat insulation performance shall meet the requirements. And the influence of the thermal conductivity varying with temperature shall be considered.
(2) Within the application temperature range, mechanical properties of material shall meet installation stress and dynamics environmental requirements during launching or landing; and the material stability shall meet installation precision of equipment.
(3) Material processing technique is fine, and its size and precision are easy to meet requirements.
(4) Thermal expansivity of materials matches with that of the thermal insulation object and connection screw, etc.
(5) Total weight of material and CVCM indexes could meet requirements.

Table 4.14 Meaning and calculation formulas of different thermal resistances

Thermal resistance	Meaning	Calculation formulas	Remarks
R_{11}	Contact thermal resistance between thermal insulation gasket 1 and ear-flake	$R_{11} = (h_{11} \cdot A_{11})^{-1}$	h_{11}—contact thermal conductivity between thermal insulation gasket 1 and ear-flake A_{11}—contact area between thermal insulation gasket 1 and ear-flake
R_{12}	Thermal conductance resistance of thermal insulation gasket 1	$R_{12} = \frac{L_1}{\lambda_1 A_1}$	L_1—thickness of thermal insulation gasket 1 λ_1—thermal conductivity of thermal insulation gasket 1 A_1—sectional area of thermal insulation gasket 1
R_{13}	Contact thermal resistance between thermal insulation gasket 1 and structure	$R_{13} = (h_{12} \cdot A_{12})^{-1}$	h_{12}—contact thermal conductivity between thermal insulation gasket 1 and structure A_{12}—contact area between thermal insulation gasket 1 and structure
R_{21}	Contact thermal resistance between thermal insulation gasket 2 and ear-flake	$R_{21} = (h_{21} \cdot A_{21})^{-1}$	h_{21}—contact thermal resistance between thermal insulation gasket 2 and ear-flake A_{21}—contact area between thermal insulation gasket 2 and ear-flake
R_{22}	Thermal conductance resistance of thermal insulation gasket 2	$R_{22} = \frac{L_2}{\lambda_2 A_2}$	L_2—thickness of thermal insulation gasket 2 λ_2—thermal conductivity of thermal insulation gasket 2 A_2—sectional area of thermal insulation gasket 2
R_{23}	Contact thermal resistance between thermal insulation gasket 2 and screw	$R_{23} = (h_{23} \cdot A_{23})^{-1}$	H_{23}—contact thermal conductivity between thermal insulation gasket 1 and screw A_{23}—contact area between thermal insulation gasket 2 and screw

(continued)

4.3 Thermal Insulation Technology

Table 4.14 (continued)

Thermal resistance	Meaning	Calculation formulas	Remarks
R_{24}	Thermal conductance resistance of screw	$R_{24} = \frac{L_3}{\lambda_3 A_3}$	L_3—length of screw λ_3—thermal conductivity of screw A_3—sectional area of stud

Table 4.15 Schedule of common thermal insulation materials

Material	Density /(kg·m^{-3})	Thermal expansivity /(ppm·°C^{-1})	Thermal conductivity /(W·m^{-1}·°C^{-1})	Reference application temperature /°C
Kapton	1,400	0.55	0.3	⩽200
Fiber-reinforced polymer	2,000	–	0.3	⩽100
Titanium alloy (TC4)	4,400	9.3	6.8	⩽350
Stainless steel (1Cr18Ni9Ti or 304L)	7,900	17.2	16.3	⩽700

2. **Application cases**

Case 1: Non-metallic thermal insulation gasket.

The commonly used non-metallic thermal insulation gasket is made of fiber-reinforced polymer and the Kapton insulation gasket. Generally, the mechanical properties and stabilities of both thermal insulation gaskets can meet the use requirements; however, they have different performances of thermal resistance. The thermal insulation gasket made of fiber-reinforced polymer is applicable to temperature under 100 °C. Kapton insulation gasket has better heat resistance than fiber-reinforced polymer gasket and can be used in temperature under 200 °C. Moreover, the latter one costs less than the former. Therefore, thermal insulation gaskets made of fiber-reinforced polymer are largely applied on installation points of electronic equipment, antenna and other parts of which temperature is less than 100 °C. Kapton thermal insulation gasket is mainly used for thermal insulation on parts with higher temperature. For example, for China communication satellite, Kapton thermal insulation gasket is used for thermal insulation of 10 N thruster installation ear-flake as well as the first generation of 490 N engine ear-flake. The temperature of these parts can be higher than 100 °C, but generally not more than 200 °C.

Case 2: Metal thermal insulation gasket.

The common metal thermal insulation gasket is TC4 titanium alloy and 304 stainless steel thermal insulation gaskets. The application temperatures of them are much higher than that of non-metallic ones, but their thermal insulation performances are much worse than that of non-metallic ones. Hence, the thermal insulation performance is usually satisfied by increasing the thickness or quantity. The thermal conductivity of TC4 titanium alloy thermal insulation gasket is lower than that of 304 stainless steel thermal insulation gasket; therefore, TC4 titanium alloy thermal insulation gasket is preferred when temperature condition is allowed. For example, for the second generation of 490 N engine in China ignites, temperature of its installation ear-flake could be as high as 260 °C. Applying Kapton thermal insulation gasket at this time is risky, so TC4 titanium alloy thermal insulation gasket is chosen and one 1-mm-thick Kapton thermal insulation gasket of the first generation of 490 N engine is replaced by two titanium alloy thermal insulation gaskets of which thickness are 1 mm, respectively.

Case 3: Special product with low thermal conductivity.

The special thermal insulation design is often required in the thermal design of spacecraft. For example, one segment of some waveguide of the communication satellite is outside the satellite, the other segments are inside the satellite, and the temperature range of the waveguide outside the satellite is usually wide and changes intensively. If measures are not taken, there will be large temperature fluctuation of equipment that relates to waveguide inside the satellite. In extreme cases, the equipment temperature may not meet requirements. For this situation, instead of using the common aluminum alloy waveguide in the whole process, one segment of the low thermal conductivity waveguide shall be used on the compartment, such as invar waveguide or titanium alloy waveguide (generally, the invar waveguide is adopted since it is applicable for silvering technique on the surface, but the surface of titanium alloy is hard to be silvered), thus reducing the influence of waveguide outside satellite on the inside of satellite. Furthermore, regarding some structure inside spacecraft, like thruster pipeline support, its entire structure is made of materials with low thermal conductivity (such as Kapton) for the purpose of meeting thermal insulation requirements of pipeline.

4.3.4 Thermal Insulation Under Gaseous Environment

4.3.4.1 Functions and Working Theory

Manned spacecraft, space station and other spacecraft are equipped with sealed cabins. Planet lander, rover or atmospheric sounder often need to work in a gaseous environment. These missions all require thermal insulation in the gaseous environment. For example, the sealed cabin of manned spacecraft requires the heat transfer to the wall as small as possible; Mars exploration requires convection heat exchange between the probe and Mars atmosphere as small as possible.

The thermal insulation measures for spacecraft in the gaseous environment are like that in the ground environment, which generally utilizes porous thermal insulation

4.3 Thermal Insulation Technology

materials with low heat conductivity to cover the deck. There should be no gas between the thermal insulation material and the deck (for example, thermal insulation materials and the deck are combined by bonding) or the influence of gas convection could be ignored. In this way, the total heat transfer is limited to the heat conduction through the low thermal conductivity material and the gas, and the heat radiation that may pass through the surface of the thermal insulation material could be absorbed and emitted with multiple times inside the thermal insulation material structure.

In addition, in some cases, if the density and heat conductivity of the ambient gas are very low and gravity is small enough, the effect of convection could be suppressed and good thermal insulation performance can be achieved by simply detaining gas inside the sealed space.

4.3.4.2 Categories and Characteristics

Porous thermal insulation materials commonly used in the gas environment include polyfoam and aerogel. Most polyfoam uses PU polyfoam and Kapton polyfoam, as shown in Fig. 4.100. Aerogel is one of the lightest solids in the world. At present, the density of the lightest monox aerogel is only 3 kg/m^3, which is known as "frozen smoke" or "blue smoke" (Fig. 4.101).

1. **Polyfoam**

Polyfoam is a kind of porous material with closed-cell loose-hole structure (hole chambers are not connected with each other) or with open-chamber loose-hole structure (hole chambers are connected). Due to small pore size (smaller than 3 mm) inside the polyfoam, the heat transfer of gas convection could be ignored. Hence, the heat transfer mode of polyfoam includes heat conductance of the internal small pore wall,

Fig. 4.100 Product of polyfoam. (a) PU polyfoam; (b) Kapton polyfoam

Fig. 4.101 Aerogel and its structure. (a) Aerogel; (b) Structure of aerogel

heat conductance of gas inside small pore and heat radiation between the pore walls and between the pores.

Generally, the performance of the closed-cell polyfoam is poor at low pressure since the residual gas pressure in the closed-cell polyfoam is high and the contribution of gas conduction is great. Study of Basotect foam shows that at the same gas pressure, the gas conduction contributes the most to heat transfer of the thermal insulation layer. When air pressure decreases, the heat conductivity of the polyfoam will drops largely since the decrease of gas pressure will weaken the heat conduction of gas inside the micropore.

Generally, the heat conductivity of polyfoam decreases with the decrease of temperature. The variation of thermal conductivity of soft polyimide foam with a density of 15 kg/m^3 under room temperature in air environment is provided. This is because the temperature drop will decrease the heat conductivity of air in polyfoam, thus decreasing the heat conductivity of the entire polyfoam. Therefore, the influence of temperature on heat conductivity shall be fully considered when using this kind of thermal insulation material at low temperature on spacecraft.

2. **Aerogel**

Like polyfoam, the heat transfer of gas convection in the aerogel can also be ignored in gas environment. There are three main ways of heat transfer inside it: heat conduction along the solid framework, gas conduction in the structure and radiation heat exchange. Aerogel is a kind of nanoporous amorphous material with low density and continuous irregular network structure. The high void age of aerogel makes the percent of solid very low, which reduces the heat conduction through the solid phase. In addition, the sinuate structure of solid phase is beneficial to reduce the heat conduction. Generally, the characteristic void dimension of aerogel is much smaller than micron dimension, which makes the gas heat transfer in aerogel smaller than that in free gas. Therefore, aerogel can achieve a total thermal conductivity less than 0.026 W/(m·K) at environmental temperature.

4.3 Thermal Insulation Technology 191

It is difficult for the aerogel to inhibit radiation because of its translucence structure. However, adding infrared radiation absorbing substances in the aerogel will weaken the radiation characteristic of it. For example, the heat conductivity of aerogel mixed with carbon at room temperature can be reduced to 0.013 W/(m·K); adding titanium dioxide in it make silica aerogel a new type of high-temperature thermal insulation material. The thermal conductivity is only 0.03 W/(m·K) when temperature is 800 K. At room temperature, the thermal conductivity of silicon dioxide aerogel with a density of 100 kg/m^3 is 0.02 W/(m·K), lower than that of air.

Aerogel is a very ideal thermal insulation material. However, its special network structure, high void age, and low density make it brittle. At high temperature, translucence aerogel is difficult to resist the effect of impedance radiation. Consequently, aerogel shall be compounded with other materials to realize the actual effect in many fields. Aerogel compound thermal insulation materials mainly include fibrous compound aerogel material and sunscreen aerogel compound, and in practice, aerogel is usually combined with opacifier and fibrous reinforcement.

4.3.4.3 Selection Principles and Application Cases

1. **Selection principles**

The application of porous thermal insulation materials shall follow the following basic principles:

(1) In determining whether the thermal insulation performance meets the thermal design, besides the performance of the material itself, the influence of temperature level, atmosphere or vacuum environment, stability and other factors on the thermal insulation performance shall be considered.
(2) The weight limit, implementation technique, adaptability to mechanical environment and other factors shall be considered comprehensively.
(3) Total weight loss and CVCM limits shall be considered. For manned spacecraft, material toxicity should also be considered.

2. **Application cases**

Case: Apply polyfoam on China's manned spaceship.

In addition to its good thermal insulation performance, PU soft polyfoam has many advantages such as ease of use, easy cutting or shaping, light weight and good stability. Therefore, it is used for the thermal insulation of the sealing cabin of China's Shenzhou spacecraft. Generally, PU polyfoam on inner surface of cabin wall is of 20 mm thick, and its density is no more than 20 kg/m^3. A polyacrylate PSA and a layer of fire-retardant fabric are applied on the side of the polyfoam facing the cabin. The performance parameters of several PU polyfoam used on Shenzhou spacecraft are listed in Table 4.16 .

Table 4.16 Performance parameters of PU polyfoam on Shenzhou spacecraft

Specification performance	C-1	X-1	C-2
Density /(kg·m^{-3})	⩽20	20–35	⩾35
Thermal conductivity /(W·m^{-1}·K^{-1})	RT, atmospheric pressure: ⩽0.04	RT, 0.5 atm: ⩽0.04 Room temperature, 0.3 atm: ⩽0.035	RT, atmospheric pressure ⩽0.05
Tensile strength /kPa	⩾70	⩾90	⩾90
Breaking elongation /%	⩾50	⩾60	⩾60
Odor grade	⩽1.5	–	⩽1.5
Toxicity /(μg·g^{-1})	Total VOC: ⩽100 CO content: ⩽25	–	Total VOC: ⩽100 CO content: ⩽25
Tear strength /(N·m^{-1})	⩾100	⩾120	⩾100
Rebound /%	20–25	25–35	35–40
Weight loss /%	⩽1	⩽1	⩽1
CVCM /%	⩽0.1	⩽0.1	⩽0.1
Hydroscopicity /%	⩽15	Confirmed as needed	Confirmed as needed
Radioresistance dose /rad (Si)	⩾10^4	⩾10^6	⩾10^4
Temperature resistance performance	Color does not change after being baked at 150 °C for 3 h	Thermal insulation performance does not change after cold and hot alterations for 300 times at −196 to 120 °C	Color does not change after being baked at 120 °C for 3 h
Application temperature zone /°C	−100 to 150	−196 to 120	−100 to 120

Note: 1 atm = 101.325 kPa

4.4 Heating Technology

4.4.1 Introduction

Generally, the thermal control of spacecraft can be achieved by heat dissipation and thermal insulation. However, due to the influence of specific spacecraft orbit, attitude and operating modes, the lowest operating temperature of the equipment cannot be achieved only by heat dissipation or thermal insulation measures. In this case, heating technology must be adopted to ensure that the equipment in low-temperature environment could work normally or meet the requirements of low-temperature survival (when the equipment does not need to operate). Heating technology is the commonest technology in thermal control system of spacecraft, which converts electric energy or solar energy into thermal energy. The full-scale temperature control system, which is composed of heating loop, thermometry sensor and thermal control electronic device, can accurately control the temperature of some special equipment.

4.4.2 Electrical Heating

4.4.2.1 Functions and Working Principles

There are many ways of heating, such as infrared heating, electrical heating and laser heating. Electrical heating is the commonest heating technology used in spacecraft thermal control system. By mounting the electrical heater on the targeting equipment and powering, the electric energy can be converted into thermal energy.

Electrical heater is generally composed of three parts: electrical heating component, electrical insulating layer and electric cable.

The materials for electrical heating components are called electrical heating materials. Theoretically, all electric conduction materials with high resistance can be used as electrical heating materials. In practice, electrical heating materials should have better mechanical property, heat resistance and oxidation resistance. The common electrical heating materials are mainly nickelin (constantan), nichrome, alchrome, simple metal with high melting point and graphite.

The main function of electrical insulation layer is to isolate electrical components from the heated equipment to get good electrical insulation. The resistance of electrical insulation materials shall be higher than 10^9 Ω cm. In some cases, except insulation performance, the electrical insulating layer shall provide mechanical support, fixation and protection of the electrical heater. Therefore, the electrical insulating layer should not only have good mechanical strength, heat resistance and high-pressure-resistant performance, but also adapt to the effect of high vacuum and outer space radiation environment in spacecraft applications. Common electrical insulation materials mainly include polyimide film, paint, fiber fabric and metal sleeve powder.

The electric cable of the heater is used for connecting power, which usually adopts a conductor with an insulating layer. Attentions should be paid to composition, structure, application temperature and adaptability to space environmental when selecting electric cables. The conductive wire core of an electric cable could be single/multi-metal wires, metal strip or metal foil, and the most common used structure is a tinning or silver copper core. The insulating layer of the conductor can be polyvinyl fluoride, polyethylene, fluorine 46, polyimide. In general, the conductors used in spacecraft are AF46 conductors with Teflon FEP as insulating layers and C55 conductors with ETFE as insulating layers.

4.4.2.2 Type and Characteristics

The commonly used electrical heater on spacecraft mainly includes film foil heater, wire strap heater, armored electrical heater.

1. **Film foil heater**

Two-layer or multi-layer polyimide films are used as insulating layer in the film foil heater, and the internal heating component is alloy foil or alloy wire electrical heater. Polyimide film-type electrical heater is generally composed of shaped foil electrical heating component with certain resistance value and insulation film by hot press. The active heating area can be tailored into desired shapes mechanically or by chemical etching. According to their shapes, they are often called patch heaters. The thickness of the patch heater can be as small as 0.2 mm. It has the advantages of flexibility, tunable resistance value, low density, low thermal inertia and strong radiation resistance, especially suitable for temperature control of precise devices. The working temperature should be controlled below 125 °C.

Film foil heater could be shaped to band, which is called wire strap heater. It can be wrapped on the surface of target product and fixed with silicone rubber when used for thermal control of spacecraft pipeline. The electrical heating components in the electrical heating band are constantan wires and the resistivity of constantan wire changes with wire diameters. In the design of electrical heating band, both the length and the designed resistance value shall be considered, and constantan wires with different wire diameters should be selected.

A typical film foil heater is shown in Fig. 4.102.

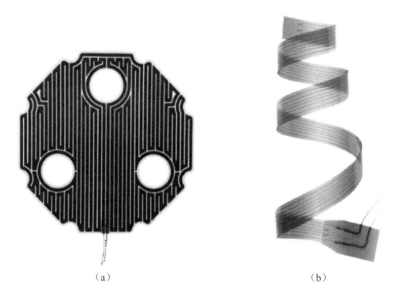

Fig. 4.102 Film foil heater. (a) Sheet-shaped film foil heater; (b) Wire strap heater

In addition to the shapes, the composite layer and heating loop numbers are the other two design parameters that meet the requirements of specific engineers. They can be one-layer single-loop heater, one-layer multi-loop heater, double-layer single-loop heater, double-layer double-loop heater, etc.

2. **Armored electrical heater**

Literally, armored electrical heater refers to a layer of "armor" attached to the heater to obtain high temperature resistance. There are two kinds of common armored electrical heaters: sleeve type and customized type.

The heating component of sleeve-type armored electrical heater is alloy wire with high resistivity, such as nichrome or alchrome. The metal sleeve outside the heat component serves as the mounting surface. Space between the sleeve and the heating component is filled with magnesium oxide or aluminum oxide ceramics to realize electronic insulation. Commercial sleeve-type armored heater can be hundreds of meters in length. For specific application, the size of the heater can be adjusted according to the design resistance value. It can be coiled or wrapped on the heated object. Armored electrical heater has the characteristics of pressure resistance, impact resistance and high-temperature resistance. Its operating temperature can be as high as 900 °C.

The customized armored electrical heater is usually customized for some special equipment, and its heating component is a heating core composed of nichrome/alchrome high-temperature alloy wire and supporting framework. The supporting framework is made of ceramics/quartz glass or other insulation materials with enough mechanical strength, excellent high-temperature insulation and thermal shock resistance. The heating core is coiled on the framework by spiral winding or other ways to obtain the resistance density much higher than that of common sleeve-type electrical heater. It can keep adequate mechanical properties when operating at 1,100 °C. The external protection and mounting shell are metals with different thickness and shapes, such as light gauge stainless pipes. Even if the framework of heating core is an insulator, it is still necessary to fill in magnesium oxide or aluminum oxide powder between the heating core and the shell to realize credible insulation. Different from sleeve-type electrical heater, the customized armored heater is an integral product with fixed resistance value and shape after production. In other words, it is usually designed for oriented applications. A typical armored heater is shown in Fig. 4.103.

3. **Ceramic-metal integrated heater**

Ceramic-metal integrated heater is a new heating technology that is developed based on nichrome electrical heater. It is made of ceramic-metal integrated material by high-temperature cofired ceramic technology, and it could definitely change the structure of the mini-type armored electrical heater, which can improve the power density and meet the requirements for small space and small power on the one hand; and on the other hand, it could improve service life and reliability of the component.

The ceramic-metal integrated heater is a ceramic-based electrical heating component that is prepared with metallic heating core and ceramic matrix by lamination

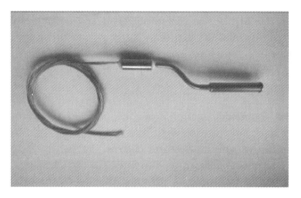

Fig. 4.103 Armored electrical heater for thrusters

modeling and cofired technology. The heating material for the conventional armored heating components is nichrome wire in very fine diameter (straight or spiral winded), and the insulation materials are quartz glass tube and magnesium oxide powder. Comparing with the conventional heating component, heating materials for the new type of heating components adopt metal or alloy (such as tungsten, molybdenum) with high resistance value and high melting point, and the heating component is modeled into thick film. The major matrix material is ceramic with high intensity, low expansion, high thermal conductivity and high-temperature resistance (such as silicon nitride, aluminum nitride), which is a good insulation material as well. The heating component made in this material combines advantages such as simple structure, high operating temperature, high mechanical strength, stable chemical performance, good insulation, anti-humidity and anti-thermal impact.

The ceramic-metal integrated heater can withstand the environmental temperature up to 1,500 °C. By choosing different heating materials, the power density can reach tens of W/cm^2, which is higher than that of current nichrome electrical heater.

The heating material of the ceramic-metal integrated heater is in planar shape, which results in a 30% to 50% decrease in surface load compared with the winding-type heating material. Since the heating component is uniform and laid distributed, insulation problems between wires can be well solved by improving ceramic matrix sinter quality; therefore, the service life of the heating component can be greatly improved. A typical ceramic-metal integrated heater is shown in Fig. 4.104.

4.4.2.3 Selection Principles and Application Cases

1. **Selection principles**

In the thermal design of spacecraft, the electrical heaters shall be selected according to the temperature, geometrical features, thermal characteristics, heater power and its application environmental testing conditions of the heated object.

4.4 Heating Technology

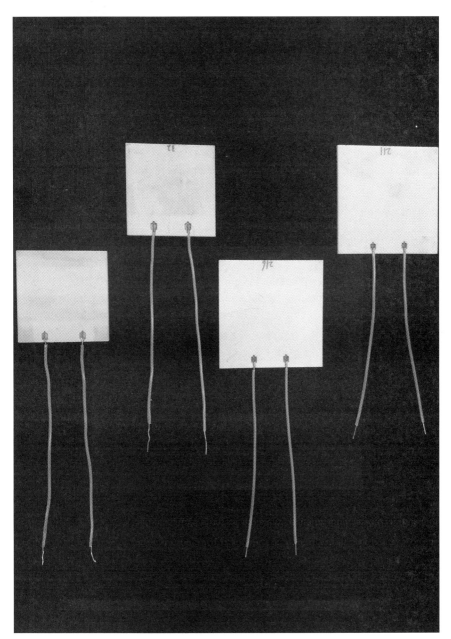

Fig. 4.104 Ceramic-metal integrated heater

(1) Film foil heater can be used for the heated object with large heating area and good temperature uniformity.
(2) The wire strap heater can be used for line or doll-shaped heated object.
(3) The wire-wound resistor-type electrical heating component can be used for equipment outside of spacecraft with small sticking area such as antenna deployed mechanism and retaining mechanism.
(4) Small thrust engines generally adopt armored electrical heater, and heater strip heater can be used for regular thruster.
(5) The application temperature of electric heater has upper limit, and electric heater shall be selected according to the requirements and limits of heating temperature control.

2. **Application cases**

Case 1: Apply film foil electrical heater in temperature control of electronic equipment.

Film foil electrical heater is the most common heater in spacecraft thermal control system, such as compensation heating and temperature control of the equipment. The application of film foil electrical heater in triple floated gyroscope is shown in Fig. 4.105.

Case 2: Apply film foil electrical heater in temperature control of optical system.

Spacecraft optical systems, such as camera, infrared Earth sensor, are very sensitive to temperature changes. Film foil electrical heater is an important way to realize temperature control. The application of film foil electrical heater in infrared Earth sensor is shown in Fig. 4.106.

Fig. 4.105 Typical application of film foil electrical heater in temperature control of electronic equipment

4.4 Heating Technology

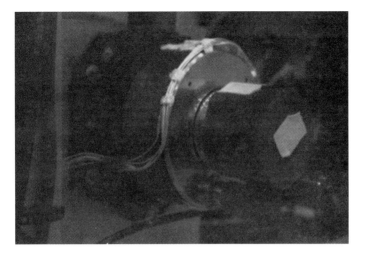

Fig. 4.106 Typical application of film-type electrical heater in infrared Earth sensor

Case 3: Apply film foil electrical heater in overall temperature control of RAMP.

In general, the film foil electrical heater can be directly pasted on the RAMP to realize unified temperature control of all equipment on one RAMP; and the heat pipe is embedded in the corresponding position of the heater, so the temperature control of the entire RAMP is realized through compensation heating and effective heat expansion of the heat pipe, as shown in Fig. 4.107. When the RAMP consists of a heat pipe network that is composed of the embedded heat pipes and surface heat pipes, a film foil electrical heater can be pasted on the surface of surface heat pipe to realize the unified heating temperature control of the entire RAMP.

Fig. 4.107 Typical application of film-type electrical heater in overall temperature control of RAMP

Case 4: Apply wire strap heater in temperature control of tube line.

The heating temperature control is necessary for the hydrazine line since liquid working fluid of the propulsion system shall be protected from low temperature. Hydrazine line is usually 6 mm in diameter and small in size, which is not convenient for the installation of film-type electrical heater. Hence, the wire strap heater is mounted by winding, as shown in Fig. 4.108.

Case 5: Apply armored electrical heater in temperature control of thruster.

The armored electrical heater is usually installed by winding, and its resistance value is related to the heater length. Its application in bi-propellant 10 N thruster is shown in Fig. 4.109.

Besides, the available space for the monopropellant small attitude control thruster is small and complicated, and the power requirement is small, which makes the sleeve-type armored heater fail to meet requirements for resistance density and

Fig. 4.108 Typical application of wire strap heater in hydrazine line

Fig. 4.109 A typical application of sleeve-type armored heater in temperature control of thruster

4.4 Heating Technology

mounting mode. Hence, a specially customized mini-type armored heater for thruster is adopted.

4.4.3 Radioisotope Heating Technology

4.4.3.1 Functions and Working Principles

Generally, the radioisotope heating technology is applied on deep space exploration to solve the energy shortage problem caused by being away from the sun or lacking of sunlight for a long time, to assist the electrical heater for temperature control and even work as the main heat supply source to provide enough energy, such as the Mars rover, the lunar lander/rover.

Radioisotope heaters such as RHU and RTG are commonly used. Among them, the working principles of RHU is: energetic particles (such as proton, α particle, β ray and γ ray) emitted by radioisotope interact with substance and they are absorbed by substance eventually. Then kinetic energy of ray transforms to thermal energy and rises temperature of the active substance to provide thermal energy to the outside. Based on RHU, RTG could transform thermal energy to electric energy by taking advantages of kinds of thermoelectric conversion modes. Application of the RTG includes electric energy and thermal energy since RTG can generate power and produce heat.

The heat emitted by radioisotope decrease progressively according to the exponent rules and does not need to control. The output thermal power of this heat source can be determined according to the following formula:

$$P(T) = P_0 e^{-\lambda t} \tag{4.29}$$

$$\lambda = \frac{0.693}{T_{1/2}} \tag{4.30}$$

where $P(T)$ refers to thermal power of time t; P_0 refers to initial thermal power; t refers to time; λ refers to decay constant; $T_{1/2}$ refers to half-time of radioisotope.

It can be seen from the above formula that the thermal power output by radioisotope heat source is related to the half-time of the radioisotope. Another important parameter of radioisotope is mass ratio power. Volume specific power is a more commonly used parameter since radioisotope compound often used as fuel.

4.4.3.2 Classification and Characteristics

The radioisotope heaters can be classified to RHU and RTG generally. While there are many types of RHU and RTG according to the adoption of different types of radioisotopes. The common radioisotopes mainly include ^{238}Pu, ^{210}Po.

Radioisotope heater has many advantages, such as long service life, compact structure, high reliability, strong survivability, no need of input of the external energy, no need of maintenance. But its disadvantage is that the calorific value attenuates with time. In the process of storage and transportation, heat rejection measures must be taken, especially the influence on human body or other creatures shall be concerned specially, and necessary protective measures shall be taken. The manufacturing technique of radioisotope heater is complicated and the cost is high.

4.4.3.3 Selection Principles and Application Cases

1. **Selection principles**

While selecting radioisotope heat/electric source, the rules below shall be generally followed:

(1) Select radioisotope heat source with higher power.
(2) It shall be ensured that the thermal power does not decrease significantly and the thermal power attenuation is not more than 10% to 20% during the reserved service life. It is regulated in the conventional selection principles that the proper half-time period is between 100 days and 100 years, and the specific selection shall meet the engineering requirements.
(3) Radioisotope's radiation toward the outside shall be as low as possible. α radioisotope is usually adopted as fuel.
(4) Verify and confirm the availability of radioisotope fuel. Generally, the applicable α radioisotope fuel is produced in reactor irradiation, so it is very expensive. Hence, its cost accounts a large proportion in the radioisotope heat (electric) source.
(5) The currently use of radioisotope fuel is limited to solid only. It shall conform to requirements such as high melting point, high thermal conductivity, oxidation resistance, corrosion resistance, insolubility, non-volatile, hard to be absorbed by creatures, hard to accumulate inside creatures. Besides, it shall have good mechanical processing performance.
(6) During the design, manufacturing and application of radioisotope fuel, it shall follow the United Nations regulations on principles of applying nuclear energy in astrospace and relevant laws and regulations of countries. Consequently, multiple protective measures such as redundancy design and so forth shall be taken in the design of radioisotope heat source so the design could be extremely reliable. Once a protective measure fails, it can still ensure safety without nuclear contamination.

2. Application cases

Case: Apply radioisotope heating technology in spacecraft of China.

Three sets of RHU are adopted on the Chang'e 3 lunar probe platform system launched on December 2, 2013, to supply heat for survival at lunar night, including one set for lander $+Y$ cabin, one for $-Y$ cabin and one for the rover.

Chang'e 3 RHU is ^{238}Pu, with the energy/weight ratio of 50 W/kg and the halftime of 87.7 years. It has stable calorific value while working on the lunar surface, and heat supply for each RHU is (125 ± 5) W. RHU is composed of cover plate, sleeve, radioisotope nuclear and radiating fin, out of which, the cover plate and sleeve are referred to as shell. The overall structure of RHU is shown in Fig. 4.110.

Among them, the RHU cover plate is used for fixing radioisotope nuclear mounted in the sleeve; the sleeve and RHU cover plate compose a chamber for containing the radioisotope nuclear which is orientated and loaded in the chamber, and a mechanical interface connected with the detector and a thermal interface connected with the two-phase liquid loop are provided; the radioisotope nuclear is a core part for generating heat; the radiating fin is used to increase the heat dissipation and control the temperature of RHU shell.

RHU is mounted on the outside of the lander and the rover and thermally couples with the inside of lunar probe through a two-phase liquid loop. During the lunar night, most of the heat of RHU could be transferred from outside of the detector to the inside of it by using two-phase liquid loop so as to ensure temperature for survival of detector at lunar night; while a small part of heat could be rejected to the astrospace by radiation of RHU. During the lunar day, the two-phase liquid loop

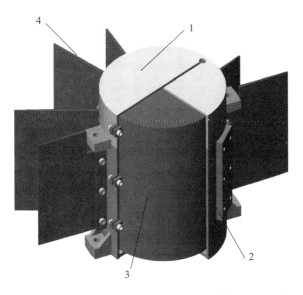

Fig. 4.110 RHU structure. 1—Coverplate; 2—Sleeve; 3—Radioisotope nuclear; 4—Radiating fin

stops operating, the RHU will no longer transfer heat to the inside of detector, and calorific value of RHU will all be rejected to the astrospace by radiation. In addition to the above basic functions, RHU temperature shall be within a reasonable range at the lunar night and the lunar day due to reliable application of RHU itself on one hand and application and management of heat, and safe and reliable application of the two-phase liquid loop on the other hand.

Figure 4.111 shows the temperature change of RHU of $\pm Y$ cabins of the lander from December 2013 to December 2014. It is seen from this figure that monthly change rules of RHU are basically the same. The highest temperature in winter (solar irradiation intensity approaches) is basically the same. The highest temperature of RHU shell in $+Y$ cabin is 226 °C, and that in $-Y$ cabin is 200 °C.

Considering the half-time (87.7 years) of RHU nuclear ^{238}Pu and thermal power attenuation (0.8%/year), RHU thermal power (which is 127 W at launching) will be reduce by 1 W in one year. In addition, because of on-orbit degradation of KS-Z white paint on RHU shell and the increase of solar absorptance, the solar absorptance increases by about 0.05–0.10 in one year, and the growth of its solar heat flux absorption is equivalent to the thermal power attenuation, and lunar dust's influence on RHU radiating surface could be ignored. Hence, RHU temperature is relatively stable. The analysis shows that the thermal power of Chang'e 3 RHU on-orbit does

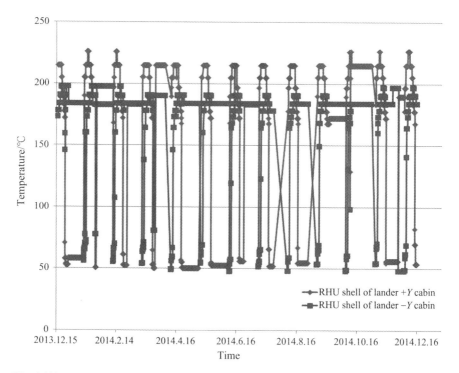

Fig. 4.111 Temperature change of lander RHU at lunar surface within one year

not decrease significantly, and its thermal supply capacity decreases by 0.8% in one year, which is in line with the attenuation prediction of radioisotope nuclear.

4.5 Temperature Measurement and Control Technology

4.5.1 Introduction

Spacecraft thermometry and control technology is the most common procedure in thermal design of spacecraft. By setting up a heating loop and a temperature sensor on the surface of the target device, the collected temperature signal is transmitted to the applicable electrical signal by the thermometry circuit and fed back to the control circuit, which drives connection/disconnection of heating loop, to get the ideal temperature. Thermometry and temperature control technologies are two essential branches of this technology.

The current thermometry and control loop are mainly composed of temperature sensor, measurement circuit, control circuit and heating loop. Most thermometry and control systems are configured with temperature control software to realize active closed-loop control.

The control modes include open-loop control and closed-loop control. The former requires the whole spacecraft to send program control commands to control connection/disconnection of heating loop, without special circuit design and software. The design is simple and reliable, suitable for short-term use. The thermometry loop is only used for temperature monitoring. For example, the compensation heating loop in the early stage of on-orbit operation operates only before the payload starts. Once the payload is started up and operates normally, the heating loop will cut off. However, the closed-loop temperature control loop could use thermometry signal and software to control connection/disconnection of the heating loop autonomously. Generally, its design is complicated and based on temperature control software. For example, the closed-loop thermometry and control system applied on some spacecraft is composed of the thermistor, the temperature control unit and the film heater loop. Among them, the temperature control unit includes secondary power module, temperature collection circuit, control circuit, heating driving circuit and other hardware as well as temperature control software. Temperature control threshold values of the heating loop can be adjusted as desired making the system reliable and flexible.

The main temperature control targets on a spacecraft are critical load, equipment with special temperature requirements and special structures, such as optical camera, huge antenna, battery, propulsion system. In practice, redundancy design of primary backup is often adopted to improve the reliability of thermometry and control system.

4.5.2 Thermometry Technology

4.5.2.1 Functions and Working Principles

Thermometry is an important branch of the thermal design in spacecraft system. It is not only an important parameter to evaluate the operation status of the spacecraft thermal control system, but also an important telemetry data of the spacecraft. Especially when the spacecraft is out of state on orbit, the temperature data has always been an important clue to determine the originator of the malfunctions. Therefore, a large number of remote channels in spacecraft need to transmit the important temperature data in real time.

According to the zeroth law of thermodynamics, temperature is an equivalent state parameter for a thermal steady system. It is an important physical quantity to describe the heat degree of an object in thermal steady state, determined by particle movement status inside objects. Different from other physical quantities, temperature cannot be measured directly. For example, the weight or length value of a measured object could be compared with relevant units, while temperature can only be derived by measuring the relevant physical quantities. Hence, temperature is a non-additive intensive quantity.

Spacecraft thermometry is based on a large amount of sensors, which are made of materials that are sensitive to temperature. The numerous thermometry technologies can be classified into different species from different perspective. It can be classified into non-electric measurement and electric measurement according to whether there is electrical signal from the sensor during measurements. It could also be classified to contact and non-contact mode according to different contact modes between sensors and measured objects. Each method can be realized by a variety of temperature sensors, such as expansion sensor, metallic thermocouple sensor, thermistor, semiconductor thermistor, optical sensor, infrared sensor.

In a spacecraft system, contact thermometry technology is adopted with the help of thermistor senor, thermal resistance sensor and thermocouple. Their working principles are introduced below.

1. **Thermometry principles of thermistor**

Thermistor is a kind of semiconductor ceramic made of multiple metals by mixing and grinding, shaping, sintering and sealed with glass, epoxy, etc. In general, NTC thermistor (negative temperature coefficient) is used as temperature sensor. The resistance decreases with the increase of temperature. The resistance temperature coefficient is usually −(1% to 6%)/°C. A typical R_t-T curve of NTC thermistor is shown in Fig. 4.112, and the mathematical expression of its resistance–temperature relation is shown as follows:

$$R_1 = R_2 \exp\left[B\left(\frac{1}{T_1} - \frac{1}{T_2}\right)\right] \quad (4.31)$$

4.5 Temperature Measurement and Control Technology

where R_1 and R_2 are resistance values of thermistors when temperatures are T_1 and T_2; B refers to the material constant of NTC thermistor.

B is a constant that describes physical characteristics of material of NTC thermistor. In general, the higher B is, the higher the absolute sensitivity will be. Technically, B is not a constant since it increases a little as temperature increases.

According to Formula (4.31), the formula of thermistor material constant B is as follows:

$$B = (\ln R_1 - \ln R_2) / \left(\frac{1}{T_1} - \frac{1}{T_2} \right) \quad (4.32)$$

As shown in Fig. 4.112, B represents the slope of the straight line passing through $(1/T_2, \ln R_2)$ and $(1/T_1, \ln R_1)$. $\ln R$ is the ordinate, and $1/T$ is the abscissa.

The temperature characteristics of thermistor are nonlinear. Therefore, its interchangeability is poor. Hence, when used for thermometry, each thermistor should be calibrated at several typical working temperature zones and fitted with Formula (4.33) or (4.34).

$$R = \exp \left[\sum_{i=0}^{n} a_i \left(\frac{1}{T} \right)^i \right] \quad (4.33)$$

$$\frac{1}{T} = \sum_{i=0}^{n} b_i (\ln R)^i \quad (4.34)$$

where a_i and b_i refer to fit coefficients, respectively.

Thermistors are usually used in the form of voltage-division circuit for spacecraft application. Thermistor R_t and fixed resistance R_N are connected in series as a loop as shown in Fig. 4.113, and its output voltage is as follows:

$$V_o = V_i \cdot R_t / (R_t + R_N) \quad (4.35)$$

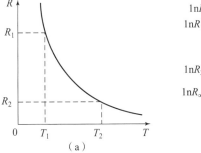
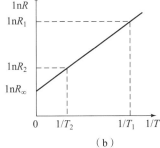

Fig. 4.112 Temperature characteristics of thermistor

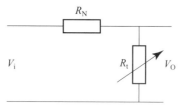

Fig. 4.113 Thermometry route of thermistor

where V_i refers to input voltage DC 5 V.

The resistance of thermistor in the working temperature range usually varies from hundreds of kilo-ohm (kΩ) to 200 Ω, which is much larger than the resistance value of lead. Therefore, the error caused by lead resistance can be ignored in a two-wire system. To minimize the measurement deviation for high-accuracy results, the deviation by lead resistance can be removed by using a four-wire constant current method.

2. **Thermometry principles of platinum resistance temperature detector**

Based on the monotonic correspondence property of conductor or semiconductor between resistance and temperature, temperature can be measured through resistance measurement, namely resistance thermometry. The thermistor mentioned above is a kind of semiconductor. Many materials possess monotonic correspondence property between resistance and temperature, but they should also have high temperature resistance, stability, suitable resistance temperature coefficient when used for thermometry. In general, metals for producing thermistor include platinum, copper, indium. Platinum is the best. The advantages of platinum are as follows:

(1) Physical and chemical properties of platinum are very stable. In other words, it is difficult to oxidize even in high temperature. Its properties do not change except in reductant medium.
(2) Resistance temperature coefficient of platinum is big, and the average of it at 0–100 °C is 3.925×10^{-3} °C.
(3) The resistivity of platinum is high, which is $0.098\ 1 \Omega \cdot mm^2/m$.
(4) Relation between platinum resistance and temperature is smooth, which could be shown with a simple equation.
(5) It is easy to purify platinum and realize a high purity.

Platinum resistance temperature detector made of platinum is divided into standard type and industrial type. Standard platinum resistance temperature detector is made of high-purity platinum filament that is wrapped on the mica framework and covered with quartz glass tube which is filled with inert gases before sealing. It is mainly used as measurement instrument for thermometric scale transfer or measurement of precise temperature, which shall be applied in the lab.

There are two main types of industrial platinum resistance temperature detectors. One is made by winding thin platinum wires of which diameter could be as small as 0.1 mm and then sealed in metal glass or ceramic protection tube; the other type uses platinum films as components, which is divided into thick-film resistor and thin-film resistor. The thick film is to print platinum ink on the alumina and bake it, then it is covered with a layer of glaze on the surface, baked again so as to form a layer of solid protective film on the surface of platinum components. The upper limit of temperature of it is about 500 °C. The thin-film platinum resistance temperature detector is made of vacuum sputtering thin-film component through photo-etched, plated with protective film and welded lead. The application temperature could be as high as 850 °C, which is suitable for industrial mass production. Thin-film industrial platinum resistance temperature detectors are often used as a thermometry sensor in spacecraft.

The industrial platinum resistance temperature detector has the advantages of good linearity, high accuracy and wide thermometry range, which can reach −196°C to 850 °C. The function relationship between its resistance and temperature is as follows:

$$R_t = R_0\left[1 + At + Bt^2 + Ct^3(t - 100)\right] \tag{4.36}$$

where R_t refers to a resistance value when temperature is t; $A = 3.9083 \times 10^{-3}\,°C^{-1}$; $B = -5.775 \times 10^{-7}\,°C^{-2}$; $C = -4.183 \times 10^{-12}\,°C^{-4}(t < 0\,°C)$, $C = 0(t > 0\,°C)$.

The resistance values of industrial platinum resistance temperature detector usually are generally 50 Ω, 100 Ω and 1,000 Ω at 0 °C. There are four allowable difference grades of them as regulated in *Industry Platinum and Copper Resistance Thermometers* JJG 229—2010 of *Measures for the Management of National Metrological Verification Regulations,* such as AA, A, B and C as shown in Table 4.17.

3. **Thermometry principles of thermocouple**

Temperature can also be measured by thermocouple thermometry based on Seebeck effect or thermoelectric effect, which is discovered by Seebeck in 1821. In a closed loop formed with two different conductors, current can be detected in the loop when a temperature gradient is applied between two ends. The driving force is afterward confirmed as electromotive force between two ends. Conductor A and conductor B

Table 4.17 Allowable difference grades and difference values of industrial platinum resistance temperature detectors

Allowable difference grades	Valid temperature ranges /°C		Allowable difference values
	Wire-wound assembly	Film-type assembly	
AA	−50 to 250	0–150	±(0.100 °C + 0.001,7 |t|)
A	−100 to 450	−30 to 300	±(0.150 °C + 0.002 |t|)
B	−196 to 600	−50 to 500	±(0.300 °C + 0.005 |t|)
C	−196 to 600	−50 to 600	±(0.600 °C + 0.010 |t|)

are called thermodes, as shown in Fig. 4.114. Joint 1 is a measurement end, and joint 2 is set up at constant temperature as required, which is called the reference end.

Fig. 4.114 Operating principles of thermocouple

When thermodes' materials are given, the thermoelectric force generated in the loop is only related to the temperature on the measurement joint and the reference joint. The total thermoelectric force is:

$$E_{AB}(t, t_0) = \int_{t_0}^{t} S_{AB} dt = E_{AB}(t) - E_{AB}(t_0) \quad (4.37)$$

where t and t_0 refer to temperatures on the measurement joint and the reference joint, respectively. A and B refer to different materials on the anode and cathode in order. S_{AB} refers to thermoelectric force ratio (Seebeck coefficient), which serves as derivative of thermoelectric force versus temperature. Its value is related to thermodes materials and joint temperature; $E_{AB}(t)$ and $E_{AB}(t_0)$ refer to separate thermoelectric forces on two joints.

If the reference end t_0 can be kept constant, for example, by putting it into a mixture of ice and water, then $E_{AB}(t_0)$ is constant. The total thermoelectric force is only related to temperature t at the measurement end and becomes a monotone function of t. The temperature on the thermometry point can be obtained by measuring the thermoelectric force. There are two necessary conditions for generating thermoelectric force: One is that the thermocouple is composed of two different thermodes, the other is that two joints of the thermocouple should have two different temperatures.

When thermocouple is used to form a measurement circle for thermometry, the following laws shall be acknowledged, with brief description as follows:

(1) Homogeneous conductor law: A closed loop consisting of a metal does not generate thermoelectric force regardless of the sectional area, the length or temperatures distribution on each part. Using homogeneous conductor law, it can be concluded that the thermocouple shall be composed of materials with two different properties. If two thermodes of the thermocouple are composed of two homogeneous conductors, the thermoelectric force of thermocouple is only related to temperatures on two joints instead of thicknesses, dimensions and geometrical shapes, or temperature distribution and its changes along thermodes. When there are temperature differences for closed loop that is composed of one material, it indicates that this material is an uneven conductor if the thermoelectric force is produced in the loop.

(2) Intermediate conductor law: After the intermediate conductor (the third conductor) is connected to the thermocouple loop, the total thermoelectric force of thermocouple loop will not be affected as long as temperatures on both ends of the intermediate conductor are the same wherever the intermediate conductor is connected to the loop. In the actual thermometry, both the connecting lead and the display instrument can be regarded as the intermediate conductor. According to the intermediate conductor law, the total thermoelectric force of thermocouple loop will not be affected as long as temperatures on both ends of the intermediate conductor are the same.

(3) Intermediate temperature law: The thermoelectric forces of two joints of the thermocouple loop whose temperatures are T and T_0 are equal to the algebraic sum of thermoelectric forces of thermocouple loop when the joints temperatures are T and T_m and the thermoelectric forces when the joints temperatures are T_m and T_0. Among them, T_m is between T and T_0 as shown in Fig. 4.115. The mathematical expression of the intermediate temperature is:

$$E_{AB}(T, T_m, T_0) = E_{AB}(T, T_m) + E_{AB}(T_m, T_0) \tag{4.38}$$

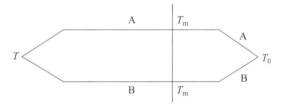

Fig. 4.115 Schematic diagram of intermediate conductor law

In the actual thermometry, the temperature on the reference end of thermocouple is difficult to set at 0 °C. It is an intermediate temperature T_m. The intermediate temperature T_m always changes. The thermoelectric force of thermocouple can be calibrated according to Formula (4.38), so as to amend the temperature of the reference end to 0 °C.

4.5.2.2 Categories and Characteristics

As mentioned above, the temperature sensors commonly used in spacecraft include thermistor, thermocouple and platinum resistance temperature detector.

In the thermal control system of spacecraft in China, a large number of thermistors are used as temperature sensors. It is possesses the advantages of strong sensitivity, large resistance temperature coefficient l, which is one or two order of magnitude as meta such as copper resistor, platinum resistance temperature detector and thermocouple. Other superiorities include simple thermometry rout, good stability, large signal, strong overload capacity, small power dissipation, wide operating temperature and small thermal capacity. Hence, thermistors are often adopted as primary thermometry components in remote thermometry condition. The disadvantages are poor interchangeability, narrow thermometry zone, nonlinear resistor temperature characteristics, which need to be calibrated.

Platinum resistance temperature detector has many advantages, such as good linearity, high accuracy, good stability, wide thermometry zone and good interchangeability. It has been widely used in industrial production and scientific research. However, due to the influence by lead resistance on the thermometry accuracy, three-wire or four-wire mode shall be used for measurement. In addition, the measurement loop of platinum resistance temperature detector shall be powered on with constant current in case of measurement error caused by self-heating. The industrial platinum resistance temperature detector is only applied for thermometry on special parts of thermal system in spacecraft of China since it needs four-wire measurement and extra signal conditioning line. Pt1000 platinum resistance temperature detector has been widely used in international spacecraft.

For thermocouple thermometry, the cold junction compensation is necessary and the output signal is as weak as millivolt level, which makes them sensitive to noise and easy to be interfered by electricity. However, it does not need signal conditioning and the electric measuring instrument can be used directly without extra consideration to the line resistance result from long measure distance. Therefore, it is widely used in ground test of spacecraft, such as thermometry during thermal vacuum or thermal balance test.

4.5.2.3 Selection Principles and Application Cases

1. **Selection principles**

For thermometry system with contact-type measurement, the following shall be considered in the selection of temperature sensor: stability of sensor in stress environment, resolution of sensor signals, thermometry accuracy and size of the sensor.

1) Thermistors

According to the actual thermometry ranges, accuracy and environmental adaptability, several alternative thermistors with different resistance–temperature characteristics are as shown in Table 4.18. Within the thermometry ranges, the thermistor with high resistance zone no more than 300 kΩ and low resistance zone not less than 500 Ω would be selected to ensure sufficient resolution and reduce remote hierarchy error. When used for temperature control, the thermistor with good stability, small thermal time constant, high B value and high resistance resolution in temperature control zone are usually recommended. For general temperature measurement, interchangeable thermistors are preferred.

4.5 Temperature Measurement and Control Technology

Table 4.18 Specification of common thermistors for spacecraft

	Specification and model	Thermometry range /°C	Range and accuracy of calibration temperature	Environmental adaptability /°C
1	MF51-3000 K ± 5% −8.2 kΩ ± 5%	−50 − 150	(0–100 °C) ± 0.3 °C Other range: ±1 °C	−55 − 150
2	MF51-3000 K ± 5% −18 kΩ ± 5%	−50 − 200	(0–200 °C) ± 0.3 °C Other range: ±2 °C	−55 − 200
3	MF51-3300 K ± 5% −30 kΩ ± 5%	−50 − 250	(0–200 °C) ± 1 °C Other range: ±2 °C	−55 − 350
4	MF61-3500 K ± 1% −3.9 kΩ ± 1.2%	−20 − 60	(−20 − 60 °C) ± 0.1 °C Interchangeable: (−20 − 60 °C) ± 0.3 °C	−55 − 125
5	MF501-4100 K ± 1% −5 kΩ ± 1%	−40 − 70	Interchangeable: (−30 − 60°C) ± 0.3°C other range: ±0.5 °C	−55 − 125
6	MF5802-2100 K ± 5% −5 kΩ ± 10%	−100 − 100	(−40 − 70 °C) ± 0.3 °C Other range: ±1 °C	−196 − 125
7	MF5802-2100 K ± 5% −3 kΩ ± 10%			−196 − 125
8	MF5604-(2.5–5.0) −50 kΩ ± 40%	−196 − −50	(77–223 K) ± 0.1 °C	−196 − 125
9	MF5408-4300 K ± 10% −10 kΩ ± 10%	100–450	(100–450 °C) ± 1 °C	−62 − 500
10	MF601-3500 K ± 1% −3.9 kΩ ± 1%	−20 − 60	(−20 − 60 °C) ± 0.1 °C	−55 − 125

2) Platinum resistance temperature detector

During selection, it should be considered whether the selected industrial platinum resistance temperature detector can meet the operating environment adaptability. The error tolerance level and value shall meet the requirements of Table 5.27.

3) Thermocouple

Theoretically, any two different conductors (or semiconductors) can be assembled into thermocouple. However, in fact, this is not the case in material selection. The following requirements shall be met:

(1) The assembled thermocouple shall have a high thermoelectric force and thermoelectric force ratio. The temperature characteristics of thermoelectric force shall have a linear univalent function relationship as far as possible.
(2) A wide working temperature is preferred. Physical and chemical stability are also important, especially for products long-term operate in temperature alternating environment.

(3) Its resistance temperature coefficient and specific heat are small.
(4) Good processability, enough toughness and good weldability are preferred.
(5) Materials are easy to get at a low price.

Table 4.19 exhibits the common thermocouple models and thermometry ranges. It can be observed that the thermometry measurement range of copper–constantan thermocouple (T-type) is relatively wide, and the thermoelectric force ratio is relatively large. In addition, it has stable temperature characteristics, good linearity, low cost and easy access, and is especially suitable for making thermocouples. Therefore, it is widely used in ground thermal test in the process of spacecraft development.

2. **Application cases**

Case 1: Application case of thermocouple thermometry.

Thermocouple thermometry is only suitable for ground test system of spacecraft due to its weak accumulated signal and poor anti-interference ability. Copper–constantan thermocouples are most used thermocouple thermometry, which is shown in Fig. 4.116.

In practice, the temperature on the reference end of the thermocouple must be known, and the reference end is usually placed in an ice–water mixture vessel to keep its temperature at 0 °C. In the spacecraft ground test, temperatures are often measured by sharing the freezing point or reference point compensator, which is convenient and economical without extra modification to measured values.

In the ground thermal vacuum test of spacecraft with many measuring points, the sharing freezing point wiring of copper–constantan thermocouple is exampled in Fig. 4.117. This method can not only largely reduce quantity of thermocouple reference points inside the freezing point vessel, but also reduce almost half of wires in the vacuum simulation chamber and vacuum sealing plugs, thus simplifying wiring and reducing negative influence of heat leakage of the conductors on measurement and test. However, in the actual installation process, the rest of the thermocouples except the common copper wire contacts, especially those between the contacts at the measuring end, must be reliably insulated.

Table 4.19 Specification of common thermocouple

No.	Material	Thermometry range /°C
1	Pt-Rh 10 Pt thermocouple (S)	0–1,600
2	Pt-Rh 13 Pt thermocouple (R)	0–1,600
3	Pt-Rh 30 Pt-Rh 6 thermocouple (B)	600–1,700
4	NiCrNi thermocouple (K)	−40 – 1,300
5	NiCrSi nisiloy thermocouple (N)	−40 – 1,300
6	Nickel–chromium/copper–nickel thermocouple (E)	−40 – 900
7	Iron/copper–nickel thermocouple (J)	−40 – 750
8	Copper-constantan thermocouple (T)	−200 – 350

4.5 Temperature Measurement and Control Technology

Fig. 4.116 Schematic diagram of thermocouple application

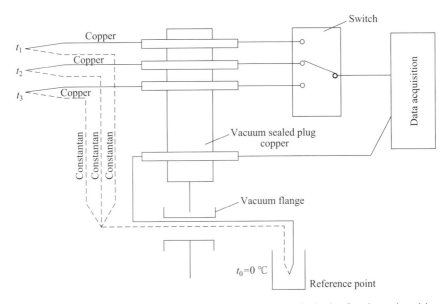

Fig. 4.117 Schematic diagram of copper-constantan thermocouple sharing freezing point wiring

Case 2: Application case of thermistor thermometry.

Different types of thermistors can be selected according to different ranges of thermometry and accuracies of the object. Take the optical camera, for example, the long-term on-orbit temperature of it is about 20 °C, which does not have high requirements for thermometry range, but do have high requirements for accuracy which shall be within ±0.1 °C. Hence, thermistor with higher B value shall be chosen. The main lens barrel of an optical camera shown in Fig. 4.118 is designed with 20-circuit temperature control loops. Temperature signals of the body tube are collected by 20 pieces of MF61 thermistors set between the clearances of patch heaters. Feedback signals drive the temperature control unit to control temperature of camera actively with the help of heater. The status of the main lens barrel pasted with heater and thermistor is shown in Fig. 4.118.

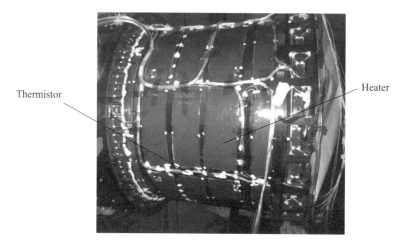

Fig. 4.118 Application case of thermistor of optical main lens barrel

For payload like external antenna, the temperature measurement accuracy requirement is not high, which is usually within ±3 °C. But it has a wider operation range, which makes it have higher requirements for the thermometry range. Therefore, B4 thermistor with thermometry range from −100 °C to 100 °C is recommended. Take the relaying antenna outside some spacecraft cabin, for example, its operating temperature is −80 °C to 90 °C, and B4 thermistor is used to collect temperature. The status of the relaying antenna waveguide assembly pasted with thermistor is shown in Fig. 4.119.

Case 3: Application case of platinum resistance temperature detector thermometry.

Platinum resistance temperature detector has many advantages, such as wide thermometry range, good linearity, high stability and simple structure. However, its sensitivity to temperature is relatively low due to the fact that the temperature measurement

Fig. 4.119 Application case of relaying antenna thermistor

4.5 Temperature Measurement and Control Technology

accuracy can be easily affected by the conductor resistance, thermoelectric force, voltage measurement error. In the field of aerospace, four-wire lead is mostly adopted to eliminate the influence of conductor resistance and on-resistance of multiple-way switch so as to get higher thermometry accuracy, while it also brings problems such as complicated design, reduction of reliability, growth of cost and so forth. Platinum resistance temperature measurement is less used in China than that abroad. In order to adapt to a wider temperature and achieve higher thermometry accuracy, the pusher cylinder of Mars soil humidity detector and the rotation shaft mechanism of reflector of NASA use platinum resistance for temperature measurement. Temperature control is realized by combining the thin-film heater or solid controller, as shown in Figs. 4.120 and 4.121.

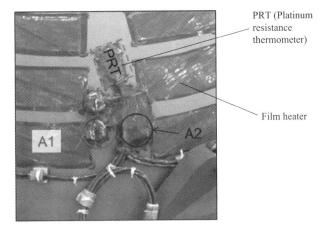

Fig. 4.120 Schematic diagram of platinum resistance temperature detector thermometry on the pusher cylinder of Mars soil humidity detector

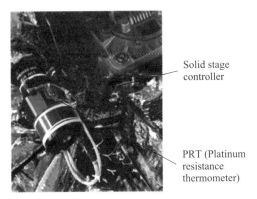

Fig. 4.121 Layout of platinum resistance temperature detector thermometry on the rotation shaft mechanism of reflector of Mars soil humidity detector

4.5.3 Temperature Control Technology

4.5.3.1 Functions and Working Principles

Temperature control technology is a kind of active thermal control technology to maintain the target object within a specified temperature range. It operates by comparing temperature of the object with the target value according to temperature signals, turning on/off the heating loop of the object, thereby conducting or not conducting heating compensation and realizing control of temperature zone. The temperature control technology is mainly applied in the thermal design of equipment with strict requirements for the temperature zone in a spacecraft, including optical/microwave payloads, batteries, propulsion systems, etc.

According to the temperature collection and control modes, the temperature control technology is divided into non-intelligent and intelligent temperature control technology. Their working principles are introduced.

1. **Working principles of non-intelligent temperature control technology**

The working principle of the non-intelligent temperature control technology is based on a feedback regulation mechanism. Temperature signals are first collected with temperature sensors, and then transmitted to inner thermometry bridge circuit to produce signal deviation by comparing with the target value. The heating loop is then on/off signal according to the deviation signal, thereby realizing temperature control in the end, as shown in Fig. 4.122.

Temperature–voltage switch is a linear balance bridge composed of resistors. One arm of the bridge circuit is a thermistor (NTC) of the controlled object, and its resistance changes with temperature. The resistance decreases with the increase

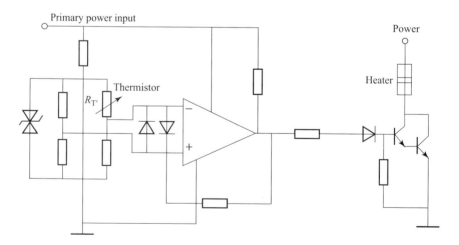

Fig. 4.122 Working principles of non-intelligent temperature control technology

of temperature. The change of resistance results in unbalance of the bridge circuit. When the temperature deviates from the setup value, a PD will be produced on the output of the bridge circuit. When the temperature is higher than the setup value, the resistance decreases and voltage on the bridge output is negative. When the temperature is lower than the setup value, the resistance increases and voltage of the bridge circuit is positive.

The input of voltage comparator is connected with the output of bridge, and a positive feedback is formed by connecting with its own output. When the input of the voltage comparator is positive voltage, its output is high level output, and when the input is negative voltage, its output is low level output.

The heating switch is composed of MOS power tube or Darlington circuit, featuring higher current load capacity to drive the heater directly. The on/off stage of heating switch, driven by the "1" or "0" electrical level sent by the previous stage (voltage comparator), can directly control the on/off stage of heater, thereby realizing temperature control.

The heating driver "switch" can be classified into negative terminal controlled switch and positive terminal controlled switch by its location. Negative terminal controlled switch refers to the heating driver located on the negative terminal of power, which is the downstream of heater, while the positive terminal controlled switch means the heating driver located on the positive terminal of power.

2. **Working principles of intelligent temperature control technology**

Comparing with the non-intelligent temperature control technology, the intelligent one mainly adopts a voltage-division-type resistance network for temperature accumulation circuit and central processing circuit. Thereby it can realize high-precision temperature accumulation, temperature control, and more flexible on-orbit remote measurement and control by software processing and the advanced temperature control algorithm.

The hardware composition block diagram of some intelligent temperature controller (which is composed of power, thermometry input, CPU control, heating output and bus interface circuit) is shown in Fig. 4.123. The reliability of independent temperature controller should be increased by warm redundant design of thermometry input and heating output and cold redundant design of CPU controller.

The temperature controller is powered with primary power of spacecraft and then converts the obtained power into secondary power using DC/DC module and finally supply each module inside it with the secondary power. The input thermometry channel with a tandem voltage-division structure of thermistor and standard resistor circuit can send the voltage signal that is transformed by the temperature/voltage circuit to the thermometry circuit. After the thermometry signal is sent to CPU microprocessing system, the thermometry data can be formed and compared with the upper and lower control threshold values as preset in the memory, thereby forming the heating control signal with combination of temperature control algorithm. The primary and backup CPU systems can receive the control data sent by 1553B bus to correct the corresponding relation between the heating circuit enabled state, the

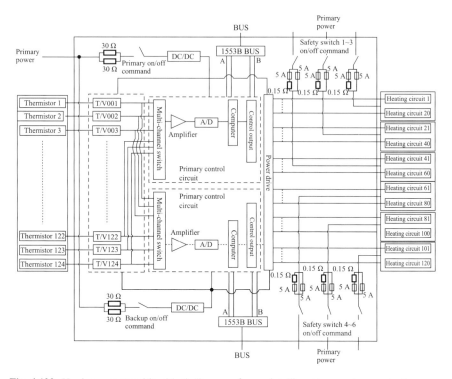

Fig. 4.123 Hardware composition block diagram of some intelligent temperature controller

forbidden state (on/off), the heater and thermistor. It can send digital telemetry data, including temperature, on/off of heating loop, temperature control threshold values and enabled the forbidden states of the heating loop, the corresponding relation between heating loop and thermistor, working state of hardware. The direct command interface can complete on/off of the safety switch and the switching of the primary and backup CPU systems.

4.5.3.2 Categories and Characteristics

As introduced in the aforesaid text, the temperature control technology can be classified into the non-intelligent and intelligent temperature control technology according to different signal processing and heating driving modes. The former can fall into the distribution-type and the integration-type temperature controls, represented by the solid-state controller and non-intelligent temperature controller representatively. The intelligent temperature controller is the representative product of the latter. Their characteristics are described as follows respectively.

1. Solid-state controller

Solid-state controller is a small independent temperature control device with small volume and light weight. It can be installed on each part of the object in distribution mode, with temperature sensors, patch heaters installed on the object nearby. The temperature control channel of solid-state controller is very limited, which is general one-circuit or several circuits. The temperature control threshold value depends on the circuit parameters of the temperature control channel. The control mode is a simple two operation statuses: on and off of the heating loop. The advantages of it are simple temperature control line and high reliability. Since all signals would not be accumulated on the temperature controller by a large amount of long cables, it can reduce measurement error and power loss caused by the long cable transportation. Satellite weight payload and complication of the cable layout would be reduced. Furthermore, it is good for thermal design layout and makes up for disadvantages of the integration-type temperature controller, and it costs low.

Since it does not have software processing capability and its temperature control fluctuation is relatively large, its temperature control resolution can be as high as ± 0.1 °C at most and the control threshold values cannot be regulated on orbit. Besides, it lacks flexibility and its control efficiency is low.

2. Non-intelligent temperature control unit

Comparing with the solid-state controller, the non-intelligent temperature control unit is equivalent to assembly of many solid-state controllers together to provide multi-circuit temperature control channel (as high as hundreds of circuits). Hence, it needs longer thermometry and control cables to connect with the temperature sensors and heaters mounted on the object, and its thermometry characteristics matches that of the solid-state controller. Because of adoption of the longer cable, thermometry signals are vulnerable to the interference by the surrounding environment.

3. Intelligent temperature control unit

The intelligent temperature control unit can realize high-precision thermometry with the help of temperature/voltage switch (resistance division network), weak signal conditioning circuit and AD converter. With PID temperature control algorithm, CPU control system software calculates the control quantity and feed it to the heating loop. The control period can be limited within 10 s, and the minimum heating time slice can be limited within 0.25 s. This type of control mode takes full advantages of the temperature control algorithm, such as small static temperature control error, small fluctuation and high efficiency, and the control resolution can be as high as ± 0.05 °C. Due to intervention of the intelligent CPU, a large amount of control parameters, such as temperature control threshold values, can be adjusted flexibly, and a large amount of telemetry data, such as temperature, can be downloaded. Its development is relatively costly because of adoption of the CPU control system.

4.5.3.3 Selection Principles and Application Cases

1. **Selection principles**

Temperature control technology is usually applied to equipment or assemblies that have strict requirements for temperature, especially the payload that have high-precision temperature measurement and control requirement. The selection principles are as follows:

(1) Different types of temperature control technology have different accuracy for temperature measurement and control, so it shall be chosen as needed.
(2) Solid-state controllers are usually used for local temperature control of the spacecraft component with little requirement for the loop. The controllers also are applied in spacecraft components with many scattering temperature control channels and temperature control cable having lightweight design command, while the temperature control parameters don't need to regulate on orbit.
(3) The non-intelligent temperature control unit is usually used for temperature control of spacecraft components, such as storage batteries and propulsion piping systems, which have low-temperature control accuracy requirements, high loop requirements and relatively concentrated components when temperature control parameter on orbit is not necessary to adjust.
(4) The intelligent temperature control unit is mainly used in the parts that require high-temperature control accuracy and many loops, especially when the temperature control parameters should be regulated on orbit. Their applications are very flexible, widely used in a variety of spacecraft, such as the optical payload of visible light/infrared/multi-spectral, microwave payload like SAR antenna, and rubidium clock.

2. **Application cases**

The development of spacecraft such as remote sensor and navigator has promoted the progress of thermal control technology in the last four decades. Many active temperature control technologies such as non-intelligent and intelligent temperature control units have come into being one after another. The temperature control precision of large-scale optical payload on-orbit is better than $\pm 0.05\ °C$.

Case 1: Solid-state controller.

China has conducted some research on the solid-state controller and mounted it on XY-1 new technical experiment satellite. Its temperature control precision reaches $\pm 0.3\ °C$.

The engineering prototype of a solid-state controller is shown in Fig. 4.124. There are two layers of this prototype structure. The control circuit and power tube are mounted on both sides of the structure. Connectors are not mounted on the structure, but led out from the inside of conductor. The product can provide two-circuit telemetry signals: One monitors the current control temperature of equipment, the other monitors the electrical level of output channel. By referring to these

4.5 Temperature Measurement and Control Technology

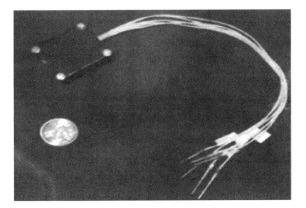

Fig. 4.124 Engineering prototype of a solid-state controller

two telemetry signals, we can judge the correctness of the operating state of the equipment.

Case 2: Non-intelligent temperature control unit.

The non-intelligent temperature control unit with high reliability is widely used in a variety of Chinese spacecraft in the early years, such as for temperature control of battery and propulsion system. The temperature control precision is better than ± 0.3 °C. Because of the inherent characteristics of the non-intelligent temperature control unit, the temperature control parameters cannot be adjusted to adapting the on-orbit changes of spacecraft in some special cases. Hence, the non-intelligent temperature control units are replaced by the intelligent ones gradually.

Case 3: Intelligent temperature control unit.

The intelligent temperature control unit is widely used in a variety of spacecraft in China, such as optical components, microwave payload, battery, propulsion system and sensor, and its temperature control precision on orbit is better than ± 0.05 °C. Figure 4.125 shows the latest generation of intelligent temperature control unit under development in China. By optimizing the layout of components, and benefiting from the latest electronic technology and thick-film technique, the integration level of this intelligent temperature controller is improved by 50% and its weight and volume are reduced dramatically, not to mention that its temperature control resolution is better than ± 0.02 °C compared with the last generation. It is equipped with on-orbit autonomous fault diagnosis and repair abilities in addition to the regulation of on-orbit temperature control parameters and transferring of telemetry data, by which on-orbit self-adaption ability of the thermal control system has been largely increased. This intelligent temperature control unit also realizes standardization of each functional module to configure quantity of modules according to spacecraft requirement.

Fig. 4.125 Intelligent temperature control unit

References

1. GB/T 14811-2008 Heat Pipe Terminology
2. H. Zhang, *Theoretical and Experimental Study of Loop Heat Pipe Two-Phase Heat Transfer Technology* (Beihang University, Beijing, 2006)
3. T. Ma, Z. Hou, W. Wu, *Heat Pipe* (Science Press, Beijing, 1983)
4. X. Xu, J. Ren, Thermal control performance analysis of spacecraft thermal control shutters, in *Proceedings of the 8th Space Thermophysics Conference*
5. Z. Hou, J. Hu, *Spacecraft Thermal Control Technology-Principles and Applications* (China Science and Technology Press, Beijing, 2007). (in Chinese)
6. Z. He, J. Jiang, *New Thermal Control Material Devices and Applications* (China Astronautics Publishing House, Bejing, 1988). (in Chinese)
7. Q. Zhong, H. Jiang, J. Huang, Discussion on equivalence of effective emission model of multilayer insulation components, in *Proceedings of the 13th Space Thermophysics Conference*, Yantai, 2017. (in Chinese)
8. J. Yuan, Z. An, J. Zhang, Lightweight functional composite materials based on hollow microspheres, in *Proceedings of the 13th Space Thermophysics Conference*, Yantai, 2017. (in Chinese)
9. D.G. Gilmore, *Spacecraft Thermal Control Handbook, Volume 1: Fundamental Technologies* (The Aerospace Corporation Press, California, 2002)
10. X. Zhou, Q. Zhao, Y. Wang, Study on the performance of a new lightweight low temperature multilayer, in *Proceedings of the 13th Space Thermophysics Conference*, Yantai, 2017. (in Chinese)
11. G. Min, Z. Zhang, Z. He, *Satellite Thermal Control Technology* (China Astronautic Publishing House, Beijing, 1991). (in Chinese)

12. W. Cao, J. Wu, Y. Yu, Development of carbon-containing polyimide film, in *Proceedings of the 6th Space Thermophysics Conference*, Guilin, 2003. (in Chinese)
13. Q. Ma, R. Fang, Y. Sun, Determination of performances of several multilayer insulation systems without interlayer, in *Proceedings of the 3rd Space Thermophysics Conference*, Yantai, 1982. (in Chinese)
14. Y. Chen, X. Ning, S. Su, et al., Experiment research on heat-transfer capability of the multi-layer insulation blankets under different vacuum degrees. Sci. Sin. Technol. 44(4), 407–416 (2014). (in Chinese)
15. W. Liu, Y. Wen, Heat transfer modeling analysis of multilayer insulation components, in *Proceedings of the 5th Space Thermophysics Conference*, Huangshan, 2000. (in Chinese)
16. S. Chen, J. Zhang, Y. Wen, Experimental study on thermal insulation performance of high-temperature multi-layer thermal insulation components, in *Proceedings of the 7th Space Thermophysics Conference*, Sanqingshan, 2005. (in Chinese)
17. T. Gu, Research on the application technology and thermal properties of multilayer insulation materials, in *Proceedings of the 4th Space Thermophysics Conference*, Chengde, 1991. (in Chinese)
18. C. Shen, Design and purification measures of multi-layer insulation vent hole of FY-1 meteorological satellite, in *Proceedings of the 4th Space Thermophysics Conference*, Chengde, 1991. (in Chinese)
19. B. Ding, S. Guo, G. Min, *Study on the Performance of Open-hole Multilayer Thermal Insulation Components. Collected Works of Academician Min Guirong* (China Astronautic Publishing House, Beijing, 2003). (in Chinese)
20. T. Sun, Q. Zhao, J. Li, Study on the adaptability of different installation methods of multilayer insulation components to the decompression process of active section, in *Proceedings of the 13th Space Thermophysics Conference*, Yantai, 2017. (in Chinese)

Chapter 5
Typical Thermal Control Design Cases of Spacecraft

5.1 Overview

The basic thermal control principles, methods and common thermal control technologies of the spacecraft have been thoroughly demonstrated in the aforesaid chapters. This chapter further introduces the applications of these principles, technologies and methods in the thermal control of the spacecraft and parts based on the aforesaid.

Generally, those spacecraft developed in China could be categorized into remote sensing satellite, communication satellite, navigation satellite, manned spacecraft, deep space exploration spacecraft and so forth. The orbit altitude of most of the remote sensing satellites is lower than 1,000 km, and that of the communication and navigation satellites is above 20,000 km. The orbit altitude of most of the manned spacecraft is lower than 400 km and with participation of astronauts, and the deep space exploration spacecraft encircles or lands on the celestial body outside the Earth. In general, the satellite and deep space probe are mainly based on the passive thermal control technology and supplemented by the active thermal control (ATC) technology; while the manned spacecraft are mainly based on ATC technology and supplemented by the passive thermal control technology. This chapter takes the typical spacecraft of the aforesaid spacecraft, for example, to introduce the thermal control system design of them.

Except for the case of system-level thermal control design, the thermal control design cases of typical spacecraft parts such as propulsion systems, battery, camera, antenna and so forth are also introduced in this chapter. What's more, the internal thermal control design of the electronic equipment has been an essential part of the development of the spacecraft electronic equipment. Therefore, this chapter also introduces the purposes, principles and steps of the thermal control design of electronic equipment, and the thermal control design cases of them are introduced as well.

5.2 Design Cases of Spacecraft Thermal Control System

5.2.1 Thermal Control System Design of Remote Sensing Satellite

5.2.1.1 Characteristics of Remote Sensing Satellite

Satellites that use remote sensors to acquire target radiation or reflection electromagnetic wave information are called remote sensing satellites, which are widely applied on meteorology, land, marine and environment and so on. The remote sensing satellite orbit covers low, middle and high orbits. Considering the observation ranges, revisiting period, spatial resolutions and other factors, the majority of the remote sensing satellites are operating on sun-synchronous orbit.

The characteristics of remote satellite's demand for thermal control design are summarized as follows:

(1) Different types of satellite payloads have different requirements for thermal control. For example, the high-precision temperature control is necessary for the optical payload to ensure the stability of on-orbit temperature; while heat dissipation requirements of the high-power part of the microwave payload need to be met. The payload thermal control is often a difficulty of thermal control design of the remote sensing satellite.
(2) The majority of the remote sensing satellite works in complicated modes, such as real-time data transmission, data record and data playback. Working status and times of payloads at different working modes are different. Neither is the heat dissipation.
(3) On-orbit attitudes of most remote sensing satellites are changeable. During operation of the payload, the whole satellite would have several attitudes offsets like side-sway, rolling, pitching, etc. Hence, the environmental heat flux not only changes with the orbital location, but also with the satellite attitude.
(4) Most payload cells of remote sensing satellites operate in short-term or periodically. The large heat dissipation during operation of the payload cells makes it necessary to consider the heat dissipation design. And the heating compensation needs to be considered since the heat dissipation is small when the payload cells are not operating.

5.2.1.2 Design Characteristics of Thermal Control System of Remote Sensing Satellite

The design characteristics of thermal control system of the remote sensing satellite are listed as follows according to the thermal control requirements:

(1) The orbit altitude of the remote sensing satellite is mostly at 500–1,000 km, in addition to the direct solar radiation heat flux, the earth infrared and earth reflection heat flux cannot be ignored.

(2) When designing the thermal control system design of the remote sensing satellite, it is necessary to comprehensively consider the satellite orbit, attitude, load operation mode, etc., and confirm the design conditions by selecting extreme environmental heat flux and extreme internal heat source. For example, for the optical imaging satellite, several on-orbit attitudes like side-sway, rolling and pitching and short-term working mode of the optical payload shall be considered; for the critical inclination orbit satellite, the change of the solar incident angle in life cycle shall be considered and the design shall be conducted by selecting the extreme cases.
(3) In general, short orbital period and short-term operation of the equipment shall be fully used in the design and heat capacity of the equipment, and the structure shall be used as far as possible to make sure the temperature could meet requirements. Relatively, the thermal analysis shall generally be transient thermal analysis, and thermal balance test shall be transient thermal balance test if necessary.
(4) Battery, power conditioning unit and other high-power long-term operating equipment shall be set up on the decks with little or less environmental heat flux. The long-term and short-term equipment shall be set up in crossover mode and thermally coupled by using heat pipes. A substitute electric heating loop shall be designed for the short-term equipment if necessary, thereby compensating heat for the equipment when it does not work.
(5) The load-bearing mechanisms of some remote sensing satellites have higher requirements for temperature stability. Thus, special thermal designs like high-precision temperature control design shall be taken to ensure the orientation precision and imaging properties of the payload.
(6) Optical payload with high-precision thermal control requirements such as camera and so on, usually needs integral design of machine, electricity, light and heat, and the thermal insulation to the surroundings shall be used as indicated in Sect. 6.3.4.
(7) When operating on lower orbit, the influence of atomic oxygen shall be considered except for the space radiation environment. Hence, atomic oxygen resistant coating is usually coated over the satellite, or the thickness of thermal control coating could be increased properly.

5.2.1.3 Design Cases of Remote Sensing Satellite

This section takes the ZY-3 satellite, for example, to illustrate the thermal control system design of remote sensing satellite.

1. **Introductions of ZY-3 satellite**

ZY-3 satellite includes payload module and service module, and its deck structure layout is indicated in Fig. 5.1. Four payloads are carried on the top of payload module of ZY-3 satellite, such as one multi-spectral camera and three three-line array cameras, as indicated in Fig. 5.2.

230 5 Typical Thermal Control Design Cases of Spacecraft

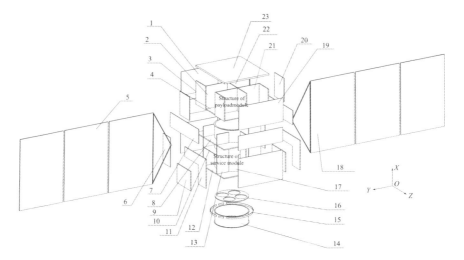

Fig. 5.1 Structural layout of ZY-3 satellite deck. 1—$-Z$ outer deck; 2—$-Z$ clapboard; 3—$+Y$ outer deck; 4, 12—Bottom deck; 5, 18—Solar array; 6—Anti-earth deck (a) with opposite to sun; 7—Middle deck; 8—Top deck; 9—Battery deck; 10—Clapboard of service module; 11—Anti-earth deck (b) with opposite to sun; 13—Shear clapboard; 14—Abutment frame to launcher; 15—Belting mechanism; 16—Hydrazine tank support; 17—Central tube; 19—$+Z$ outer deck; 20—$-Y$ outer deck; 21—$+Z$ clapboard; 22—Middle clapboard; 23—Top structure of payload module

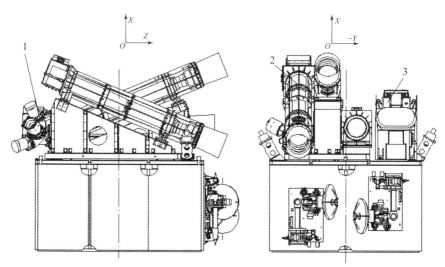

Fig. 5.2 Payload layout of ZY-3 satellite. 1—Star sensor; 2—Three-line array camera; 3—Multispectral camera

5.2 Design Cases of Spacecraft Thermal Control System

The satellite orbit is a sun-synchronous circular orbit at 10:30 a.m. and with orbit attitude of 500 km. The flight attitude mainly includes the flight attitude at sub-satellite point, the side-sway one and emergency one, etc.

The working modes could be classified into normal working mode, intensive working mode and idle working mode. The normal working mode refers to the regular operation mode of satellite, the longest imaging time per orbit is 15 min, and the total time for imaging in one day is no more than 50 min. The longest imaging time per orbit in the intensive working mode is 15 min, and the total time for imaging in one day is no more than 75 min. While the idle working mode is that there are few or no satellite missions and the payload does not work at all in one day. The long-term heat dissipation of the whole satellite is about 625 W, and the short-term heat dissipation is about 730 W, as indicated in Table 5.1.

2. Introductions of thermal control design of ZY-3 satellite

1) Thermal design of service module

Generally, the equipment in the service module of ZY-3 satellite operates in long term. According to the actual distribution of heat dissipation and change laws of the space environmental heat flux, heat radiating surfaces are set up on many surfaces of decks, as the dark shadow areas indicated in Fig. 5.3. Isothermal measures are taken as follows to intensify the radiation inside the service module: All surfaces inside the service module, except the mounting areas of equipment, shall be coated with high-emittance thermal control paint. All equipment inside the service module

Table 5.1 Heat dissipation statistics of ZY-3 satellite

Module	Long-term heat dissipation/W	Short-term heat dissipation/W	Remarks
Service module	400		Include battery, shunt, discharge regulator, DC/DC, etc.
	120	–	Include gyroscope, star sensor electronics, SADM, etc.
	35	–	Include telecontrol unit, telemetry units, etc.
	70	–	Include GPS receiver, USB responder, etc.
Payload module	–	320	Include signal processor, camera controller, etc.
	–	250	Include compression encoder, data processor, solid state amplifier, etc.
		40	Include solid state memorize, etc.
		120	Include DC/DC, etc.

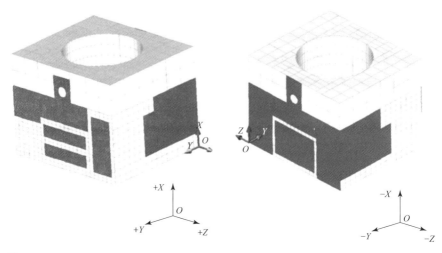

Fig. 5.3 Radiating surfaces distributions in service module of ZY-3 satellite

shall be coated with high-emittance thermal control paint or processed by black anodic oxidation treatment. Silicon grease shall be filled at the mounting surfaces for equipment with high heat dissipation and heat pipes are used for heat expansion. For equipment with little heat dissipation, keep dry contact or use thermal isolators between the equipment and the mounting surface to reduce heat conduction.

Special thermal control measures are taken on the following units, excluding common electronic equipment.

(1) Thermal designs of tanks and propulsion pipelines.

The range of operating temperature of the propulsion tanks and pipelines is 5–60 °C. Therefore, the thermal control design and active temperature control design as introduced in Sect. 6.3.1 shall be taken to ensure the normal operation of the propulsion system on orbit.

(2) Thermal designs of cadmium–nickel battery.

Cadmium–nickel batteries with narrow range of operating temperature of 0–15 °C are applied on the ZY-3 satellite, and the temperature difference between the battery modules and that between the inner cells in the same module shall be no more than 5 °C. Therefore, thermal insulation measures as follows are taken between batteries and the surrounding: MLI shall be covered over the outer surfaces of batteries; single aluminized polymide film shall be pasted over inner surfaces of small modules of the battery except the mounting plate. For the purpose of reducing temperature difference between the battery modules and that between cells in the same module, heat pipes shall be embedded inside the battery mounting plate. Since electric heating is used to actively control the temperature of the batteries, heating plates are specially designed and installed between batteries and their mounting plates, and the polymide film

5.2 Design Cases of Spacecraft Thermal Control System

heater is pasted on the heating plate. The thermal design of battery is indicated in Fig. 5.4.

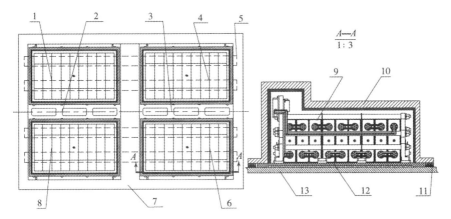

Fig. 5.4 Schema diagram of thermal design of battery. 1—Battery (a); 2—Heating plate (1); 3, 12—Heating plate (2); 4—Battery (b); 5—Embedded heat pipe; 6, 9—Battery (c); 7, 13—Battery mounting plate; 8—Battery (d); 10—Multi-layer thermal insulation shield; 11—Nylon clasp

(3) Thermal design of gyroscopes.

Gyroscope assembly which has higher requirements for start-up temperature (which shall be higher than 10 °C) is adopted on ZY-3 satellite. The inside of the gyroscope head is sensitive to temperature. Hence, thermal isolators shall be used for heat insulation between the gyroscope head and the mounting plate of it and between the mounting plate and the mounting deck. Meanwhile, the operational heater is set up on the outer surface of the gyroscope head, and the thermistor is set up on the top flange (since temperature on this part is closest to the inside oil temperature of gyroscope). The requirement of start-up temperature of the gyroscope assembly is guaranteed by taking the aforesaid measures. The thermal design of gyroscope assembly is indicated in Fig. 5.5.

2) Thermal design of payload module

The equipment in the payload module of ZY-3 satellite usually works in short term. Short-term working equipment with high heat dissipation includes camera, data transmission, etc. In addition to making full use of heat capacity of equipment and structure, many measures, such as application of heat pipe for heat expansion, application of radiating surfaces and so forth, shall be taken to make sure heat generated by the equipment could be dissipated in time during operation. To ensure that temperature is above the requested minimal value while the equipment is not operating, the electric heater is used.

As indicated in Fig. 5.2, the camera is mounted on the top of payload module. Higher requirements have been made for three-line array camera and multi-spectral camera in regard to the uniformity and the stability of temperature field of the camera

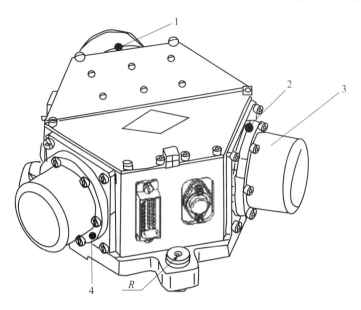

Fig. 5.5 Schematic diagram of thermal design of gyroscope assembly. 1—Primary thermistor for temperature control; 2—Redundant thermistor for temperature control; 3—Mounting locations of operational heater (three locations); 4—Thermistor for temperature measure

bearing structure, which requires that the temperature of the top structure shall be kept at 20 °C ± 2 °C. Hence, MLI shall be used to cover the upper and lower surfaces of top structure, thereby reducing the influence of equipment temperature fluctuation inside the payload module on temperature field of the top structure; meanwhile, electric heaters are designed on surface of the top structure so as to ensure the stability and uniformity of temperature. Temperature distribution of the top structure at some moment is indicated in Fig. 5.6, which could meet the requirement of 20 °C ± 2 °C

5.2.2 Thermal Control Design of Communication Satellite

5.2.2.1 Characteristics of Communication Satellite Missions

Communication satellite refers to a kind of satellite that could realize communication on ground, water and atmosphere as a relay station. Currently, almost all communication satellites are active satellites loaded with a large amount of electronic equipment for signal transferring, which could operate at low/intermediate/high orbit and Molniya orbit.

Compared with other spacecraft, the requirements of the communication satellite in regard to the thermal control design are listed as follows:

5.2 Design Cases of Spacecraft Thermal Control System

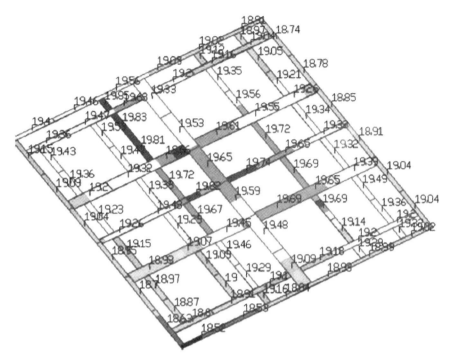

Fig. 5.6 Temperature field of the top structure of ZY-3 satellite

(1) Service life of it is very long. The design life of most communication satellites is longer than 15 years.
(2) Power of the satellite is very high. Power of some communication satellites is over 10 kW, while that of ultra-large communication satellite is higher than 20 kW. As a result, the heat dissipation of high-power communication satellite cable and waveguide is as high as hundreds of watts, which cannot be ignored.
(3) Heat dissipation of the whole satellite is very high. For example, heat dissipation of the payload module of DFH-4 communication satellite is as high as 4,500 W, and that of DFH-5 communication satellite is as high as 9,000 W. Heat dissipation of a single equipment is very high, for example, that of a 20 kW power conditioning unit made in China could be as high as 800 W. Besides, heat flux on equipment baseplate is quite high as well.
(4) A considerable number of communication satellites that operate on geosynchronous orbit (GEO) could enter GEO through the geostationary transfer orbit with their own orbit maneuvers. Generally, on the geostationary transfer orbit, payloads such as repeaters are not switched on or generate heat until they arrive at GEO.
(5) Generally speaking, payloads are in a long-term charging mode once the GEO communication satellite is orientated. However, the heat dissipation varies greatly

under different operating modes of repeaters (such as saturation mode, withdrawal mode and static state mode). Meanwhile, cycled backup is often applied on the repeater of communication satellite, which leads heterogeneous distribution of heat dissipation.
(6) Space radiation environment of GEO are much more severe than that at low earth orbit whose altitude is lower than 1000 km. Hence, the influence of space radiation shall be considered in the thermal control system of GEO long-life satellite.

5.2.2.2 Design Characteristics of Thermal Control System of Communication Satellite

Regarding regular hexahedron communication satellite that adopts stable tri-axial attitude control modes at static geosynchronous orbit, characteristics of its thermal design are simply concluded as follows:

(1) Generally, the earth reflection and infrared heat flux will not be considered since the satellite operates on GEO.
(2) Because the orbit periodicity (24 h) is long and the $+X, -Z, -X$ and $+Z$ decks of the satellite are alternatively shining by the sun for a longer time, they are unfit for working as heat radiating surfaces. $+Y$ and $-Y$ plates with small environmental heat flux, stable orbit periodicity and changing with seasons, are usually chosen as heat radiating surfaces.
(3) For satellite with high heat dissipation, since the $+Y$ and $-Y$ plates are alternatively shining at different times, heat pipes can be utilized to realize thermal conduction coupling between $+Y$ and $-Y$ radiating surfaces, thus, improving the heat dissipation capacity if necessary. In the case that $+Y$ and $-Y$ plates fail to provide sufficient radiating surfaces, the deployable radiator could be adopted.
(4) The equipment with high heat dissipation shall be set up on $+Y$ and $-Y$ radiating surfaces as much as possible, and conductive interface fillers shall be adopted to intensify thermal coupling between the equipment with high heat dissipation and the radiating surface. For the circumstance that a large number of high heat dissipation equipment is set up on the inner structural panels of satellite (for example, equipment with total heat dissipation of up to kilowatt are set up on the horizontal plate of E3000 platform), it is generally necessary to realize intensified thermal coupling between the inner structural panels of satellite and the radiating surfaces by using heat pipes.
(5) Generally, heat pipes are used in radiating surfaces for distributing heat. In most cases, the orthogonal heat pipe network is adopted to uniform the heat dissipation distribution and improve efficiency of radiating surfaces and reduce demand for electric power.
(6) Active thermal control measures like using electrical heater shall be taken to adapt to changes of heat dissipations or environmental heat flux at different stages, different operation modes from launching to the end of life. Backup design of heaters is necessary to meet requirements for 15-year operation.

(7) For the electric propulsion satellite, it is necessary to carry out adaptive design for intermittent large heat dissipation requirements of electronic equipment related to electric propulsion.
(8) Products or materials used in thermal control system shall be capable of adapting to no lower than 15-year GEO space radiation. For example, using the second surface mirror of glass as the coating on the radiating surface is preferred. Antistatic accumulation measure is taken for thermal control on the surface of satellite. For example, adopt anti-static thermal control coating or use conductive film as MLI surface film and connect ground.

What's more, since the environmental heat flux on the radiating surface changes with seasons, and the total heat dissipation of equipment and heaters in satellite is stable, typical cases of thermal designs are classified as: cold cases refer to transfer orbit case and equinox case at BOL (including earth shadow), and hot cases refer to summer solstice and winter solstice at EOL. The thermal analysis could be steady-state thermal analysis based on the same reasons.

5.2.2.3 Thermal Design Cases of Communication Satellites

By taking DFH-4 platform communication satellite of China, for example, the thermal control system design of communication satellite is introduced in this part.

1. Introductions of DFH-4 platform communication satellite

According to the function and equipment composition, DFH-4 platform communication satellite could be categorized to propulsion module, service module, payload (communication) module and solar array. Propulsion module, service module and solar array compose DFH-4 bus platform. DFH-4 platform satellite enters GEO through its own orbit maneuvers. On GEO, stable tri-axial attitude control mode is adopted, and the on-orbit deployable status of it is indicated in Fig. 5.7 and each module of satellite is indicated in Fig. 5.8.

DFH-4 platform has a large number of communication satellite devices. The majority of high dissipating equipment is mounted on $+Y$ and $-Y$ decks. Some equipment (like momentum wheel, gyroscope, etc.) is mounted on the inner structural panel. The equipment layouts on $+Y$ and $-Y$ decks of the service module and the payload module are indicated in Figs. 5.9, 5.10, 5.11 and 5.12.

Heat dissipation distributions of each module of satellite are indicated in Table 5.2.

2. Introductions of thermal design of DFH-4 platform communication satellite

As introduced in the aforesaid text, the heat radiating surfaces of DFH-4 platform communication satellite are, respectively, arranged on the outer surfaces of $+Y$ and $-Y$ plates of service module and the payload module. MLI are used to cover the outer surfaces of satellite except the radiating surfaces. The second surface mirror of glass is applied as the radiating surface coating and it is pasted on the surface of the structural plate with conductive adhesive.

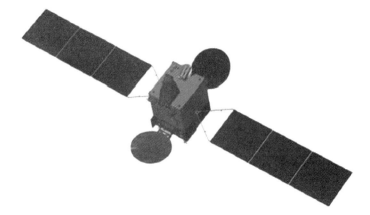

Fig. 5.7 Schematic diagram of on-orbit deployable DFH-4 platform satellite

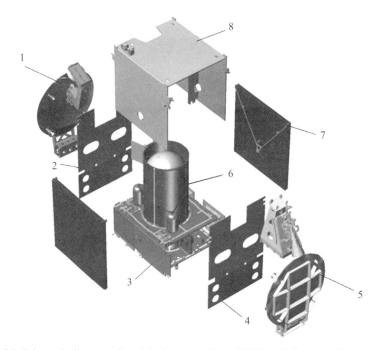

Fig. 5.8 Schematic diagram of module decomposition of DFH-4 platform satellite. 1—Ku band antenna; 2—+X deck; 3—Service module; 4——X deck; 5—C band antenna; 6—Propulsion module; 7—Solar array; 8—Communication module

1) Thermal design of propulsion module

Heat dissipation of electronic equipment in the propulsion module is mainly transferred to the $+Y$ and $-Y$ radiating surfaces by radiation. Hence, the equipment

5.2 Design Cases of Spacecraft Thermal Control System 239

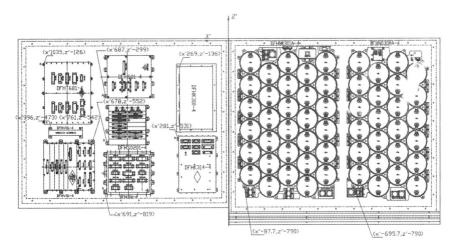

Fig. 5.9 Layout of devices in service module $-Y$ deck of DFH-4 series communication satellite

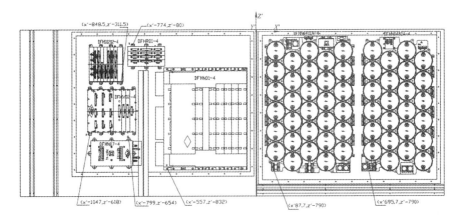

Fig. 5.10 Layout of devices in service module $+Y$ deck of DFH-4 series communication satellite

surface is coated with high-emissivity thermal control paint or black anodic oxidation. For equipment with higher temperature only radiated by its surface area is not enough, such as gyroscope, a doubler is set up for it.

Tanks of DFH-4 platform are mounted inside the central cylinder and temperature difference between the two tanks shall be no more than 5 °C. Thermal control measures taken for both tanks are indicated in Fig. 5.13. The thermal control paint with high-emissivity shall be coated over opposite hemisphere surfaces of both tanks, and MLI shall be covered over other surfaces of tanks; thermal isolators shall be used between tanks and cylinders and aluminizing films shall be pasted over the inner surfaces of cylinders; electric heaters shall be mounted on the hemisphere surfaces

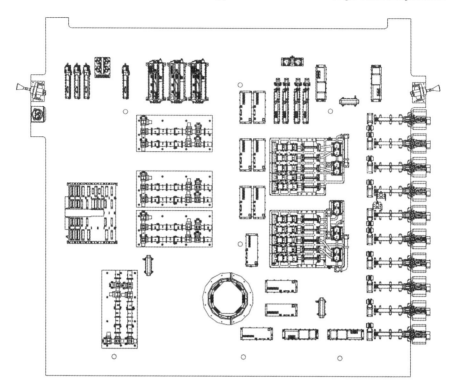

Fig. 5.11 Layout of devices in payload module $+Y$ deck of DFH-4 series communication satellite

on $-Z$ sides of both tanks so as to ensure the temperature and temperature difference of tanks meet the requirements.

The thermal control measures of the liquid lines and the cylinders of the propulsion system are introduced in Sect. 5.3.1 in detail, out of which, heaters on the surface of cylinders are bonded as indicated in Fig. 5.14.

2) Thermal design of service module

For the purpose of enhancing the radiation exchange, surfaces of all equipment in the service module shall be coated with high-emissivity thermal control paint or black anodic oxidation coating, and areas, except the mounting surfaces of $+Y$ and $-Y$ decks in service module, shall be coated with high-emissivity thermal control paint. Interface fillers are used to reduce thermal contact resistance between the dissipating equipment and the deck.

For the early DFH-4 platform satellite, each set of solar array has four panels and only the outermost plate would deploy at the geostationary transfer orbit. All panels will deploy after the satellite arrives on GEO. During geostationary orbit transfer, the undeployed solar wings will shield parts of the radiating surfaces of the service module. Meanwhile, the heat dissipation of each plate in the service module will vary greatly depending on whether the gyroscope works or not and the operation of

5.2 Design Cases of Spacecraft Thermal Control System   241

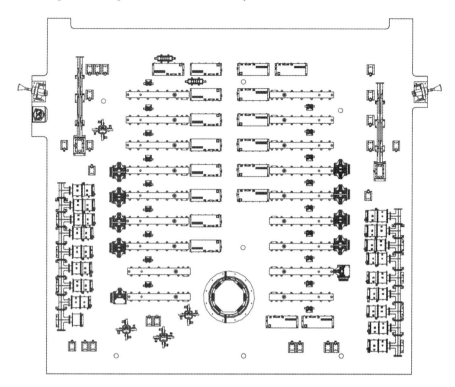

Fig. 5.12 Layout of devices in payload module $-Y$ deck of DFH-4 series communication satellite

Table 5.2 Heat dissipation of DFH-4 communication satellite

Name	Heat dissipation on transfer orbit/W	Heat dissipation on geosynchronous orbit/W	Remarks
Communication module	15	1450	Include 120 W heat dissipation of cable and waveguide
	118	1550	Include 150 W heat dissipation of cable and waveguide
Service module	363	272	Heat dissipation of cable is 35 W, excluding the one of battery
	141	150	Heat dissipation of cable is 20 W, excluding the one of battery
Propulsion module	91	135	

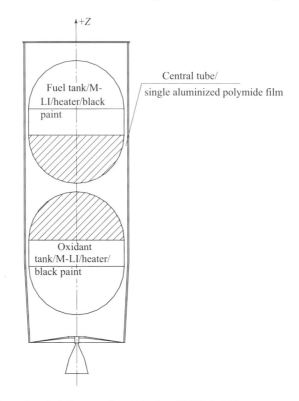

Fig. 5.13 Thermal control diagram of central tube of DFH-4 platform

different momentum wheel assembly (wheel assembly is 3:2 backup). To adapt to the aforesaid cases, orthogonal heat pipe network/heat pipes are applied on service module $+Y/-Y$ plates, respectively, as indicated in Figs. 5.15 and 5.16. Due to high heat dissipation (150–300 W) of PCU and high heat flux in the local area of equipment baseplate (which could be as high as 4 W/cm^2), the quantity and locations of heat pipes beneath this equipment depend on the suggestions given by the PCU manufacturer. Furthermore, mounting holes are set up in the middle of PCU baseplate to further reduce thermal contact resistance between PCU and structural panels.

Thermal control of Ni–H battery for DFH-4 platform is indicated in Fig. 5.17. To meet isothermal requirement for cells, the orthogonal heat pipe network shall be adopted to minish temperature difference between the cells in the module and between two modules.

In general, a lower storage temperature is favorable for keeping the performance of the long-life Ni–H battery. Hence, the temperature of Ni–H battery shall be kept at a lower level even in hot condition (temperature of Ni–H battery on DFH-4 platform is usually kept lower than 5 °C in summer solstice or winter solstice at the end of life), which makes it necessary to heat the battery in cold conditions, otherwise, its temperature will be lower than the allowable lowest temperature. Heaters of the

Fig. 5.14 Layout of helium tank heater

battery shall be pasted on the cell sleeve directly. Deck surfaces shall be pasted with aluminizing film as indicated in Fig. 5.17 for the purpose of reducing deck's thermal radiation on battery. In order to enhance radiation exchange between the battery and the equipment mounted on the decks in the propulsion module, the surfaces of battery cells shall be coated with high-emissivity thermal control paint.

3) Thermal design of payload module

Payload equipment is mounted on the communication satellites. In a way, the layout design of payload module equipment has been a part of thermal design. From the perspective of thermal control, the goals of layout are: The distributions of heat dissipation on $+Y$ and $-Y$ plates of the payload module are approximately equivalent in order to avoid insufficient radiating area on the single plate; heat dissipation on the single plate shall be as uniform as possible; equipment with equivalent temperature requirement shall be arranged in the same area for the convenience of designing heat dissipation capacity according to temperature zones; the mounting areas for heaters and heat pipes shall be reserved. However, the requirements for electricity, mechanism and heat shall be considered in the layout of payload module comprehensively, and requirements for electricity are usually considered in priority. Therefore, it is difficult to realize uniform distribution of heat dissipation on the single plate, especially considering that the cycled backup is often applied on the repeater of communication satellite and heat dissipation of the independent equipment and heat flux on baseboard are very high. As a result, the orthogonal heat pipe networks are

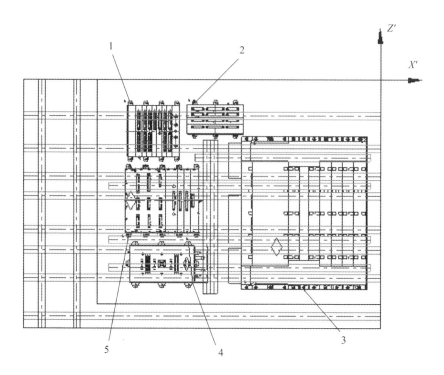

Fig. 5.15 Heat pipes layout of service module $+Y$ plate of DFH-4 platform. 1—RTU of service module; 2—10 N heating controller; 3—PCU; 4—Battery connected relay box; 5—Power distribution unit of service module

widely applied on the payload module of communication satellite. The orthogonal heat pipe networks that are particularly designed for equipment layout as indicated in Figs. 5.11 and 5.12 are indicated in Fig. 5.18.

Outer surfaces of payload equipment are also coated with high-emissivity thermal control coating. Generally, the conductive fillers are also used to reduce thermal contact resistance between the equipment and the panel. In addition to the mounting areas of equipment, other areas on the $+Y$ and $-Y$ plates of payload module shall be coated with high-emissivity thermal control paint.

In the phase of geostationary transfer orbit, most of the payload equipment is not powered on, and heat dissipation in the payload module is very low. Electric heaters are used inside the payload module to keep equipment and radiating surfaces at an appropriate temperature. These heaters mainly operate in the phase of geostationary transfer orbit. At GEO stage, these heaters would operate as well when the total heat dissipation drops because of repeater withdrawn or static operation. Heaters are not backed up one by one since orthogonal heat pipe networks are adopted.

5.2 Design Cases of Spacecraft Thermal Control System

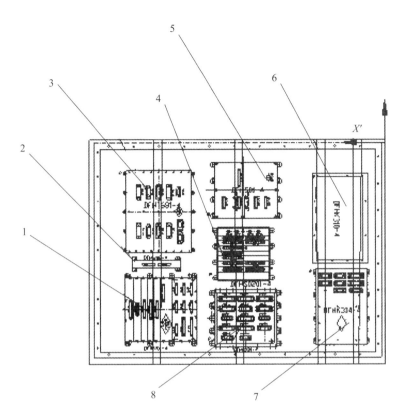

Fig. 5.16 Heat pipes layout of service module $-Y$ plate of DFH-4 platform. 1—Power distribution unit of service module; 2—APS for TMTC; 3—Propulsion electronics; 4—RTU of service module; 5—CDMU; 6—AOCE; 7—Power converter; 8—OBC

5.2.3 Thermal Control System Design of Lunar Probe

5.2.3.1 Characteristics of Lunar Exploration

The moon is the only satellite of the Earth, which is the primary goal for human beings to explore extraterrestrial objects. The lunar probe includes the platform and the payload. The former is developed for the purpose of realizing engineering and scientific goals, and new platforms shall be developed aiming at different exploration missions. The latter is configured for the purpose of realizing scientific goal, and different scientific exploration goals require different configurations of scientific payload.

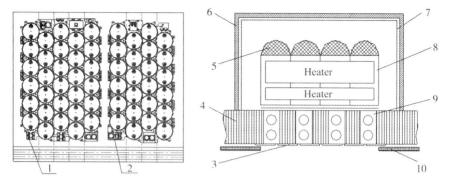

Fig. 5.17 Schematic diagram of battery thermal control of DFH-4 platform. 1—Battery A; 2—Battery B; 3—OSR; 4—Battery panel; 5—Black paint; 6—Satellite panel; 7—Aluminized film; 8—Battery; 9—Heat pipe; 10—MLI

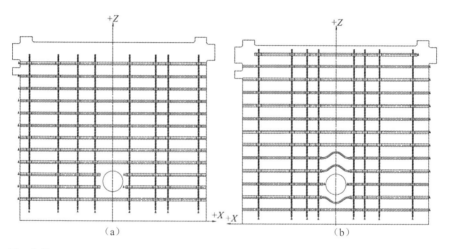

Fig. 5.18 Heat pipes layouts of payload module $+Y$ and $-Y$ plates of DFH-4 communication satellite (a) $+Y$ plate; (b) $-Y$ plate

Compared with the Earth orbit space mission, characteristics of the lunar exploration are listed as follows:

(1) During the lunar exploration, the configuration of the detector may have changes greatly. The detector may have many working modes and various attitudes. As the payload varies, working modes and the power demand also change.
(2) The lunar infrared radiation is strong, and it is extremely hot on the lunar surface during the day and very cold in the night. Radioisotope nuclear heat source is an essential and effective measure to maintain equipment temperature above the allowable lowest limit.
(3) For landing and cruise missions, landform on the lunar surface would affect landing attitude and environmental heat flux on the heat radiating surface of the

probe. Meanwhile, lunar dust on the lunar surface may cause pollution on the thermal control coating. Furthermore, the acceleration of gravity is 1/6 g on the moon, which would limit application of heat pipe.

(4) Thermal protection design shall be taken into account, since the thermal influence of plume from the high thrust engine is serious.

5.2.3.2 Design Characteristics of Lunar Probe Thermal Control System

Characteristics of the lunar exploration determine that thermal control system of the lunar probe is different from that of GEO thermal control system, which could be simply summarized as follows:

(1) Generally, separate module thermal design is taken and isothermal design is applied on each single module. For the detector working on the lunar surface, the influence of the lunar gravitational field on the operation performances of heat pipes shall be considered in the isothermal design.
(2) Since the lunar infrared radiation is strong, the location of the radiating surface of the detector shall try to avoid where it could be affected by the lunar infrared radiation, or it shall be set up where there is little influence of lunar infrared radiation. For the detector working on lunar surface, the radiating surface of it is often set up on the side facing to sky.
(3) Sometimes, the variable-conductance heat pipes shall be used to adjust heating capacity actively for the purpose of saving compensation power.
(4) Regarding detector that operates on the lunar surface and experiences lunar night, thermal insulation design and heat acquisition design are required. For the thermal insulation design, inner and outer surfaces of the deck are covered with MLI. For the heat acquisition design, radioisotope heat source could be used.
(5) For the detector that operates on the lunar surface and experiences lunar night, heat leakage caused by structure connection cannot be ignored, measures as follows shall be taken to achieve the maximum thermal insulation effect from aspects of reducing contact area, adding thermal isolators and using low-conductivity materials, and so forth.
(6) Protection designs shall be made for the thermal radiation effect and plume thermal effect of high thrust engine. Generally, the design combines high-temperature, middle-temperature and low-temperature MLI. Sometimes, aluminum foil is used to replace nickel foil in high-temperature MLI to reduce weight.
(7) Because of the increase of the relevancy between the thermal control system and the probe system as well as other systems and the increase of coordination between them, sometimes, other systems are required to cooperate to realize thermal control (for example, CE-3 rover needs to fold $+Y$ solar wing for thermal insulation at lunar night).
(8) During the thermal analysis of the thermal control system, the full-task procedures of different launching windows shall be analyzed. Apart from considering changes of coating parameters and solar radiation intensity, landform, detector

attitude and orbit changes shall be considered as well. Besides, astronomical phenomena such as lunar eclipse, etc., shall be considered as well. For probes running in lunar orbit, the thermal analysis shall focus on transient conditions; for probes working on the lunar surface, thermal analysis shall focus on stable conditions.

5.2.3.3 Thermal Design Cases of Lunar Probe

This part takes CE-3 lunar probe lander of China, for example, to introduce the thermal control system design of the lunar probe.

1. **Introductions of CE-3 lunar probe lander**

CE-3 lunar probe includes landing detector and rover. Before separation of the landing detector and rover, the rover is mounted on $+X$ side of the landing detector so as to compose CE-3 lunar probe (as indicated in Figs. 5.19 and 5.20) and to complete launching and landing on lunar surface. When the CE-3 lunar probe lands on the lunar surface, the release and separation mechanism that is mounted on the landing

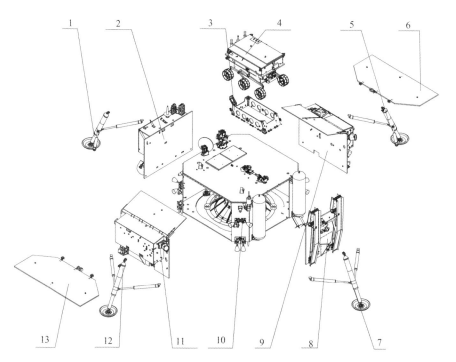

Fig. 5.19 Disassembling schematic diagram of CE-3 lunar detector module. 1, 5, 7, 12—Landing buffer mechanism; 2—$-Z$ module; 3—Rover connection and release mechanism; 4—Rover; 6—$-Y$ solar array; 8—Transfer mechanism; 9—$-Y$ module; 10—Center module; 11—$+Y$ module; 13—$+Y$ solar array

5.2 Design Cases of Spacecraft Thermal Control System 249

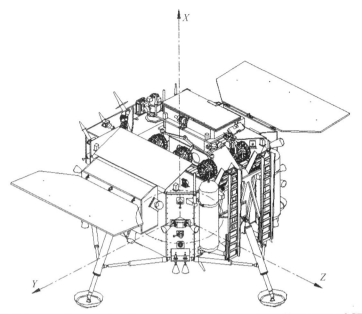

Fig. 5.20 Schematic diagram of earth-moon transfer and circumlunar flight status of CE-3 lunar probe

detector would release the rover on the lunar surface, and then the landing detector and rover could start exploration missions, respectively. The flight configuration of CE-3 lunar probe at all stages is indicated in Figs. 5.20, 5.21 and 5.22.

CE-3 lunar probe is launched and enters the earth-moon transfer orbit directly with flight attitude of $+X$ facing to the sun. After experiencing braking near the moon, it enters the polar circular lunar orbit with an incline angle of 90°. It would land at N44° at lunar surface when there is a chance after flying for five days and its $-Z$-axis would point to the lunar equator.

The equipment layout design of CE-3 probe lander is classified and centralized according to the equipment working stage. For equipment that do not experience the lunar night, they shall be centrally set up in the center module and $-Z$ module; for those would experience the lunar night, they shall be centrally arranged in $-Y$ and $+Y$ modules n, as indicated in Figs. 5.23, 5.24, 5.25 and 5.26.

The center module is mainly equipped with propulsion tank, propulsion pipeline, propulsion electronics and gyro assembly of GNC sub-system. A 7500 N engine is set up in the middle of $-X$ deck. $-Z$ module is mainly set up with most GNC equipment and landing sensor, etc. $\pm Y$ modules are equipped with TMTC instruments, data management equipment, electric instruments, payload, etc.

The heat dissipations in each module of CE-3 lander vary greatly at different mission phases. The heat dissipations in each phase of $+Y$ module are listed in Table 5.3.

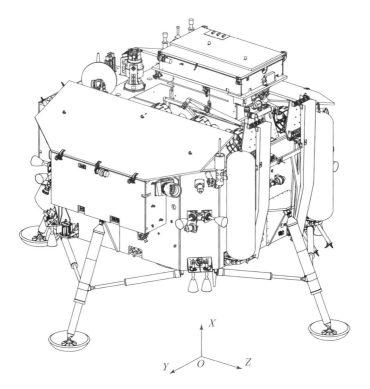

Fig. 5.21 Schematic diagram of CE-3 lunar probe landing on the moon

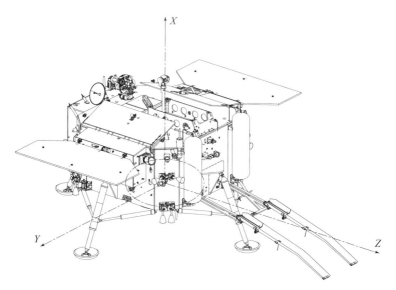

Fig. 5.22 Schematic diagram of working status of CE-3 lander on lunar surface

5.2 Design Cases of Spacecraft Thermal Control System 251

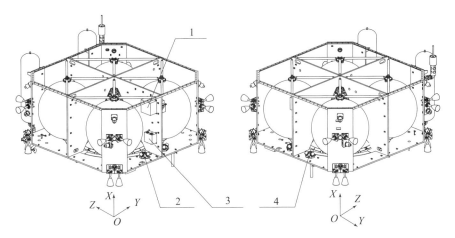

Fig. 5.23 Layout plan of center module of CE-3 lander. 1—Propulsion electronics; 2—Gyroscope assembly 1; 3—Propulsion power distribution unit; 4—Gyroscope assembly 2

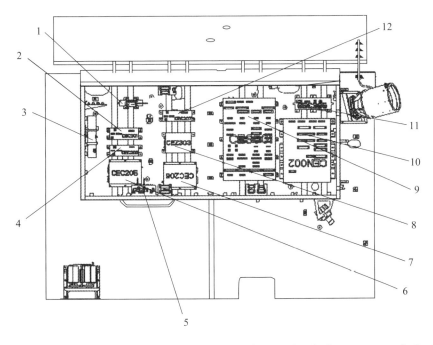

Fig. 5.24 Equipment layout of +Y module of CE-3 lander. 1—X band microwave network; 2—X band responder 1A; 3—Multiplexer A; 4—X band responder 2A; 5—S band microwave network; 6—X band solid state amplifier 1A; 7—X band solid state amplifier 2A; 8—Electronics of lunar dust gauge; 9—System management unit; 10—PCU; 11—UHF receiver; 12—S band responder

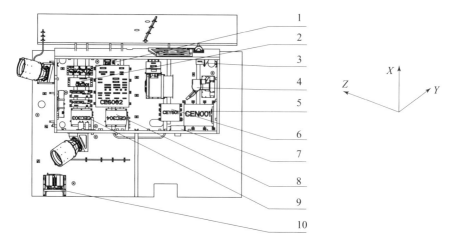

Fig. 5.25 Equipment layout of −Y module of CE-3 lander. 1—Data transmission modulator; 2—Unified frequency source; 3—SQCM probe; 4—Lunar optical telescope; 5—Battery; 6—Payload electronics; 7—Descending camera; 8—X band solid state amplifier 2B; 9—X band solid state amplifier 1B; 10——Y radioisotope heat source

2. Introductions of lander thermal design of CE-3 lunar probe

1) Thermal design of center module

Four tanks, pipes and valves of the propulsion system, propulsion electronics, gyro assembly and so on, are settled in the center module. Heat dissipation is mainly caused by gyro assembly and propulsion electronics. The total power consumption is 45 W. Gyro assembly is set up in the middle, which has higher requirements for heat dissipation. Hence, two heat pipes are laid in the deck beneath the gyro assembly, thus, expanding heat to the entire deck to dissipate.

The thermal capacity of the propulsion tank is large. Considering the short flight time, the surface of the tank is covered with MLI, and the temperature of the tank is maintained only by its own thermal capacity. Thermal control measures for propulsion pipelines and the tank are introduced in Sect. 5.3.1 in detail, out of which, the heater on the surface of cylinder is installed by suspension. The heat radiation and plume thermal effect caused by the operation of 7500 N engine are introduced in Sect. 5.3.1 in detail.

2) Thermal design of ±Y module

Thermal control functions and designs of ±Y modules are similar, and only +Y module is taken, for example, to introduce the thermal control design. +Y module needs to adapt to the requirements for radiating at the lunar day and thermal insulation at the lunar night, thus, realizing effective heat rejection of equipment during the lunar day, heat acquisition at the lunar night and thermal insulation design of system, etc.

Regarding heat dissipation at the lunar day, heating equipment in +Y module is set up on the side plate of it. Variable-conductance heat pipes (please refer to

5.2 Design Cases of Spacecraft Thermal Control System 253

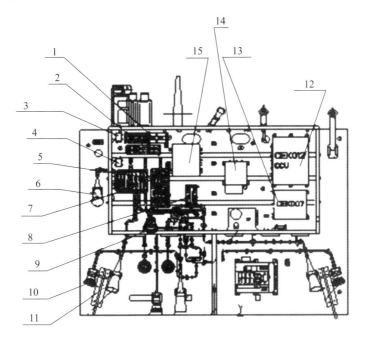

Fig. 5.26 Equipment layout of $-Z$ module of CE-3 lander. 1—Signal processor for distance and speed measurement; 2—Frequency synthesizer; 3—Redundant crystal oscillator for distance and speed measurement; 4—Primary crystal oscillator for distance and speed measurement; 5—Power distributor for distance and speed measurement; 6—Receiving antenna and its support for speed measurement 1; 7—Receiver for distance and speed measurement; 8—Transmitter for distance and speed measurement; 9—Transmitting antenna 1 and its support for speed measurement; 10—Transmitting antenna 3 and its support for speed measurement; 11—Clapboard and its support 2; 12—Central control unit; 13—Laser imaging sensor electronics; 14—Accelerometer; 15—IMU electronics

Fig. 5.27 for heat pipe layout) are embedded on the side plate, so they could transfer heat of equipment to the radiator to reject. To allocate heat load on each variable-conductance heat pipe reasonably, the U-shaped heat pipe is incorporated between two variable-conductance heat pipes.

To reduce the influence of intensive infrared radiation on the lunar surface as well as realize thermal insulation with $+Y$ module at lunar night, an independent radiator (which is not on the side plate where heating equipment of $+Y$ module locates) is provided for $+Y$ module and set up above the lander roof, as indicated in Fig. 5.27. For the purpose of making full use of the radiation area, heat pipes are embedded in the radiators and coupled with the condensation section of the variable-conductance heat pipe (VCHP).

Regarding energy acquisition at lunar night, radioisotope heat source is used to provide heat for equipment. Furthermore, the two-phase fluid loop is used to transfer heat of radioisotope heat source to $+Y$ module controllably with assistance of the gravity. The schematic diagram of layout of gravity assisted two-phase fluid loop is

Table 5.3 Heat dissipation distributions of +Y module of CE-3 lander

Equipment name	Heat dissipation/W						
	Earth-Moon transfer	Circumlunar stage	Landing stage	Lunar night	Lunar night		
					Sleep	Awaken	Lunar night
PCU	65	87/44	109	36	18	16	0
S band responder	18	0	0	0	0	0	0
Electronics of lunar dust gauge	0	0	0	5	0	0	0
System management unit	28	28	28	28	28	0	0
Multiplexer A	0	0	0	0	0	0	0
X band solid state amplifier 1A	32	32	32	32	32	32	0
X band solid state amplifier 2A on	0	0	0	0	0	0	0
X band responder 1A	10	10	10	10	10	10	0
X band responder 2A	7	7	7	7	7	7	0
UHF receiver	0	0	0	18	0	0	0
Star sensor electronics b	16	16	16	0	0	0	0
Total	176	180/137	202	136	95	65	0

indicated in Fig. 5.28. The condenser of two-phase fluid loop is embedded in +Y side plate of +Y module. The inner surface of +Y side plate is coated with high-emittance coating so as to dissipate heat to the inside of module. The radioisotope heat source is mounted on the bottom of +Y module.

Regarding thermal insulation, the most important way is to use VCHP. During lunar night, the condensation section of the VCHP would be blocked by the non-condensable gas, thereby blocking thermal transfer from the equipment inside module to the radiator. Thermal isolators shall be added between +Y module and other structures, and the outer surfaces of +Y side plate of +Y module as well as the inner and outer surfaces of other decks shall be covered with MLI.

3) Thermal design of −Z module

The mission of equipment in −Z module ends as soon as it lands on lunar surface. Hence, the thermal design of −Z module mainly aims at maintaining temperature

5.2 Design Cases of Spacecraft Thermal Control System

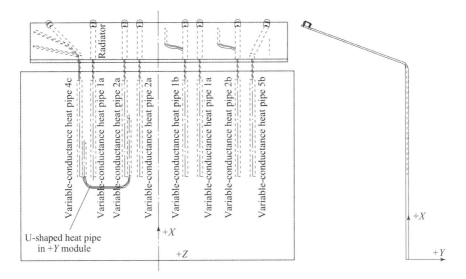

Fig. 5.27 Schematic diagram of heat pipes in +Y module of CE-3 lander

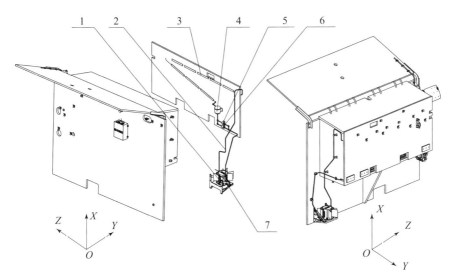

Fig. 5.28 Schematic diagram of layout of gravity assisted two-phase liquid loop. 1—Evaporator; 2—Pipeline; 3—Condenser; 4—Compensation chamber; 5—Control valve 1; 6—Control valve 2; 7—RHU

during the circumlunar stage and the landing stage. The isothermal thermal design is applied on $-Z$ module and heat pipes are embedded inside $-Z$ side plate, as indicated in Fig. 5.26. The OSR radiating surface is opened on the $-Z$ side plate of $-Z$ module to dissipate heat of equipment by radiation. All parts of $-Z$ module, except the radiating surface, are covered with MLI.

5.2.4 Thermal Control System Design of Manned Spacecraft

5.2.4.1 Mission Characteristics of Manned Spacecraft

The manned spacecraft of China operates on low earth orbit. Spacecraft with the participation of man in China includes Shenzou series manned spacecraft, Tiangong series spacelab and space station, etc.

Compared with other spacecraft like satellite, etc., the mission characteristics of the manned spacecraft are listed as follows:

(1) Provide comfortable living conditions for astronauts: High-scalability capsules are necessary since the manned spacecraft has been inhabited for a long time. Comfortable living conditions inside the capsules shall be provided to astronauts, including proper temperature, humidity, wind speed, atmospheric pressure, oxygen concentration, microorganism, concentration of CO_2, etc.
(2) The materials shall be highly non-toxic: Astronauts need to work and live in the capsules for a long time, so non-toxic materials should be applied in the capsules.
(3) Larger weight, size and power dissipation: The maximum weight of single manned spacecraft is 22t; the longest axial length is 18 m and the maximal diameter is 4.1 m; the maximal power dissipation is above 25 kW.
(4) Complicated configuration: The space station has many modules, consisting of core module, laboratory module I, laboratory module II, optical module, cargo spacecraft and manned spacecraft, and they are assembled on orbit by mechanical arms. The space station configuration is complicated and varied during the assembly.
(5) Multiple mission modes and multiple attitudes: The modes include autonomous flight, rendezvous and docking, combination mode and so forth; the attitudes include three-axial stabilization-oriented, yaw maneuver, solar-orientated, inertia flight attitude, etc.
(6) Long life: On-orbit service life of the space station is 15 years.
(7) High requirements for reliability and safety: Guaranteeing the safety of astronauts is the top priority in the design of the manned spacecraft. Once a huge fault happens, the manned spacecraft shall be able to return to the Earth autonomously or manually, and to guarantee life security of astronauts with supports of other systems as well as participations of astronauts.

5.2 Design Cases of Spacecraft Thermal Control System 257

(8) Strict requirements for equipment temperature: To meet spacecraft's requirements for long service life and high reliability, the range of equipment temperature is narrower compared with that of satellite. The range of temperature of most equipment shall be $-5\,°C$ to $40\,°C$.
(9) Many test loads with high-power dissipation and varying working modes: Most on-orbit tests shall have participation of astronauts.
(10) Ergonomics design: Ergonomics designs shall be carried out for equipment operated by astronauts, the living area and sleeping area of astronauts.
(11) Maintenance design: The on-orbit maintenance design shall be made to meet 15-year service life of the space station.

5.2.4.2 Design Characteristics of Thermal Control System of Manned Spacecraft

TCS of the manned spacecraft shall not only ensure temperatures of the equipment and the structure, but also create a comfortable temperature and humidity environment for astronauts in the capsule. The design characteristics of the manned spacecraft are simply summarized as follows:

(1) Application of mechanically pumped fluid loop (MPFL): Aiming at temperature and humidity controls of the capsule, MPFL is used to control them both by coupling with the ventilation system. Cold plates connected in MPFL are applied for equipment with high-power dissipation when the passive thermal control method and ventilation system are ineffective.
(2) Application of ventilation system: It could intensively gather air in certain area of capsule, and then distribute it organizationally. By using this system, hot air is transferred to the condensation dryer to decrease air temperature and remove moisture in the air. The generated cold air is transferred to the equipment with high-power dissipation. Air is also transferred to the human living area and the sleeping area, thereby controlling the air temperature and wind speed in the most comfortable scope that human would feel.
(3) Thermal management for integration of separate spacecraft: MPFL and ventilation system are used to transfer heat or air from the hot area to the radiator or to the cold area. Thus, heat from high-power equipment can be utilized to warm those equipment with little dissipation, or to prevent condensation of moist air in the cold area. For the space station composed of many modules, the thermal coupling design through MPFL and ventilation systems among different modules are essential, thereby realizing unified distribution and comprehensive utilization of heat in multiple modules, and optimizing the thermal control design. Backup of thermal control design between modules shall be made to improve the reliability of TCS. This can also reduce the weight of TCS.
(4) Strong robustness of thermal control design and high conditioning capability: The manned spacecraft usually adopts MPFL, ventilation system and other active thermal control methods. These methods have strong adaptability to the spacecraft attitude and to the drastic changes of heat dissipation caused

by frequent on/off of test payload. Besides, it could adjust temperature and humidity independently. And its capacity to regulate equipment temperature is also very strong.

(5) Thermal insulation of capsule and non-capsule, independent thermal control design of separate modules: Heat transferring modes, control objects and targets of the manned spacecraft capsule and non-capsule are different. The non-capsule only needs to ensure temperatures of the equipment and the structure, while the capsule shall not only ensure temperature of equipment and structure, but also ensure air temperature and humidity as well as wind speed. Based on the aforesaid, the capsule and non-capsule shall use independent thermal control design of separate modules.

(6) Maintenance design: For the purpose of improving reliability of the thermal control design and the independent health management, the maintenance design of thermal control equipment has been strengthened. Cost and risk of redundancy and maintenance shall be balanced reasonably. When partial fault happens, it shall ensure normal operation of the platform equipment and support the implementation of maintenance. When maintenance is done, TCS should have the capability to support normal operation of the payload.

(7) Man-Machine ergonomics: Regarding thermal control equipment that needs to be operated by astronauts and replaced on orbit, the man-machine ergonomics design is necessary and it shall be assessed as qualified by astronaut ergonomics.

(8) Universality: Thermal control designs and frameworks of each modules of space station shall be consistent. The universality of thermal control software and products is essential to reduce workload, increase the accumulation of product reliability data, improve on-orbit interchangeability and reduce the variety of maintenance space parts.

(9) Long life: Aiming at 15-year long-life design of the space station, the working fluid of MPFL shall be compatible with the contact material. On-orbit maintenance design for equipment with moving parts or for key equipment is essential. The exposed thermal control coating and material should have strong adaptability to the space environment, especially be able to endure erosion by atomic oxygen with high concentration.

(10) On-orbit autonomous management: The manned spacecraft usually adopts active thermal control design, such as MPFL, ventilation system and electric heater. The computer of thermal control system shall have the ability of autonomous fault detect, isolator and recovery (FDIR).

5.2.4.3 Thermal Design Case of Manned Spacecraft

This section takes Shenzhou series manned spacecraft of China, for example, to introduce the thermal control system design of the manned spacecraft.

1. **Introductions of Shenzhou series manned spacecraft**

5.2 Design Cases of Spacecraft Thermal Control System

Shenzhou series manned spacecraft could execute short-term independent flight mission as well as execute rendezvous and docking space flight mission to realize berthing and splitting with the spacelab. The configuration of the manned spacecraft, as indicated in Fig. 5.29, takes three-module layout: orbital module in the front, re-entry module in the middle and propulsion module on the back. The manned spacecraft could support 1–3 astronauts.

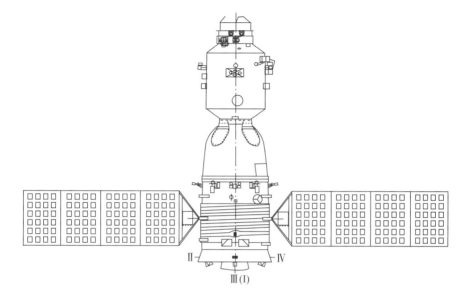

Fig. 5.29 Overall configuration of Shenzhou series manned spacecraft

2. Introductions of thermal design of Shenzhou series manned spacecraft

1) Active thermal control technology

(1) MPFL.

The mechanically pumped fluid loop is the core of thermal control system of the manned spacecraft. MPFL is an active thermal control system integrated with transferring, distribution and dissipation of heat. MPFL has strong ability to adjust temperature and humidity. This is the key role to control temperature and humidity of the spacecraft precisely, especially in multi-mission and multi-working mode of spacecraft. The major configurations of MPFL are indicated in Fig. 5.30. According to functions of MPFL, it could be classified as inner loop and outer loop.

The inner loop runs through three modules of the manned spacecraft, which gathers heat from air and devices in the orbital module and re-entry module by using the condensation dryer or the cold plate, thereby transferring heat by the circulation of working medium. The inner loop could exchange heat with the outer loop by using an intermediate heat exchanger. Therefore, heat is transferred from the capsule to the outer loop. The outer loop is composed of heat exchanger, cold plate, radiator,

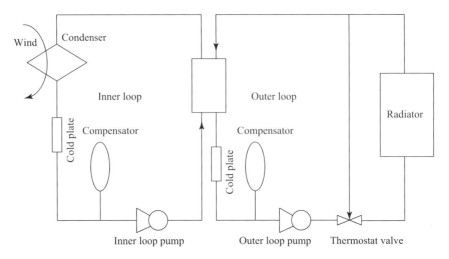

Fig. 5.30 Schematic diagram of mechanical pumped liquid loop

pump, valves, pipeline, working fluid and the compensator sensors and so on. All parts of the outer loop are installed in the propulsion module. The intermediate heat exchanger transfers heat from the inner loop to the outer loop, while cold plates gather heat of equipment in the propulsion module. Eventually, heat would be dissipated to the outer space by using the radiator mounted on the outer surface of the propulsion module. The working mode of the outer loop is controlled by thermal control units in propulsion module so as to realize fuzzy temperature control, i.e., determining the flow rate to radiator autonomously.

Since the inner loop is installed in the capsule, the working fluid of it is required to be non-toxic and nonflammable. Hence, glycol aqueous solution is chosen as working fluid for the inner loop. Since the outer loop is connected to the radiator, working fluid of it shall have a low freezing point and a wider range of operating temperature. Thus, the fluid will not freeze when heat load of the inner and outer loops is the minimum. Therefore, perfluorotriethylamine is chosen as working fluid of the outer loop.

To ensure working ability of the outer loop during the lifetime of spacecraft, the loop system is designed based on the maximum heat load. Usually, it is good for heat transfer when the flow rate is large in the pipe, heat exchangers and cold plates. However, this will increase the flow resistance in the loop and power loss of the pump. Meanwhile, it will become more difficult to develop a pump meeting the requirements. In circumstance that flow is fixed, the heat exchanger and the radiator shall be designed according to the maximum heat load. Thermal control paint with high emissivity and low solar absorbance shall be coated over the surface of radiator.

(2) Ventilation system for heat exchange.

The fan is used in the capsule to generate airflow to enhance convection heat transfer in the equipment area and the man living area. In the stage of combining

5.2 Design Cases of Spacecraft Thermal Control System

with the spacelab, thermal hose in the equipment zone of spacelab shall be pulled back to the re-entry module to realize ventilation convection between the spacelab and the spacecraft.

(3) Electric heater.

Electric heater is applied for each module, to heat air in capsule or in cold structure. Equipment with special temperature requirements or mounted outside of the module usually need electric heaters.

2) Passive thermal control technology

(1) Heat transfer measures.

High-emissivity thermal control paint or black anodic oxidation coating shall be sprayed over the outer surface of devices inside the capsule, device mounting plates and supports, to enhance radiation inside capsule.

The outer surface of the re-entry module shall be coated with S781-C thermal control paint so as to improve the absorption of the environmental heat flux and reduce infrared radiation. A radiating surface is designed on the outer surface of the propulsion module and coated with S971 thermal control paint in order to reject heat of devices mounted on the disk.

Interface fillers shall be used between the mounting surfaces of some equipment with high heat dissipation.

(2) Thermal insulation measures.

Heat dissipation in the orbital module is small, so its outer surfaces shall be covered with MLI to reduce heat leakage from the outer surface of the module, and to isolate influence of intense change of external heat flux on the module temperature as far as possible. In order to prevent atomic oxygen's erosion toward multi-layer materials, outer surfaces of MLI shall be covered with a layer of composite film which is resistant to atomic oxygen erosion.

The outer wall of non-radiating surfaces of propulsion module shall be covered with MLI. High-temperature MLI (of which outer film is stainless steel foil with high-temperature resistance oxidation coating) shall be mounted on the backwash shield of propulsion module so as to reduce influence from the propulsion engine.

Inner surfaces of the orbital module and the re-entry module shall be covered with foam to reduce convective heat transfer between the air and the wall. Besides, thermal isolators shall be added between the device with little dissipation and mounting plates, or the equipment and the structure outside of the module.

5.3 Thermal Design Cases of Spacecraft Assembly

5.3.1 Thermal Design of Propulsion System

Almost all spacecraft are equipped with propulsion system for large orbital maneuver, small orbital correction, attitude control, etc. Currently, the propulsion system applied on the spacecraft includes monopropellant chemical propulsion system, bipropellant chemical propulsion system and electric propulsion system, etc. Typical parts of the propulsion system include thruster, tanks, pipelines, valves, etc. For spacecraft needing orbit transfer or braking, etc., there is an orbit maneuver thruster with higher thrust, such as 490 N, 2,000 N, 3,000 N and 7,500 N engines.

The purposes of thermal control of the propulsion system include preventing freezing of liquid propellant and liquidation of gaseous propellant, controlling pressure ranges of gas tanks, preventing large temperature difference between fuel and oxidant for bipropellant system and large temperature gradient for solid propellant, ensuring proper temperature of thruster/engine before operation and so forth. Thermal control designs of the typical parts are introduced as follows.

5.3.1.1 Thermal Design of Tanks

1. **Thermal design of propellant tanks**

In general, propellant tanks on satellite are spherical and filled with propellant. Propellant tanks are usually isolated thermally from the surrounding by using isolators and MLI. Film electric heaters pasted on their surfaces are often used to assist temperature control, as indicated in Fig. 5.31.

Tank shells are mostly made of titanium alloys and the propellant conductivity is low. Propellant quantity in the tank changes at different flight stages of the spacecraft. Heat flux on the surface of tank is sometimes uneven. All these factors should be taken into consideration when designing the layout of heaters. And heaters could be set up in zones respectively to adapt to heating requirements at different stages if necessary. For GEO satellite entering the final orbit with its engine, most of its propellants are consumed at the stage of orbital transfer and only a little is used at the stage of geosynchronous orbit. At the orbital transfer stage, the heat capacity of propellants is high, thus, the propellant temperature could meet requirements with inertia. The remaining propellant is distributed in the hemisphere where the liquid outlet of the tank is located, under the function of the liquid management device. Therefore, heaters are only set up on the outer surface of the hemisphere where the liquid outlet is located to improve the heating efficiency.

For the planet detector, the propellants on it are mainly used for capturing planet orbit and landing on the planet surface. Thus, tanks are basically full at the long-term flight stage before landing on the explored planet. In this case, heaters shall be set up on the outer surfaces of the entire tank. The heaters are divided into independent

5.3 Thermal Design Cases of Spacecraft Assembly 263

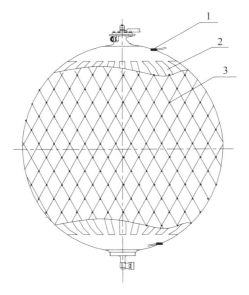

Fig. 5.31 Schematic diagram of gas tank thermal control. 1—Thermistor; 2—Film heater; 3—MLI

loops. Each loop covers certain surface area of the tank. Thus, the local temperatures of different parts of the tank can be controlled independently if necessary.

Usually, the temperature difference is required among tanks. Generally, heater loops could be designed for the single tank, respectively. All heating loops shall be set up with the same threshold value of temperature control, which is required to be narrow and with higher target temperature. Autonomic tracking mode of temperature control can also be adopted to heat the cold tanks automatically and to reduce temperature difference between tanks. Regarding special cases indicated in Fig. 5.13, radiation heat transfer measures could be enhanced additionally.

2. **Thermal design of gas tanks**

Gas tanks include the helium tank in the bipropellant chemical propulsion system and the xenon tank in the electric propulsion system. Thermal control of helium tank is mainly used to control the temperature in an appropriate range to ensure the operating pressure. Since the gas exhausting process from helium tank is an adiabatic expansion process, helium temperature would drop. To improve the minimum temperature of the helium cylinder during a single operation and in case of temperature of downstream parts of gas circuit (like reducing valve) lower than the allowable lower limit, temperature of the helium tank shall be preheated to a reasonable high level every time before it exhausts. Since the extraction flow of xenon gas from the tank is very low, the temperature of the xenon tank does not change obviously during the exhaust. However, one characteristic of the xenon is that its boiling point is close to room temperature under high pressure. For example, the boiling point at 15 MPa is close to 20 °C. Hence, the thermal control purpose of xenon tank is mainly to avoid xenon liquidation.

Similar to that of the propellant tank thermal control, gas tanks are usually isolated thermally from the surrounding by using isolators and MLI. Film electric heaters are mounted on the surface. Generally, heaters could be pasted on the surface with silicone rubber, as indicated in Fig. 5.32. However, the following situation should be considered.

The tank will expand when it is filled with high-pressure gas. In contrast, it will recover when the gas is exhausted. To adapt to this change of the gas tank, film heaters are suspended on the outer surface of the gas tank. Otherwise, the heater may be damaged when it is pasted directly on the tank. The film temperature control assembly is composed of film heaters and the polymide film, suspended on the tank surface with pins, as indicated in Fig. 5.32.

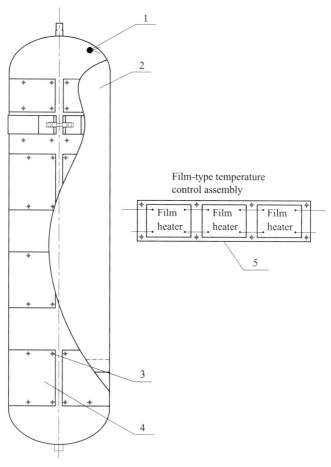

Fig. 5.32 Schematic diagram of thermal control of helium tank. 1—Thermistor; 2—MLI; 3—Pin; 4—Film-type temperature control assembly; 5—Polymide film

5.3.1.2 Thermal Design of Pipeline and Valve

There is zero heat dissipation on the pipeline. Valves produce little transient heat dissipation only when it is powered on. The combination of the active control heating and the passive thermal insulation is usually taken to ensure the temperature of pipeline, valve and inside propellant meet requirements.

The pipeline heaters are mostly heater strips or heater bands that are wrapped over the outer surface of the lines directly. During the design of heating power, line diameters and changes of the environmental temperature shall be considered. Therefore, in some cases, the density of the heating power (W/m) shall be changed along the line to make sure that the line temperature could be kept in the allowable range while it passes through the "hot zone" and the "cold zone" of the spacecraft. Generally, patch heaters shall be pasted on the regular part of the valve surface and they shall be connected with heater bands of the nearby lines to form a heating loop, thereby reducing requirements for the heating loop of the spacecraft. Furthermore, since there is a big difference between the thermal environment of the inner loop and that of the outer loop in the module, the inner and outer heating loops shall be set up as different loops, respectively, if necessary, in order to prevent line temperatures on different parts from being too high or too low, due to the asynchronous inside and outside heating requirements when the same loop is adopted.

Thermal insulation mounting shall be adopted between lines/valves and the satellite or between their supports and valves. Or, low heat conductivity materials could be used to make supports. MLI shall be covered on the outer surfaces of lines and valves in order to isolate radiation heat exchange.

For lines and valves inside the module, if the entire module temperature is higher in life cycle, which is above the propellant freezing or liquidation temperature, thermal coupling with the whole module could be considered. Outer surfaces of lines and valves are designed to have high emissivity, such as blackening, thus, intensifying the radiation heat exchange. Heat conduction mounting shall be adopted between lines/valves and the satellite.

5.3.1.3 Thermal Design of Thruster

1. **Monopropellant thruster**

The thermal control of the monopropellant thruster is to guarantee the temperature of catalyst bed, which is required to be higher than 120 °C at the beginning of operation. Hence, electric heaters and thermal insulation designs shall be taken. Except for the limited space for arranging heaters, the major problem encountered in the monopropellant thruster is that the catalyst bed temperature is very high, which is over 800 °C when the thruster operates. Therefore, heaters and thermistors that could resist high temperature shall be adopted.

The schematic diagram of thermal control of the monopropellant thruster of China is indicated in Fig. 5.33. Minitype armored electric heater and thermistor are used.

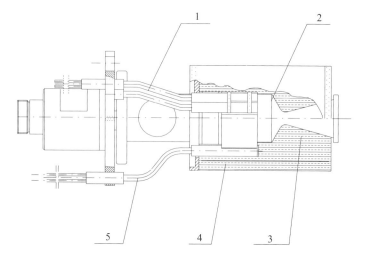

Fig. 5.33 Schematic diagram of thermal control of monopropellant thruster. 1—Minitype armored electric heater; 2—Engine; 3—High-temperature MLI; 4—Thermal-protective sleeve; 5—Minitype armored thermistor

Armored electric heater, thermistor and engine catalyst bed are connected with a fixed collar. The catalyst bed is covered with high-temperature MLI. A thermal-protective sleeve is on the outer surface of MLI.

In order to reduce the thermal influence of catalyst bed on the magnetic valve, the catalyst bed is connected to the thruster flange with stainless steel made thin-wall support, as indicated in Fig. 5.34. The support has many holes to reduce heat conduction path.

2. **Bipropellant thruster/engine**

Thermal controls of bipropellant thruster/engine are mainly used to keep injector temperature higher than the freezing point of the propellant before ignition. Similar to the idea of thermal control of monopropellant thruster, electric heater and thermal insulation design is mainly used. Since the temperature of the injector is higher than

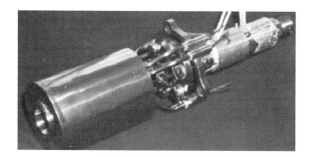

Fig. 5.34 Appearance of monopropellant thruster equipped with thermal controller

that of the conventional film heater, the heater adopts armored heating wire with stainless steel shell. A China-made 10 N bipropellant thruster with thermal control device is indicated in Fig. 5.35. The heating wire is wound on the outer surface of the injector, as indicated in Fig. 4.109. In order to reduce heat loss of the heater, the armored heating wire shall be mounted first and then the entire outer surface of the injector shall be covered with high-temperature MLI.

3. **Xenon-Ion thruster**

There are two kinds of widely applied matured electric propulsion systems: Hall thruster and electrostatic thruster of Xenon-Ion. Xenon-Ion thruster is the earliest one that is applied on engineering in China. LIPS-200 Xenon-Ion thruster that is applied on DFH-4S platform is indicated in Fig. 5.36.

Different from the requirement that the chemical thruster mainly ensures that the temperature before ignition is not low, the thermal control of Xenon-Ion thruster could not only ensure that temperature of inner parts is no lower than the lower limit of the allowable temperature when the thruster does not operate, but also ensure that

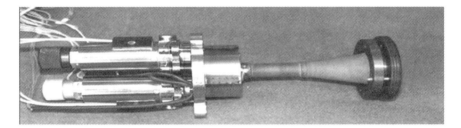

Fig. 5.35 Appearance of bipropellant 10 N thruster of China

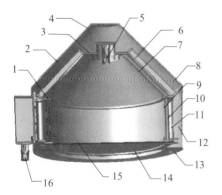

Fig. 5.36 LIPS-200 Xenon-Ion thruster. 1—Middle electrode boots; 2—Outer shell of cone segment; 3—Bottom electrode boots; 4—Outer shell of rear; 5—Primary cathode; 6—Bottom magnetic; 7—Bottom anode cylinder; 8—Bottom screen-grid cylinder; 9—Top anode cylinder; 10—Top screengrid cylinder; 11—Top magnetic steel; 12—Outer shell of cylindrical segment; 13—Outer shell of forepart; 14—Accelerating grid; 15—Screen grid; 16—Neutralizer cathode

the temperature does not exceed the upper limit of the allowable temperature during ignition when a lot of heat is produced. Take LIPS-200, for example, the applied thermal control measures are listed as follows:

(1) By taking anodic oxidation coating on the outer surface of anodic cylinder, inner and outer surfaces of the screen-grid cylinder and inner surface of the shell, the emissivity of these surfaces could be increased to enhance the radiation heat exchange between these surfaces. Thus, heat dissipated on the anodic cylinder during ignition could be transferred to the shell. Outer surface of the shell shall take thermal coating with low absorbance and high emissivity, which could reduce absorption of sunlight and good for radiating the internal heat to the space.
(2) Armored strap-type electric heater is set up on the cylinder area of post-shell of Xenon-Ion thruster. The cathode inside the Xenon-Ion thruster can be heated radiatively by increasing the post-shell temperature of thruster.
(3) The outer surface of post-shell of Xenon-Ion thruster is covered with middle-temperature MLI to reduce thermal radiation of post-shell on the pointing regulator that is connected to it.

5.3.1.4 Thermal Protection of Engine Plume and Radiation

Thermal effect of the plume and the radiation on the spacecraft deck and outer surface of equipment shall be taken into consideration. Small attitude control engine could take the metal canister for thermal protection. Thermal protection of large engine needs to take high-temperature heat shield, and the deck and the outer surface of equipment could add high-temperature and middle-temperature MLI on their low-temperature multi-layers as thermal protections.

High-temperature heat shield is a special multi-layer structure. Take high-temperature heat shield of 7,500 N engine of CE-3 spacecraft, for example, its structure is indicated in Fig. 5.37. Thermal radiation and plume of engine could be prevented from damaging the spacecraft deck and equipment by mounting a support between the engine and the module. To increase radiation to space as possible as it could, the heat shield takes a segmented frustum-shaped configuration. The configuration of each assembly from inside (which faces to the engine) to the outside (which faces to the module) includes high-temperature covering layer, high-temperature multi-layer, middle-temperature multi-layer and low-temperature multi-layer. High-temperature/middle-temperature/low-temperature multi-layers are, respectively, formed by alternatively overlapped multi-layer reflecting screens and spacing materials that apply to corresponding temperatures.

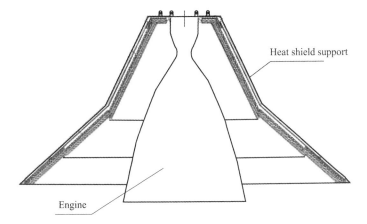

Fig. 5.37 Schematic diagram of high-temperature heat shield

5.3.2 Thermal Design of Battery

5.3.2.1 Characteristics of Battery

Battery is an important device of the spacecraft energy system. The regular battery types include lithium–ion battery, nickel–hydrogen battery and cadmium–nickel battery.

Battery temperature is an important factor to ensure its service life and property. Depending on the battery type and the spacecraft mission, different ranges of temperature are needed for battery. In general, temperature of cadmium–nickel battery shall be within 0–15 °C; and that of nickel–hydrogen battery shall be within −10 °C to 20 °C; and that of lithium–ion battery shall be within 10–30 °C. To ensure that all cells have the same charge and discharge rates, temperature differences among all battery modules shall be smaller than 5 °C. Cell-to-cell temperature difference in one battery set shall also be smaller than 5 °C. In some circumstance, the cell-to-cell temperature difference shall be smaller than 3 °C. There is one more requirement for lithium–ion battery: Its temperature shall be kept as 0–10 °C when it is laid aside for a long time.

The battery would absorb heat in the early stage of charge. In contrast, the battery would dissipate heat during discharge or in later stage of discharge. Heat dissipations would change with charge status, temperature, charge rate and discharge load, etc. Regarding nickel–hydrogen battery and cadmium–nickel battery with trickle charge during the break of charge and discharge, trickle charge would generate heat as well.

Since the heat dissipation for battery discharge is high, the required range of operation temperature is narrow and there are strict requirements for temperature difference between batteries, thermal insulation design is usually taken and independent radiators are set up. Besides, heat pipes would be taken for equalizing the temperature. Meanwhile, heaters are used to control temperature actively. Thermal design cases of nickel–hydrogen battery and cadmium–nickel battery are indicated in Sects. 5.2.1 and 5.2.2. This section introduces thermal design cases of lithium–ion batteries on spacecraft.

5.3.2.2 Thermal Design of Lithium–Ion Battery

Lithium–ion battery set is assembled by several independent cells, as indicated in Fig. 5.38. The cells are mounted on the sleeve to compose cell unit, and a certain number of cell units are assembled as battery sets.

The heat dissipation of cells shall be conducted to the mounting plate through the sleeve and then radiated to the space through the outer surface of mounting plate. Therefore, heat resistance between cells and structural panels are reduced as far as possible. For this reason, as the intensity allowed, cell shells and sleeves made of aluminum alloy are preferred. Adhesive shall be filled in the entire bonding area between cell and its sleeve to enhance heat conduction. Meanwhile, a certain contact area between sleeves and structural panels shall be ensured, and interface fillers shall be used between the sleeves and panels.

To ensure uniform heating, the heating power of battery set shall be distributed on each cell. Film heaters with the same heating power shall be pasted on the surface of each sleeve, as indicated in Fig. 5.39.

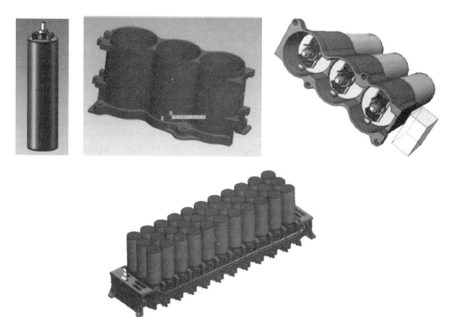

Fig. 5.38 Cylindrical lithium–ion battery set

5.3 Thermal Design Cases of Spacecraft Assembly

To even up the temperature between battery sets or among cells, heat pipes or the orthogonal heat pipe network shall be embedded in the structural panel beneath the battery set, as indicated in Fig. 5.40. OSR shall be pasted on the outer surface of the structural panel as a radiating surface. Without affecting heat dissipation of the surrounding equipment, a heat shield of MLI shall be set up for the battery set to further reduce temperature difference or to optimize thermal designs such as heating power and radiating area, etc.

Fig. 5.39 Heaters pasted on sleeve surface

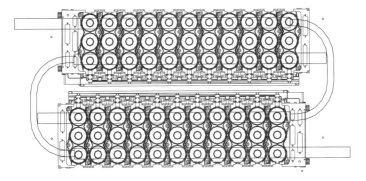

Fig. 5.40 Schematic diagram of embedded heat pipes beneath battery set

5.3.3 Thermal Design of Electrical Equipment

5.3.3.1 Purposes of Thermal Design of Electrical Equipment

In general, the thermal design of electronic equipment refers to the thermal designs of internal components and circuit boards, and under the thermal boundary conditions, the operation temperature of components and circuit boards could meet the specified requirements by adjusting heat dissipation path and heat resistance of components.

In most cases, thermal designers of electronic equipment focus on whether components temperatures could meet the derating value at hot condition. Although reducing temperature and stress caused by temperature variation are the cores of thermal design, in fact, thermal design of electronic equipment is much wider. The thermal design of electronic equipment shall cover the followings:

(1) At operation temperature, the temperature of components shall not exceed the derated value and shall be higher than required minimum temperature.
(2) At hot qualification temperature limit, the temperature of components shall not exceed its rated temperature.
(3) At hot operation temperature limit and designed fault mode, the temperature of components that work normally shall not exceed its rated temperature.
(4) Reduce temperature variation ranges of components.
(5) Satisfy temperature requirement of printed circuit board.
(6) Design thermal interface of electronic equipment to satisfy the IDS requirements.

The component temperature of electronic equipment refers to the junction temperature of components. For components without PN junction, the temperature could refer to shell temperature or hotspot temperature. The designed fault mode refers to the failure of a module or a function area in electrical equipment. Except the fault parts, other modules or functional areas shall ensure that the electrical equipment maintains the same or similar performances, such as single-stage failure mode of multi-stage module instead of primary and backup switch mode.

What needs to be emphasized is that the thermal interface of electrical equipment is obtained by thermal design. Take temperature reference point of electrical equipment, for example, from the perspective of spacecraft system, and this point could refer to the temperature of mounting interface of the electrical equipment and the spacecraft. Hence, whether the temperature on the reference point of electrical equipment could represent the temperature of the mounting interface of spacecraft must be verified via thermal analysis or thermal balance test of electrical equipment.

The thermal design of electrical equipment shall be incorporated with the mechanical and electric design, to ensure function, performance, reliability, safety and environmental adaptability of electrical equipment. Thermal design of electrical equipment shall obey the following principles:

(1) Take mature methods, materials, components and techniques.
(2) Try to satisfy temperature requirements according to the selection and the layout (high rated temperature, small heat dissipation and small contact heat resistance of mounting structure) of components.
(3) Try to adopt passive thermal control technology to intensify thermal conductance design between components, printed circuit board and cabinet.
(4) Do not weaken electronic property of electric box.
(5) Do not affect EMC of electric box.
(6) Do not add much weight and power.

5.3 Thermal Design Cases of Spacecraft Assembly

(7) Do not reduce intensity and rigidity of box.
(8) Life of the adopted thermal control materials shall not be lower than the design life of electrical equipment. It shall not affect electric properties and keep it safe, nonflammable, non-toxic, little vacuum degassing, pollution free or little pollution.
(9) Have good maintainability and testability.

5.3.3.2 Thermal Design Procedures of Electric Boxes

As a part of electrical equipment design, the thermal design shall be made with electricity property and function design, mechanical design, EMC design and space environmental adaptability design together or alternatively, which runs through the entire procedure of electrical equipment design. Specific steps of thermal design of electrical equipment are listed as follows:

(1) Circuit module division and panel design, mounting locations of high-power electronic components, heat dissipation distribution of printed circuit board shall be decided firstly.
(2) Materials for printed circuit board, thickness and area of copper shall be designed appropriately for multi-layer printed circuit board with higher heat dissipation.
(3) For wire arrangement of printed circuit board, dissipation distribution and heat transfer route to the housing of the box shall be considered.
(4) For the technological design of the printed circuit board, installation modes of components with heat dissipation shall be proper.
(5) Primarily analyzing the temperature of printed circuit board to see if partial hotspot exists. If the partial hotspot cannot be eliminated, the mounting locations of relative components shall be changed or special heat conduction measures shall be taken.
(6) Designing mounting mode of printed circuit board to the box housing. The connection mode shall be good for heat conduction.
(7) For structural design of high heat dissipation electrical equipment, it shall make sure that there is sufficient mounting contact area and mounting locations, and it shall provide sufficient mounting contact area for high heat dissipation printed circuit board and component so as to reduce heat resistance inside equipment.
(8) Carry out thermal analysis of the whole machine to obtain the temperature distribution of components and determine whether it meets the requirements. If the component temperature fails to meet requirements, it shall be solved by changing mounting locations and mounting modes of components as well as by taking special thermal conduction measures, such as changing structure of printed circuit board or changing connection mode of printed circuit board and the housing. Then, thermal analysis could be redone.
(9) Thermal balance test shall be taken if necessary, thus, obtaining temperatures of the printed circuit and high-power components and verifying if they meet requirements. If not, necessary corrections shall be done until relevant temperatures could meet requirements.

In general, when results from thermal analysis or thermal balance test of electrical equipment show that thermal design requirements are not satisfied, the housing design of the electrical equipment shall be improved. If needed, mounting modes of the components could be regulated partially or heat conduction measures could be adopted additionally. Regarding circumstances that fail to meet thermal design requirements of electrical equipment even if the aforesaid improvements have been taken, circuit shall be redesigned by optimizing circuit or reducing power to meet requirements eventually.

5.3.3.3 Thermal Design Cases of Electrical Equipment

This part takes power condition unit (PCU) of some spacecraft, for example, to demonstrate thermal design of electrical equipment.

1. **Overview**

PCU of some spacecraft is indicated in Fig. 5.41. The range of its operation temperature is −15 °C to 50 °C and the qualification temperature range is −35 °C to 70 °C. PCU could be classified to power distribution module, discharge module, MEA module, charge module, etc., as indicated in Fig. 5.42. Each module structure is processed as "I"/"C"-shaped cavity by using one bulk of aluminum for the benefit of inner heat transfer.

2. **Temperature requirements for components**

Rated temperatures and grade-I derating temperatures of PCU components are listed in Table 5.4.

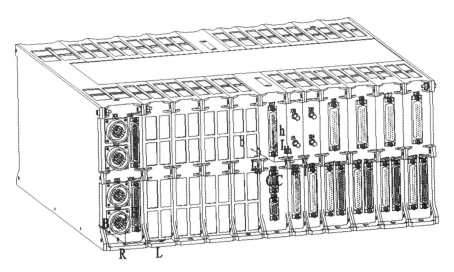

Fig. 5.41 Model of PCU

Fig. 5.42 Explosive view of power controller module

3. **Thermal design of typical components**

High-power components of PCU mainly include MOS transistor for shunt regulator, charge regulator, isolating diode, transformer for discharge, energy-storage inducer, etc. The following measures shall be in the thermal design of this type of high-power component.

1) TO/F packaging power components

TO/F packaging power components such as high-power MOS transistor, isolating diode and so on, are fixed on the baseboard of PCU with thumbscrews directly, and use 250–500 μm thick insulating film which also acts as a thermal pad, as indicated in Figs. 5.43 and 5.44.

2) Energy-storage inducer and transformer

Energy-storage inducer and transformer are directly mounted on the inner structure and filled in with GD480 heat-conducting glue in the mounting gaps to reduce thermal conductance resistance, as indicated in Fig. 5.45.

3) Power resistor

The ceramic-type packaging power resistor shall be mounted on the inner structure directly and filled in with GD480 heat-conducting glue in the mounting gaps to reduce thermal conductance resistance, as indicated in Fig. 5.46.

4) Power relay

The power relay shall be mounted on the inner structure with thumbscrews directly and use 500 μm thick insulating film which also acts as a thermal pad, as indicated in Fig. 5.47. Thus, dissipation is rejected from the mounting surface to the structure.

Table 5.4 Range of operating temperature of the components

No.	Name	Model	Heat resistance/(°C·W^{-1})	Maximum rated temperature/°C	Grade-I derating temperature/°C
1	Diode	1N7043	0.83	150	90
2	Diode	1N7043	0.83	150	90
3	Diode	1N7043	0.83	150	90
4	Diode	1N7043	0.83	150	90
5	Diode	1N7043	0.83	150	90
6	MOS power transistor	2N7224	0.83	150	85
7	MOS power transistor	2N7224	0.83	150	85
8	MOS power transistor	2N7224	0.83	150	85
9	MOS power transistor	2N7224	0.83	150	85
10	MOS power transistor	IRF5M5210	1	150	85
11	MOS power transistor	2N7225	0.83	150	85
12	MOS power transistor	IRF5M5210	1	150	85
13	Linear power	LM117H	21	150	85
14	Linear power	LM117H	21	150	85
15	Linear power	LM117H	21	150	85
16	Linear power	LM117K	1.9	150	85
17	Linear power	JW7805	3.5	150	85
18	Relay	1JB-75	–	85	65
19	Sampling resistor	RQCG7W0.005Ω	–	125	Temperature derating curve

(continued)

5.3 Thermal Design Cases of Spacecraft Assembly

Table 5.4 (continued)

No.	Name	Model	Heat resistance/(°C·W^{-1})	Maximum rated temperature/°C	Grade-I derating temperature/°C
20	Sampling resistor	RQCG7W0.005Ω	–	125	Temperature derating curve
21	Sampling resistor	RQCG7W0.005Ω	–	125	Temperature derating curve
22	Sampling resistor	RQCG7W0.005Ω	–	125	Temperature derating curve
23	Sampling resistor	RQCG7W0.005Ω	–	125	Temperature derating curve
24	Sampling resistore	RX9063W0.01Ω	–	275	Temperature derating curve
25	Sampling resistor	RX9063W0.01Ω	–	275	Temperature derating curve
26	Integrated circuit	AD1674	23.5	125	85
27	DC/DC	B65170	6.8	125	85

Fig. 5.43 Mounting schematic diagram of TO power transistor

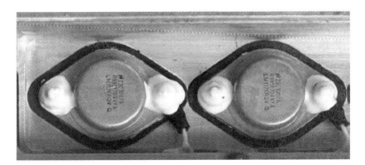

Fig. 5.44 Mounting schematic diagram of F-shaped power transistor

Fig. 5.45 Mounting diagram of inducer and transformer

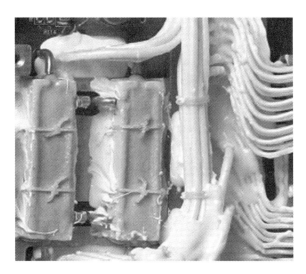

Fig. 5.46 Schematic diagram of ceramic-type power resistor

5.3.4 Thermal Design of Camera

5.3.4.1 Object of Thermal Design of Camera

The thermal design of camera is related to optical properties and optical-mechanism system of the camera, which is mainly used to ensure that thermal distortion of the camera optical system, is acceptable. Purposes of thermal design of camera are listed as follows:

5.3 Thermal Design Cases of Spacecraft Assembly

Fig. 5.47 Mounting schematic diagram of power relay

(1) Control temperatures of optics lens, major structure and focal plane assembly.
(2) Control radial temperature gradient, circumferential temperature gradient and axial temperature gradient of optics lens.
(3) Control temperature and temperature difference of the major structure.
(4) Satisfy requirements of the camera assembly in regard to space distribution and time domain distribution of temperature.
(5) Reduce temperature fluctuation when CCD works.
(6) Design thermal interface between camera and spacecraft, including interface of its radiator and spacecraft.

5.3.4.2 Characteristics of Thermal Design of Camera

Generally, main parts such as optics lens, focal plane assembly and the major structure have higher requirements, including slight temperature fluctuation and tiny temperature difference (temperature gradient). Hence, high-precision temperature control technology is applied for camera.

Regarding thermal design of charge-coupled device (CCD) camera assembly, heat transfer enhancement in small space near the mounting area of heating components is also one characteristic of thermal design of camera. Heat conduction rope, microheat pipe and loop heat pipe and so on are commonly used measures.

5.3.4.3 Thermal Design Case of Camera

1. **Overview**

The camera of some spacecraft is composed of front lens assembly, post-lens assembly, light shield and focal plane assembly. The optical lens of camera includes primary lens assembly (hereinafter referred as primary lens) and its supports, secondary lens assembly (hereinafter referred as secondary lens) and its supports, and tertiary lens assembly (hereinafter referred as tertiary lens) and its supports. The schematic diagram of camera structure is indicated in Fig. 5.48.

To meet optical requirements of camera, it shall ensure that the temperature of each optical part on the primary optical axis is uniform. The temperature gradient should be as small as possible. Besides, CCD temperature could be kept in the required range. Typical temperature indexes of the camera are indicated in Table 5.5.

2. **Thermal design status**

1) Light shield of camera

The light shield of camera is a structural assembly made of cylindrical CFRP, which is mounted on the satellite without connecting to the camera. The main thermal control measures of light shield are listed as follows:

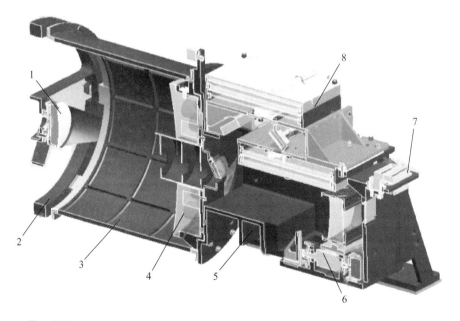

Fig. 5.48 Schematic diagram of a type of camera structure. 1—Secondary lens assembly; 2—Secondary lens supporting framework; 3—Front lens barrel; 4—Primary lens assembly; 5—Primary lens supporting framework; 6—Tertiary lens assembly; 7—Thermal control plug; 8—Focal plane assembly

5.3 Thermal Design Cases of Spacecraft Assembly

Table 5.5 Typical temperature indexes of camera

Item	Temperature indexes
Secondary lens assembly	20 °C ± 1.5 °C
Primary lens assembly	20 °C ± 1.5 °C
Circumferential temperature gradient of primary lens assembly	⩽1 °C
Tertiary lens assembly	20 °C ± 2 °C
Front lens barrel	20 °C ± 1.5 °C
Temperature difference of front lens cone	Circumferential difference ⩽1 °C, axial difference ⩽1.5 °C
Triple-rod	19 °C ± 2 °C
Temperature difference of different rods of triple-rod at the same location	⩽1 °C
Primary supporting framework	20 °C ± 3 °C
Axial temperature gradient of primary framework	⩽2.5 °C
TDICCD component	0–12 °C

(1) Paint black paint on the inner surface of the camera light shield to eliminate light scattering.
(2) MLI shall be used on the outer surface of the light shield.
(3) Fiberglass isolator shall be used on the joint of the light shield and other structures.

2) Primary camera lens

The primary camera lens is made of pyroceram. The thermal environment inside and outside the primary lens is very different. Temperature level and its stability of the primary lens would directly affect the imaging quality of the camera. Hence, the thermal insulation combined with active thermal control shall be applied. The main thermal control measures are listed as follows:

(1) The primary lens is only connected with lens frame with nine joints to increase heat resistance between them.
(2) The primary lens framework uses titanium alloy and heaters are set up along the circumferential direction of the outer surface.
(3) Heaters are installed on the back of primary lens framework to ensure temperature stability and uniformity.
(4) The primary lens framework is covered with MLI.

3) Secondary camera lens and triple-rod of the front lens

The secondary camera lens is mounted on the triple-rod which is made of titanium alloy. It is in the front of the primary lens and its space environment is worse than that of the primary lens. The main thermal control measures of the secondary lens are listed as follows:

282 5 Typical Thermal Control Design Cases of Spacecraft

(1) Blacken the surface of the triple-rod of the front lens.
(2) Set up heaters on the back of secondary lens base to keep its temperature stable.
(3) Set up heaters on the triple-rod of the front lens, respectively, to keep them from affecting temperature stability of the secondary lens and to maintain temperature within the allowable limits.

4) Tertiary lens of camera

The tertiary lens of camera is mounted on the tertiary supporting frameworks of post-lens and is located in the center of the rear of the camera. The environmental conditions of it are better environment than those of the primary lens and the secondary lens. Its temperature is mainly influenced by the post-lens and mounting support of camera. Therefore, thermal insulation is applied by combining with the heater. The main thermal control measures of the tertiary lens are listed as follows:

(1) Choose titanium alloy to make tertiary lens supporting framework.
(2) The back of tertiary lens is only connected with lens frame with nine joints to increase heat resistance between both.
(3) Heaters are set up on the back of tertiary lens along the circumferential direction.
(4) The tertiary lens supporting framework is covered with MLI.

5) Focal plane

Owing to the configuration of this camera, there is no space to set up heaters inside the focal structure. Thus, the coupling thermal design of camera focal plane and its surrounding parts is employed to achieve thermal stability of the camera focal plane. The emphasis is on the design of heat transfer channel on the focal plane (including heat transfer path and heat storage) and the design of system temperature stability.

The lower focal plane assembly is installed in the primary supporting framework. Considering temperature stability of the optical system, the heat dissipation during operation shall only be rejected to the upper focal plane assembly. Due to mounting precision, there shall be no thermal conduction fillers between the electrical equipment of focal plane assembly, so heat could only be transferred by surface mounted heat pipe on the upper and lower focal planes. Heat transfer channel of focal electrical equipment is composed of the lower focal plane assembly, upper focal plane assembly, heat pipe and radiator, as indicated in Fig. 5.49.

5.3.5 Thermal Design of Antenna

5.3.5.1 Purposes of Thermal Design of Antenna

Generally, the antenna is mounted on the outer surface of spacecraft. Compared with those devices installed inside of the spacecraft, the antenna would be affected severely by deep cold space and sunlight. Transmitting antenna usually has some

5.3 Thermal Design Cases of Spacecraft Assembly

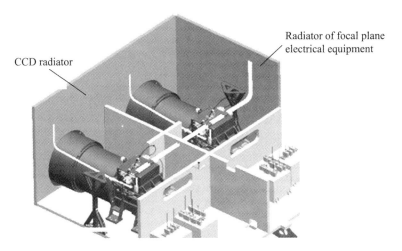

Fig. 5.49 Schematic diagram of camera heat pipe and the satellite

heat dissipation. Heat dissipation increases with the transmitted power, while heat flux increases with the operating frequency of antenna.

Antenna thermal design plays an important role. The thermal control requirements of the majority of antennas could be categorized into two kinds: One has equivalent requirements as the electrical equipment does, such as locking and release device, active revolving mechanism (including deployment mechanism, pointing drive mechanism), which usually has a narrow range of the operating temperature; the other is a structural part with a wider allowable temperature range. Purposes of thermal design for these structural parts are to keep antenna temperature in the allowable temperature range of the materials, and to reduce temperature difference of each part. Predicted temperatures are references for analyzing antenna thermal distortion and determining temperature conditions of the thermal vacuum test.

A growing number of spacecraft has used the active phased array antenna currently, such as active Synthetic Aperture Radar (SAR). Thermal control requirements of this type of antenna are equivalent to those of electrical equipment. In most cases, the temperature difference among antenna parts shall be controlled as required.

5.3.5.2 Thermal Control Measures of Common Antennas

1. White paint

White paint with low solar absorbance and high infrared emissivity is widely used on the antenna surface, especially on the solid reflecting antenna. The white paint is usually applied on the operation surface of reflector, transmitting waveguide and feed or surfaces of other radiating parts. It is used to reduce thermal influence of sunlight and avoid over-high temperature. For the solid reflector, the application of white paint is good for reducing temperature difference between the sun-illuminated area and the shielded area of the reflector. Regarding spacecraft with anti-static requirements, the anti-static white paint of which volume resistivity is 1×10^6 $\Omega \cdot m$ to 1×10^9 $\Omega \cdot m$ shall be used. Generally, it is forbidden to paint on the solid reflector frontage after it is metallized. What's more, for large solid reflectors, thermal effect caused by specular reflected sunlight from white paint shall be assessed sometimes.

2. **MLI**

MLI is also a popular measure of antenna thermal control, which is mainly applied on the back of the solid reflector, supporting structure and surfaces of parts with little heat. It is forbidden to cover MLI on the operation surface of antenna reflector or microwave channel since microwave cannot penetrate through the conductive aluminum coating.

3. **Germanium-coated polymide film**

Germanium-coated polymide film is usually applied on the operation surface of solid reflector, horn feed and active phased array antenna. When it is used on the opening surface of solid reflector, it is covered as a tent over the opening surface. The tent is usually called as "sun shield," which is used for shielding sunlight and preventing sunlight shining on the reflecting surface directly. This could reduce temperature difference and the highest temperature level on the reflecting surface. When the reflecting surface is not illuminated by sun shining, the tent could raise the minimum temperature level to a certain extent. Germanium coating is used for reducing static accumulation on the polymide film. When there are none anti-static requirements, polymide film with white paint could be used instead of germanium-coated film.

4. **Heat pipes/loop heat pipes**

Regarding active phased array antenna, heat pipes/loop heat pipes would be applied in most cases so as to transfer heat dissipation inside the antenna to the radiator or even up temperature of each part.

5. **Heaters**

For the majority of antennas, heaters are mainly applied on the locking and release device or active revolving mechanisms. For active phased array antenna, the application of heaters is one of the important thermal control measures, which could improve temperature level as well as reduce temperature difference of relevant parts.

5.3.5.3 Thermal Design Cases of Antennas

1. **Deployable solid reflectors**

The deployable solid reflector on DFH-4 platform satellite in folded and deployed state is indicated in Fig. 5.50.

5.3 Thermal Design Cases of Spacecraft Assembly

Fig. 5.50 Folded and deployed status of deployable solid reflector on some DFH-4 spacecraft. (a) Folded status; (b) Deployed status

The back surface of the primary reflector is covered with MLI and sprayed with white paint on the front. Antenna supporting/deployable arm and deployable mechanism is covered with MLI as well. Outer surfaces of antenna feed and sub-reflector are sprayed with white paint. Because of high heat dissipation of feed assembly, radiating fins that are processed along with feed assembly are used, as indicated in Fig. 5.51. The locking and release device of antenna has been equipped with an electric heater and covered with MLI, as indicated in Fig. 5.52. The deployable mechanism has been mounted with an electric heater as well. These heaters would operate when the antenna is folded and does not operate any more when antenna is deployed.

Fig. 5.51 Thermal control of feed source and sub-reflector

Fig. 5.52 Heater for locking and release device

2. **Variable-pointing beam antenna**

The variable-pointing beam antenna of some spacecraft is indicated in Fig. 5.53, including primary reflector, sub-reflector, feed, drive mechanism, rotation joint of waveguide, locking and release device, etc.

As indicated in Fig. 5.53, the front of primary reflector is coated with a germanium-coated sun shield, which covers the sub-reflector and its supporting rods and feed. The back of the primary reflector is covered with MLI. Overall thermal design measures are taken on the drive mechanism and rotation joints on the back of primary reflector. A MLI tent is used, which extends from the back of primary reflector to

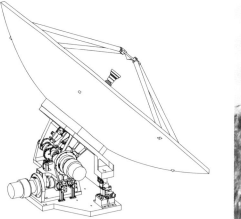

Fig. 5.53 Variable-point beam antenna and its thermal control

5.3 Thermal Design Cases of Spacecraft Assembly 287

the mounting deck, thereby shielding the drive mechanism, rotation joint, locking and release device, waveguide, etc. Besides, surfaces of these assemblies shall be sprayed with thermal control paint with high emissivity and heaters are set up for the drive mechanism and locking and release device, respectively.

5.3.6 Thermal Design of Drive Mechanism

5.3.6.1 Thermal Design Purposes of Drive Mechanism

The drive mechanism is applied on the deployable assembly, swing assembly, orientation mechanism or servo-tracking mechanism. It is also employed to drive other loads such as antenna, camera, mechanical arm and so on. The drive mechanism includes structural parts, drive assembly, etc. While the structural parts of the drive mechanism include the mounting support, etc. The drive assembly is the core part of the drive mechanism, including motor, reducer, encoder, etc.

The drive mechanism is usually mounted outside the spacecraft module and insulated from the module deck. Thermal control of the structural parts is to keep its temperature in the allowable range of materials. Encoder of drive mechanism has more stringent temperature level. When the motor is working, high dissipation would increase the difficulty to control temperature of the drive assembly. However, lubricant in the reducer, especially liquid lubricant, has narrow applicable temperature range. Liquid lubricant usually works above $-70\ °C$.

As a principle, thermal control measures cannot interfere movement of the mechanism. Generally, designs and installations of MLI shall be made according to movement modes, and the maximum envelope of the drive mechanism. Anti-friction thermal control coating shall be used on parts that have relative movement. Besides, the installability of thermal control materials and products shall be taken into consideration. For example, whether there is enough space for installation and fixation of heaters and MLI shall be considered.

5.3.6.2 Common Thermal Control Measures of Drive Mechanism

Since structural parts of the drive mechanism have wide applicable temperature range, the common thermal control measure is to spray thermal control paint with low solar absorbance and high infrared emissivity, or to be covered with MLI.

The thermal design emphases of drive assembly are heat rejection at hot conditions and heating at cold conditions. Generally, heaters are pasted on the surface of drive assembly. For the drive assembly that operates in long time and with high heat dissipation, methods of spraying thermal control paint with low solar absorbance and high infrared emissivity or pasting thermal control film are often used for heat rejection. Sometimes, a specialized radiator could be designed for heat rejection of the dissipating part. For the drive assembly that operates in short time and with low heat dissipation, thermal measure of being covered with MLI is usually taken.

5.3.6.3 Thermal Design Cases of Drive Mechanism

The typical two-axial drive mechanism of some spacecraft antenna is indicated in Fig. 5.54. Status of both shafts is the same. Both shafts will work all the time.

The primary thermal analysis shows that the second surface mirror of glass is to be pasted on the outer surface of motor and encoder, to radiate heat of motor and encoder effectively. However, technique of pasting glass mirrors on the mechanism surface is unreliable. Hence, a sleeve-type radiator has been designed as indicated in Fig. 5.54: OSR is pasted on the outer surface of the sleeve and the radiator is mounted on the cylinder part of motor; conductive interface fillers shall be used between radiator and motor. The inner surface of sleeve and the outer surface of encoder adopt high-emissivity coatings. Heat of motor and encoder could be transferred to the sleeve by thermal conduction and radiated to the space.

At cold conditions, heaters shall be used to keep temperature of the drive mechanism above minimal limit. Heaters shall be set up on the inner surface of sleeve, outer surfaces of reducer and encoder, respectively. This could obtain a uniform-temperature distribution on each part of mechanism, since the shell of this drive mechanism is made of titanium alloy.

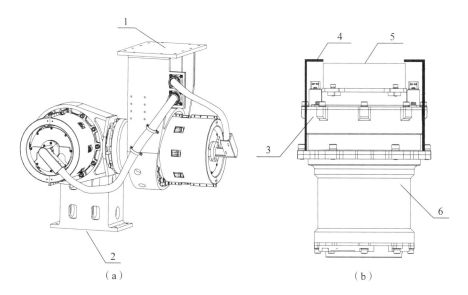

Fig. 5.54 Schematic diagram of a drive mechanism assembly. (a) Assembly status; (b) Status of single shaft. 1—Support connecting with antenna; 2—Support connecting with satellite; 3—Motor; 4—Sleeve; 5—Encoder; 6—Reducer

Chapter 6
Thermal Analysis Technology

6.1 Overview

Thermal analysis can be regarded as virtual test or numerical test, which plays an important role in spacecraft thermal control design, ground test verification and in-orbit technical support. In the design phase, thermal analysis is the most important means to determine the configuration (i.e., functional and physical characteristics). In the design verification stage, the thermal model is correlated with the results of ground thermal balance test, and the flight performance is further predicted. While being verified by test, the thermal analysis also validates the thermal design. In some cases, due to the complexity of the thermal environment in the mission, the boundary conditions in the test can only be set to a mission independent state. In this case, the purpose of the test is purely to verify the thermal analysis, and the design is mainly demonstrated by the analysis predication of the correlated thermal models. In a word, the thermal balance test requires the in-depth participation of thermal analysis to constitute the verification of thermal design. A good case study should cover extreme cases of a mission profile. Although the cases selected in the design phase are sufficient, there may be cases not covered by the previous design during real flight, when spacecraft thermal behavior can solely predicated by analysis. In addition, at the end of mission, thermal analysis is one of assertions to predict the life of spacecraft by simulating the trend of temperature evolution.

The aerodynamic heating process during ascent or atmospheric entry phase of spacecraft, the combustion and plume process when the engine/thruster fires both belong to the field of space thermophysics. However, due to the complexity of the related processes, professional analysis is usually required and provided to the thermal analysts as input conditions, so it is not described in this chapter.

Even so, the realm of spacecraft thermal analysis still covers a wide range, such as the calculation of radiation heat exchange coefficient, the calculation of orbital parameters and external heat flux, the screenings of design/analysis cases, the discretization of mathematical physics equations of heat transfer and fluid flow, the solution of algebraic equations, the simulation of function of particular thermal products or those of special heat transfer processes, correlation of thermal analysis models. These are all thermal analysis techniques. Most modern spacecraft thermal analysis uses widely accepted industrial software for modeling and simulation. Considering this, this chapter only describes common methods or concepts that are not directly used in most cases but are important for proper thermal analysis.

6.2 Space Energy Conservation Equation

6.2.1 Thermal Network Equation

The energy conservation equation of spacecraft in orbit described by *G-C* (thermal conductance heat capacity) network is

$$m_i c_i \frac{dT_i}{dt} = Q_i + q_i + \sum_{j=1}^{N} D_{ji}(T_j - T_i) + \sum_{j=1}^{N} G_{ji}(T_j^4 - T_i^4) \quad (6.1)$$

where subscripts i, j stand for nodes; T for temperature; m for mass; c for specific heat; t for time; Q for the heating rate of external heat flux; q for the internal heat source power; D_{ji} and G_{ji}, respectively, for linear thermal conductance (overall heat transfer coefficient) and radiative conductance between nodes j and i.

In Eq. (6.1), the left side is the internal energy change rate of the node, and on the right are the external heat flux absorbed by the node, heat dissipation, heat flowing rate into node i through all linear conductance, heat flowing rate into node i via all radiative conductance.

Linear conductance has many forms, such as $\frac{kA}{L}$ based on Fourier's law of heat conduction, contact heat transfer coefficient $\frac{A}{R_c}$, convective heat transfer coefficient hA, and (upstream to downstream) fluid transport items $\rho u A C_p$. Linear thermal conductance may be simple or complex. A typical example of the latter in thermal conductance is the heat transfer between upstream and downstream fluids in three-dimensional flow. Its determination often needs to solve continuity equation and momentum equation, because it is related to velocity field.

Generally, it is more difficult to determine radiative thermal conductance, which usually requires numerous complicated calculations.

6.2 Space Energy Conservation Equation

The following forms of thermal network equations are often used in engineering:

1. **Periodically transient equation**

$$\begin{cases} m_i c_i \frac{dT_i}{dt} = Q_i + q_i + \sum_{j=1}^{N} D_{ji}(T_j - T_i) + \sum_{j=1}^{N} G_{ji}\left(T_j^4 - T_i^4\right) \\ Q_i(t + \tau_0) = Q_i(t) \end{cases} \quad (6.2)$$

where τ_0 is the orbital period.

The periodically transient equation corresponds to the most commonly used design analysis state in engineering; that is, the internal heat source and external heat flux are regarded as periodic variables in continuous orbits. When the calculation time is long enough, the temperature results will be periodic; that is, periodical stability or equilibrium will be achieved. The period of temperature is not necessarily equal to the orbital period, but also affected by fluctuation period of internal heat dissipation.

2. **Quasi-steady-state equation**

$$\begin{cases} m_i c_i \frac{dT_i}{dt} = \overline{Q_i} + q_i + \sum_{j=1}^{N} D_{ji}(T_j - T_i) + \sum_{j=1}^{N} G_{ji}\left(T_j^4 - T_i^4\right) \\ \overline{Q_i} = \frac{\int_0^{\tau_0} Q_i dt}{\tau_0} \end{cases} \quad (6.3)$$

The quasi-steady-state equation corresponds to some thermal balance test state of spacecraft, in which not transient, but periodic average orbital heating is simulated.

3. **Steady-state equation**

$$\begin{cases} 0 = \bar{Q}_i + q_i + \sum_{j=1}^{N} D_{ji}(T_j - T_i) + \sum_{j=1}^{N} G_{ji}\left(T_j^4 - T_i^4\right) \\ q_i = \text{const}(i) \end{cases} \quad (6.4)$$

The steady-state equation corresponds to a small number of thermal balance test state of spacecraft, in which the periodic average orbital heating is simulated, and the internal heat source is set as a constant.

It is worth noting that even if the solutions of periodically transient or quasi-steady equations are periodic stable, the periodic arithmetic mean of the solutions of the two equations is not necessarily equal to the result of the steady-state equation [1–3], as in general $\overline{T^4} \neq (\overline{T})^4 \left(\overline{T} = \frac{\int_0^{\tau_0} T dt}{\tau_0}, \overline{T^4} = \frac{\int_0^{\tau_0} T^4 dt}{\tau_0}\right)$.

6.2.2 Computational Domain and Boundary Conditions

It is the basis of correct modeling to reasonably determine the computational domain and the relevant boundary conditions of a problem to be solved.

1. **Spacecraft in orbit**

When the spacecraft runs in orbit, 4 K cold black background is the unique first kind of boundary condition for temperature. External heat flux absorbed by spacecraft surface belongs to the second kind of boundary condition. Equipment heat dissipation can also be generally regarded as the second kind boundary conditions of equipment surface. These two boundary conditions are usually applied in the form of surface heat source.

2. **Vehicle landing on celestial body**

When a vehicle stays or moves on the surface of a celestial body, similar to in-orbit flight, direct solar irradiation and celestial albedo are still the second kind boundary conditions, but the infrared radiation of the celestial body's surface of the can be treated as either the second or the first kind of boundary conditions. With the view factors to celestial body being affected by distance, the infrared radiation from celestial body received by the spacecraft surface is not uniform. The celestial body surface near the spacecraft is divided into grids, and the temperature of these grids is included in the analysis model as the first kind of boundary nodes. This method is usually more convenient.

If the celestial body on which the spacecraft resides has atmosphere, the above-mentioned direct solar irradiation and albedo may be affected by atmospheric attenuation and scattering. In addition, cloud layer radiation participates, which can be regarded as the first or second kind boundary conditions. Another significant difference is the need to add a third kind of boundary conditions to reflect the convection with the atmosphere.

3. **Cabin segment**

If the object to be solved is a single cabin segment of spacecraft, compared with the complete spacecraft mentioned above, an additional first kind of boundary nodes should be considered: the temperatures of all the parts that do not belong to this cabin segment but have heat conduction or radiation heat exchange with it act henceforth as the first kind conditions of the segment-level model. These first kind boundary conditions need to be iterated with the rest cabin modules. This situation is equivalent to dividing the whole spacecraft computational domain into several sub-domains at the cabin segment level. If simply ignoring the rest cabin segments may cause false external heat flux or radiative factors to the space, the technique of converting the effects of rest cabin segments into equivalent boundary conditions will be explained in Sect. 6.6.

4. Internal Equipment

Only considering the temperature solution of the equipment inside the cabin, the surface of the equipment does not exposed to external environmental heating, nor does it radiate heat to the space, but there are many radiation exchanges between this surface and other surfaces of the spacecraft. Although the detailed analysis of the equipment can be embedded into system-level integrated analysis of the spacecraft, a large amount of internal detailed information of the equipment needs to be transferred to the system analyst, which is not convenient for design iterations. It is more effective to separate the detailed analysis of equipment from the system-level analysis.

In this case, there are generally two ways to set the boundary for equipment analysis: The temperature of its shell is taken as the first kind boundary condition; or besides assigning the temperature of the mounting interface as the first kind of boundary condition, all surrounding surfaces can be regarded as an enclosure at the average temperature, thus becoming the first kind of boundary condition connected to equipment through radiative heat exchange. In fact, these radiative boundary temperatures are affected by both the components and the system, so the values also need design iteration cycle.

Although the heat flow through the installation interface can also be used as a boundary condition, it is practically seldom due to the difficulty of direct determination. Even if used, it is not the only boundary condition. Another definite first kind of boundary conditions is still necessary for solution.

5. Additional heating

In addition to the external orbital heating and the internal heat source, there is sometimes other space or induced thermal environmental heating, which needs to be set as boundary conditions or source items.

The free molecular heating is generally neglectable when the perigee altitude is more than 180 km. It needs to be evaluated in some cases, such as the ascent phase. The free molecular heating is generally given by Formula (2.1), which can be used as a heat source without complicated CFD analysis.

When the spacecraft surface is heated by high-speed airflow, the surface heating rate is usually obtained by aerodynamic calculation and applied to the surface as a heat source item. This method is simpler than Newton's cooling law provided that gas temperature and heat transfer coefficient are known. Actually, neither is easy to be obtained in high-speed airflow problem. However, the aerodynamic calculation is usually carried out on the assumption that the wall is adiabatic or kept at certain fixed temperature, so it needs to be modified by gas enthalpy and the actual temperature of the wall. However, the wall temperature is unknown and needs to be solved, so the modified aerodynamic heating is actually the second kind of boundary condition, which needs to be updated continuously in the analysis. If ablation also exists on the surface, it is necessary to deduct the material ablation heat from the calculated aerodynamic heating. In summary, it is necessary to ensure that the applied surface heating is net heating flowing into the surface caused by aerothermodynamic effect.

The heating of engine operation to the surroundings includes rarefied gas plume heating, high-temperature gas radiation and nozzle's surface radiation. The heating of rarefied gas plume is mainly due to the energy transferred by the molecular motion, which is related to the temperature of the heated surface and the adaptation coefficient. Gas radiation is body radiation. These two heating items generally need plume analysis. Nozzle radiation is surface radiation, which can be obtained from plume analysis. But more common way is to export the nozzle surface temperature from engine combustion analysis or bench test and to use the temperature as a first kind of boundary condition. And both rarefied gas plume and gas radiation heating are applied as heat source simultaneously. But if radiation heating of the nozzle is taken as heat source, its temperature should not be taken as the boundary condition.

6. **Gas flowing**

To calculate the temperature field of three-dimensional flowing, the velocity field must be solved first, so the boundary conditions of velocity and pressure are necessary. The velocity boundary condition includes the flow rate through the crossing section of the fan, and no-slip assumption; that is, the velocity of airflow on the solid surface is zero. For a closed cavity, the pressure boundary can be set to the static pressure of the gas before the flow starts, which represents the static pressure near the non-flow area in the closed cavity; as for the venting system, ambient pressures at vents need to be defined.

Thus, for the aforementioned flying cabin segment-level or unit-level analysis, if three-dimensional gas flow is involved, in addition to the above-mentioned boundary conditions, particular boundary conditions related to flow need to be defined: When the gas flow crosses the cabin segments, it is difficult to determine the gas velocity and pressure conditions on the cabin–cabin crossing section. Generally, the velocity and pressure given by the multi-cabins combined model should be taken as the boundary conditions for single cabin analysis. For the aforementioned analysis of units inside a cabin, if gas flows by its surface, velocity and temperature of the surrounding gas, or gas temperature and convection coefficient on the surface, are generally extracted from system-level analysis and used as the first and third kind of boundary conditions correspondingly in the unit model. If a fan is used on the unit's wall (obviously rare in aerospace), the velocity, pressure and temperature of the surrounding gas, obtained from the system-level analysis, are the boundary conditions on the cross section of the vent.

6.2.3 Discretization

Numerical discretization refers to the methodology of transforming partial differential equations into algebraic equations. At present, the discretization schemes of time differential items are still nothing more than explicit, implicit and Crank–Nicolson method. Therefore, the general discretization methods focus on the treatment of

6.2 Space Energy Conservation Equation

spatial differential items. Note that Eq. (6.1) does not actually contain the differential items of spatial variables. Therefore, the general thermal analysts do not directly face the discretization of spatial differential items. The only problem they face is meshing, which is related to discretization. However, a rough insight to the discretization method used in thermal analysis software is helpful for a correct understanding of the settings of some boundary conditions, as well as the output or mapping rules of the analysis results.

According to the representing positions of the variables in question, two types of grids are mainly used in thermal analysis [4–6], as shown in Fig. 6.1.

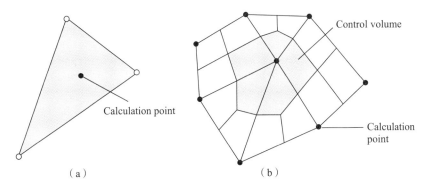

Fig. 6.1 Discrete grids. (a) Calculation point located at the center of the grid; (b) Calculation points located at grid intersection points

The calculation point in Fig. 6.1a is located in the center of the grid, corresponding to the finite difference method and the control volume method. The two methods are similar in form, but different in fact: The finite difference method is to directly expand the differential items in the differential equation by Taylor series to obtain algebraic equations, and the truncation errors directly depends on the mesh size, so a relatively refined grid is usually required; the control volume method takes the calculation point as the representative of the control volume. Firstly, the distribution function of the dependent variables is assumed, and then the distribution is adopted in the control equations, integrating over the control volume then obtaining the algebraic equation describing the relationship of variables between the calculation point and the adjacent ones. The discrete equation always satisfies the conservation on the control volume, so the mesh size selection is more flexible. The common point between the control volume and the finite difference method based on Taylor series expansion is that the calculation point's value is used as the representative of the control volume, while the difference is that the latter does not need to assume the distribution of dependent variables. For the control volume method using the grid in Fig. 6.1 (a), the grid is the control volume.

The calculation points in Fig. 6.1 (b) are located at the grid intersections, which can be considered as the finite element method, but strictly speaking, it is the standard finite element method only in the pure heat conduction problem without radiation. For radiation-involved problems, most software use finite difference method or the above-mentioned control volume method similar to finite difference method. For the flow problem, most software uses the control volume method based on finite element method.

The standard finite element methods are mainly variational principles method and weighted residuals method. The variational principles method is based on variational principle; that is, the problem of differential equation with boundary is converted into the solution of corresponding functional extreme value, which is also known as Rayleigh–Ritz method. The weighted residuals method substitutes approximation for strict value and makes inner integrals of the residual errors and selected weighting function, provided that the weighing functions make the integral equal to zero. Although the variational principles and weighted residuals method, although the equations in question are different (the former is the Euler equation corresponding to functional, the latter is original differential equation), they all have the same core idea; that is, the domain of the problem is divided into finite sub-domains and interpolate the function to be solved at finite number of points on each sub-domain approximately. Over the single sub-domain, the equation to be satisfied by the function value at interpolating points and external conditions is established, according to the relevant physical analysis or mathematical analysis. Following systematic integration establishes the overall equation over the entire domain about the function values at all interpolating points and boundaries. Once the value of the interpolating point is solved, any value at any points over the domain can be approximated with interpolating function. The finite element method cannot directly guarantee the conservation on the control body, but it has a strong geometric flexibility [7].

The control volume method can also be regarded as a special weighted residuals method, setting the weighting function to 1 only on the control volume being studied and 0 everywhere else. An important difference between the control volume method and the finite element method is that the finite element method needs a definite distribution of variables, and the relation of any point's variable values to that of calculation point is definite over the solution domain; in the control volume method, the distribution law of variables must be set before integration, but after the discretization equation is established, the distribution law of variables between calculation points is no longer needed.

When the control volume method shown in Fig. 6.1 (b) is adopted, the calculation points are grid intersection points (i.e., nodes of finite elements), and the control volume is formed by some grids (elements) around the element nodes.

6.2.4 Thermal Model Construction and Solution Process

Referring to Fig. 6.2, the construction and solution of the thermal network model used to calculate the temperature field mainly includes the following steps:

(1) Determine the range represented by nodes i and j in Formula (6.1), i.e., meshing the geometries.

6.2 Space Energy Conservation Equation

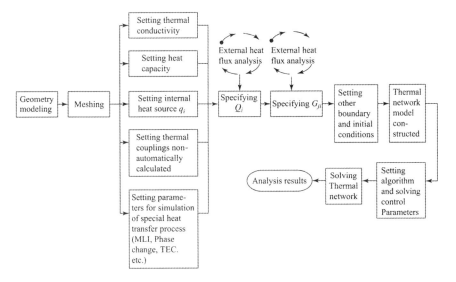

Fig. 6.2 Thermal analysis procedure

(2) Heat capacity $m_i c_i$ (or mass and specific heat) of the node is usually a known input parameter for thermal analysis, the assignment of its value is easy in model construction.

(3) Among the heat source items, the internal heat dissipation q_i is usually input as a known parameter. However, it is generally necessary to calculate the external heat flux Q_i specially, and the results are exported to the thermal model as boundary conditions.

(4) The linear conductance D_{ij} in Eq. (6.1) needs to be determined:

A. A zone of a solid continuous medium may be divided into multiple nodes, and the heat conduction between the nodes satisfies Fourier's law. After the thermal conductivity of the material is set, the thermal analysis modeling software can automatically calculate the linear conductance.

B. The linear conductance in the fluid flow energy governing equation may arise from thermal conduction and flow transport. Therefore, except for one-dimensional flow, it is almost impossible to determine it manually by analysts, which generally requires computational fluid dynamics (CFD) analysis via particular software.

C. Contact thermal conductance, and thermal coupling via entities not built geometrically need to be input by the analysts.

D. Some thermal products such as heat pipes and semiconductor cooler require the analysts to determine reasonable linear conductance to represent their thermal performances.

(5) In most cases, a special calculation of radiation heat transfer coefficient is required to determine the radiative conductance G_{ji}, and the result is used as input parameters of the thermal networks.
(6) To set up the iterating algorithm and convergence criteria necessary for numerical computation, as well as the boundary and initial conditions not covered by the above steps.

Figure 6.2 shows radiative conductance analysis, external heat flux analysis and thermal network solution as three processes, which do not mean that all thermal analyses must be carried out separately. Instead, depending on the problems to be analyzed and the software's ability, it is possible to combine these three processes into one modeling. If it is split into three analyses, the three meshes may overlap completely or partially, or they may not overlap at all. Generally, the mesh of the thermal network is the most complete. For meshes that do not overlap with the other two, the only restriction is to ensure that the data between the meshes is mapped correctly.

The above process is applicable to the situation that the design status, input parameters and case settings are determined. For the thermal analysis in the whole design cycle, the thermal analysis may be iterated many times. For example when extreme/worst cases are uncertain, it is sometimes necessary to carry out multiple trial analysis of external heat flux, screening formal cases-based possible combination of external heat flux and operating modes of equipments with heat dissipation, and sometimes even doing trial temperature computing aiming to different states to determine formal cases. Similarly, the above process does not show the iterative process of analysis–test (or flight)–thermal model correlation, but it should be noted that thermal model correlation is also an important task of analysis.

The steps of the above process are not strict in order, and many steps can be exchanged.

Sections 6.4–6.6 describe the external heat flux analysis involved in step (3), the radiation analysis involved in step (4), and the simulation of particular thermal products involved in step (5).

6.3 External Heat Flux Analysis

Most of the problems in the analysis of the external heat flux are focused on spacecraft flying around celestial bodies. The analysis process is shown in Fig. 6.3. Referring to "6.4 Radiation Analysis", for a surface A_i on spacecraft exposed to heat source E_R, corresponding external heat flux absorbed by it is $\varepsilon_{\lambda,i} A_i B_{ij} E_R$. Therefore, it is necessary to know the surface's thermo-optical properties $\varepsilon_{\lambda,i}$, which is characterized by the absorptive characteristics of the radiation source, the geometric information required to determine the orientation and area A_i, and the thermal environment parameters describing the radiation intensity E_R of the heating source; while there are numbers of factors to determine the absorption factor B_{ij} of the surface to the

6.3 External Heat Flux Analysis

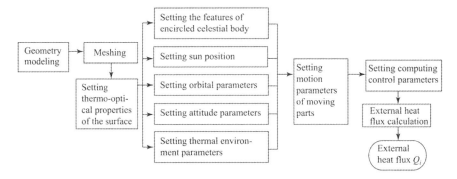

Fig. 6.3 Procedure of analysis of external heat flux

radiation source, including geometric and thermo-optical information of the surface itself and other related surfaces, as well as the positional relationship of the surface to celestial bodies. For the motion around celestial bodies, there is usually more than one celestial body involved, so it also needs the positional relationship among celestial bodies; thus, multiple parameters, such as the features of the encircled celestial bodies, the position of the sun, orbital parameters, attitude parameters, are further required.

Generally, external heat flux analysis software allows a variety of orbit definition, such as Keplerian orbit definition, β angle definition, geographic latitude and longitude definition, and heat source vector definition. Parameters required vary with the type of definition, as shown in Table 6.1, where $\sqrt{}$ indicates mandatory and $\times$ indicates unnecessary.

Table 6.1 Input parameters required for orbit definitions

Parameter category	Keplerian orbit definition	β angle definition	Geographical latitude and longitude definition	Heat source vector definition
Characteristics of encircled celestial body	$\sqrt{}$	$\sqrt{}$	$\sqrt{}$	$\sqrt{}$
Sun position	$\sqrt{}$	$\sqrt{}$	$\sqrt{}$	$\times$
Orbit parameters	Use 5[a]	Use height only	Use height only	$\times$
Attitude parameters	$\sqrt{}$	$\sqrt{}$	$\sqrt{}$	$\times$
Thermal environment parameters	$\sqrt{}$	$\sqrt{}$	$\sqrt{}$	$\sqrt{}$
Control parameters	$\sqrt{}$	$\sqrt{}$	$\sqrt{}$	$\sqrt{}$

[a] See Table 6.2

Keplerian orbit definition is the most widely accepted method, and its specific input parameters used are shown in Table 6.2.

Some of the parameters in Table 6.2 are determined by mission, orbit and other factors. They are input to the external orbital heat flux analysis as known parameters. The following mainly describes some parameters that need to be chosen or even calculated by thermal analysts.

6.3.1 Sun Position

The position of the sun is generally described by the solar day angle Φ_S, which can be easily obtained through the date:

$$\Phi_S = \frac{\text{Date} - \text{Date0}}{365.25} \times 360 \tag{6.5}$$

where Date represents the date to be calculated, Date0 represents the vernal equinox date.

Sometimes it is also necessary to use the solar right ascension Ω_S or declination δ_S to express the sun's position. The relation between these two parameters and the solar day angle is as follows:

$$\tan \Omega_S = \cos I \cdot \tan \Phi_S \tag{6.6}$$

$$\sin \delta_S = \sin I \cdot \sin \Phi_S \tag{6.7}$$

where I is the inclination of the ecliptic plane to the equatorial plane of the encircled celestial body (planet).

For spacecraft flying around the Earth or the moon, it is generally considered that the solar right ascension is approximately equal to the solar day angle, i.e., $\Omega_S \approx \Phi_S$.

The angle β is the minimum angle between the orbit plane and the solar vector. This parameter is very important for understanding the exposure of the spacecraft to the sunlight. When defining orbit by β angle type, the analyst must specify it as a known condition. Common orbit analysis software can compute β. Once the solar day angle is known, β can also be calculated according to the following formula:

$$\sin \beta = \cos i \cdot \sin I \cdot \sin \Phi_S + \sin i \cdot \sin \Omega \cdot \cos \Phi_S \sin i \cdot \cos I \cdot \cos \Omega \cdot \sin \Phi_S \tag{6.8}$$

Some software can also determine the sun's position by specifying the date.

6.3 External Heat Flux Analysis 301

Table 6.2 Input parameters for external heat flux analysis

Parameter category	Parameter name	Determined by mission	Determined by orbit design	Determined by other subsystems	Defined by thermal analysis	Remarks
Characteristics of encircled celestial body	Mean radius	•				
	Inclination of the equatorial plane to the ecliptic plane I	•				
	Orbit period	•				
Sun position	Solar day angle Φ_S/Solar right ascension Ω_S/solar declination δ_S				•	Only 1 of 3
Orbit parameters	Orbital inclination i		•			
	Perigee altitude		•			Only 2 of 4 required
	Apogee altitude		•			
	Eccentricity					
	Orbital period					
	Ascending node right ascension Ω		•			
	Argument of perigee (apogee)		•			
Attitude parameters	Orientation of spacecraft coordinate system	•				
Thermal environment parameters	Direct solar flux magnitude				•	
	Albedo				•	

(continued)

Table 6.2 (continued)

Parameter category	Parameter name	Determined by mission	Determined by orbit design	Determined by other subsystems	Defined by thermal analysis	Remarks
	Planetary IR flux magnitude				•	
Control parameters	Starting point				•	
	Number of calculation positions				•	
	Motion parameters of moving parts			•		

6.3.2 Orbital Parameters

Orbit parameters are generally determined by orbit design and are used as input parameters for external heat flux analysis. However, for quasi-sun-synchronous orbits, the right ascension of ascending node can also be obtained according to the following formula based on the solar right ascension and local time at descending node:

$$\Omega = \Omega_S - 15 \times (12 - \text{Hrs}) + 180 \quad (6.9)$$

where Hrs is local time at descending node.

6.3.3 Thermal Environment Parameters

Thermal environment parameters refer to solar flux intensity (magnitude), infrared emission intensity and albedo of encircled celestial body (for simplicity, denoted as planet in the following paragraphs of this section).

1. **Solar flux**

The solar flux intensity depends on the distance between the spacecraft and the sun, which is equivalent to the distance between the encircled planet and the sun. With a in astronomical unit AU standing for the average distance between planet and the sun, e for the eccentricity of the planet's orbit around the sun, f for the true anomaly of the sun in the celestial coordinate system, S for the solar intensity 1,367 W/m² at 1AU from the sun, the solar flux intensity around planet is

$$q_s = \left[\frac{1 + e \cos f}{a(1 - e^2)} \right]^2 S \quad (6.10)$$

For the Earth or the moon, $a = 1$, $e = 0.016,7$, $f = \Phi_S + 78.59$; for Mars, $a = 1.523,69$, $e = 0.093,377$, $f = \Phi_S - 248$ [8]. The relationship of the solar flux intensity to the solar day angle is shown in Figs. 6.4 and 6.5, respectively.

Most thermal analysis software can calculate the solar flux intensity automatically according to the position of the sun. These parameters can be specified by date, or solar day angle, or solar right ascension and declination.

2. **Infrared Radiation of Planet**

Referring to Eq. (2.5), the annual average of the Earth's infrared radiation is related to the solar intensity around Earth and the Earth's albedo. Once the solar intensity value is known, the Earth's infrared radiation intensity can be determined by Eq. (2.5). If the selected albedo is between 0.30 and 0.35, the infrared radiation intensity of the Earth is 214–247 W/m².

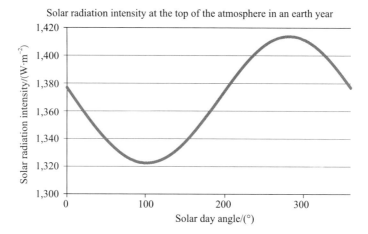

Fig. 6.4 Solar radiation intensity around the Earth and the moon

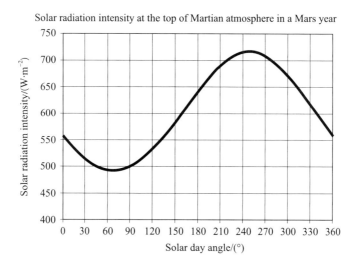

Fig. 6.5 Solar radiation intensity near Mars

For most of the Earth orbit vehicles, there is no need to consider the variation of the Earth's infrared emission intensity with latitude and longitude. If this is to be considered, a more accurate method is recommended by Marshall Flight Center: looking up a table according to three factors: orbit inclination, estimated thermal time constant, and object's sensitivity to infrared or albedo. This table is based on Marshall Flight Center's statistical measurement data of several low-orbit spacecraft.

In contrast, the infrared radiation of the lunar surface varies greatly with latitude and longitude, which should not be ignored in general. Due to the low thermal

6.3 External Heat Flux Analysis

conductivity of lunar soil and the absence of atmosphere, the temperature of the lunar surface is mainly affected by the absorbed solar flux. Thus, according to the spherical triangle rule, the temperature at given latitude and longitude can be expressed as:

$$T(\phi, \varphi) = \cos^{\frac{1}{4}} \varphi \cdot T(\phi, 0) \tag{6.11}$$

where

$$T(\phi, 0) = \begin{cases} [T_{ss}^9 \cos \phi + T_{270}^9]^{\frac{1}{9}}, & \phi \in [-90, 0) \\ [T_{ss}^7 \cos \phi + T_{90}^7]^{\frac{1}{7}}, & \phi \in [0, 90) \\ [\frac{3\sigma\varepsilon}{C}(\phi - 270) + T_{270}^{-3}]^{-\frac{1}{3}}, & \phi \in [90, 270) \end{cases}$$

$$T_{ss} = \left[\frac{S(1-\rho)}{\sigma\varepsilon}\right]^{\frac{1}{4}}$$

$$T_{90} = \left[-180 \times \frac{3\sigma\varepsilon}{C} + T_{270}^{-3}\right]^{-\frac{1}{3}}$$

where S is the solar flux magnitude in W/m^2; ρ is the average albedo of the moon to the sun; ε is the infrared emissivity of the moon; σ is Stefan–Boltzmann constant; ϕ is the longitude from the subsolar point in (°); φ is latitude in (°); T is the surface temperature of the moon in Kelvin (K); C is a fitting constant.

Using the above temperature expression and the lunar surface infrared emissivity, the lunar infrared radiation intensity can be determined.

In most cases, the variation of Mars surface temperature with latitude and longitude must to be considered, which is smaller than that of the moon but larger than that of the Earth. Due to the existence of Martian atmosphere, it is difficult to establish a formula similar to Eq. (6.11). At present, there are only insufficient observation data that is available as input for analysis.

3. **Albedo**

As mentioned before, in the problem of the Earth centered flight, the fixed albedo of the Earth is generally taken in the range of 0.30–0.35. In fact, the albedo also varies with the geographic latitude and longitude, and it is related to the infrared radiation intensity of the Earth's surface. If these problems are to be considered, it is similar to the above consideration of the infrared variation with latitude and longitude and orbit inclination, the estimated thermal time constant and the sensitivity of the object to Earth's infrared emission or albedo are used to query the recommended values for worst hot/cold case from the table. In addition, albedo shall be corrected according to the angle from subsolar point and then the short-term albedo correction shall be added, or the orbit average albedo shall be corrected according to the β angle.

The lunar surface has quite low albedo, with a fixed value of 0.073 generally taken. The average albedo of Mars is about 0.29.

6.3.4 Staying on Celestial Body

The above discussion mainly focuses on vehicles flying around a Planet. Some of them are applicable to the environment heating to a vehicle staying on the celestial body. Obviously, there are also some special problems. Here we mainly discuss the moon and the Mars.

There is no atmosphere on the moon, which makes the direct solar irradiation onto lunar surface same as the orbit around the moon. For vehicles landing at the lunar surface, the effective albedo and the lunar infrared emission only come from nearby surface. If following the way of defining the orbit around the moon, meshes refined enough to depict the variation of lunar surface infrared radiation with longitude and latitude need to be built. For the albedo, this approach cannot reflect the fact that the shadow areas of the spacecraft does not actually reflect. Therefore, a simpler method is to include a part of the near lunar surface in the model and mesh it (although the lunar geometric characteristics are used in the problem of flight around it, in fact, the moon is regarded as a geometry without mesh), thus avoiding the above two difficulties.

For infrared radiation and albedo on Mars, the same treatment method as that of the lunar surface is generally used. The radiation top layer of the Martian atmosphere is very close to the surface of Mars. In the absence of dust storms, it can be considered that the infrared radiation intensity of the top layer of the atmosphere and the surface of Mars are approximately the same.

In addition to the infrared and albedo of the Mars surface, the thermal environmental factors on Mars surface also include the infrared radiation of the atmosphere. On the one hand, the absorption and scattering of sunlight by the atmosphere cause the direct solar radiation preaching the surface of Mars being attenuated; on the other hand, it increased by diffusion. Atmospheric infrared, attenuated direct and diffused solar radiations are coupled. Atmospheric transparency affects these three at the same time. Low transparency causes direct solar radiation decreasing, and diffused solar radiation and atmospheric infrared emission increasing, but the quantitative relationship among them remains unknown so far. This make it complicated to determine thermal environment parameters of Mars surface.

The infrared radiation of Martian atmosphere is generally described by equivalent atmospheric radiation. For the case of high sky transparency, the equivalent atmospheric temperature of Mars is $-128.2\ °C$, corresponding to the infrared radiation heat flux of 25 W/m^2; for the case of low sky transparency, for example, during the dust storm, the equivalent atmospheric temperature of Mars is $-58.7\ °C$, corresponding to the infrared radiation heat flux of 120 W/m^2. These treatments only qualitatively reflect the impact of atmospheric transparency, but do not establish a quantitative relationship of direct and diffuse sunlight on the atmospheric transparency.

According to the ability of the software tool, there are two ways to deal with solar radiation composed of the direct and the diffuse component.

One is that analysts need to specify two parameters, respectively: attenuated direct solar irradiation intensity G_{bh} and diffuse solar irradiation intensity G_{dh}.

Assuming that local solar elevation angle is $\theta_h(°)$, the atmospheric optical depth is τ, and the local albedo is al, G_{bh} and G_{dh} are given by Formulas (6.12) and (6.13), respectively [8].

$$G_{bh} = G_{ob} \cdot \cos(90° - \theta_h) \exp\left[-\tau/\cos(90° - \theta_h)\right] \quad (6.12)$$

$$G_{dh} = G_h - G_{bh} \quad (6.13)$$

$$G_h = G_{ob} \cos(90° - \theta_h) \frac{f(\theta_h, \tau, al)}{1 - al} \quad (6.14)$$

where G_{ob} is the solar radiation at the top of Martian atmosphere; $f(\theta_h, \tau, al)$ is a normalized net heat flux function.

This method requires the optical depth to be specified. When there is no dust storm around the globe, the atmospheric optical depth of Mars can be assumed to be 0.5; the extreme value of heavy dust storm can be assumed to be 4. The smaller the optical depth, the higher the atmospheric transparency. Therefore, the specified optical depth should be consistent with the infrared radiation intensity of the equivalent atmosphere used in the same analysis

The second method does not need to specify the values of G_{bh} and G_{dh}; but in addition to the solar radiation at the top of Martian atmosphere, atmospheric extinction coefficient and diffuse sky radiative factor are needed.

For the Mars lander, compared to the orbiter, the selection of Mars surface Albedo may differ, and the global average albedo is no longer used. Instead, it should be selected in a wider range according to the region.

The wind on the surface of Mars can also be regarded as a thermal environment, which is generally introduced into the model as a convection boundary condition. The temperature of the wind is usually the same as temperature of the local ground surface, and the speed needs to be specified by the analyst.

6.4 Radiation Computing

Most of the radiation computings in aerospace are faced to radiation of surfaces separated by the nonparticipating medium which does not emit, absorb or scatter radiation. The vacuum strictly meets these requirements, and the gas in the pressurized cabin also approximately satisfies such assumption.

6.4.1 View Factor

View factor, also known as shape factor, geometrical or black body view factor, is defined under the assumption of uniform effective radiation and diffuse surface (both emission and reflection of surface are diffuse), as shown in Fig. 6.6. The view factor F_{ij} is defined as the fraction of the energy diffusely emitted by face i that is directly incident on face j, expressed mathematically as (6.15) [9]

$$F_{ij} = \frac{1}{\pi A_i} \int_{A_i} \int_{\omega_{j-i}} \cos\theta_i \, d\omega_{j-i} dA_i = \frac{1}{A_i} \int_{A_i} \int_{A_j} \frac{\cos\theta_i \cos\theta_j}{\pi r_{ij}^2} dA_i dA_j \qquad (6.15)$$

where dA_i and dA_j are the elements on the surfaces i and j; r_{ij} is the distance between element dA_i and dA_j, θ_i and θ_j are the angles between line linking elements and the normal of surface i and j, respectively. $d\omega_{j-i}$ is the solid angle of dA_j to dA_i, and A_i and A_j are the areas of surfaces i and j.

If spherical coordinates are used, $d\omega_{j-i} = \sin\theta_i d\theta_i d\beta_i$, and the view factor can also be expressed as

$$F_{ij} = \frac{1}{\pi A_i} \int_{A_i} \int_{\theta_i} \int_{\beta_i} \cos\theta_i \sin\theta_i d\theta_i d\beta_i dA_i \qquad (6.16)$$

The view factor F_{ij} has the properties of unity, reciprocity and superposition, which can be easily derived according to its definition.

The calculation of view factor is the basis for other radiation analysis such as radiative exchange factor and orbital heating. The main calculation methods of view factor

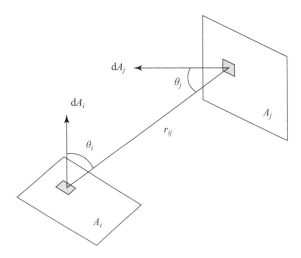

Fig. 6.6 View factor

6.4 Radiation Computing

include contour integral technique, Nusselt sphere technique, hemicube method and Monte Carlo technique.

1. **Contour integral technique**

Formula (6.15) is a double integral over surface, that is, a quadruple integral over lines in a general sense. It can be converted into loop integral according to Stokes' theorem, thus yielding:

$$F_{ij} = \frac{1}{2\pi A_i} \oint_{C_i} \oint_{C_j} \ln R \, \mathrm{d}C_i \, \mathrm{d}C_j \tag{6.17}$$

where C_i and C_j are the paths around the perimeters of surface A_i and A_j, respectively; R is the length of a line between two points on the perimeters of A_i and A_j [10].

2. **Nusselt sphere technique** [11]

As shown in Fig. 6.7, I_1 represents the radiation emitting intensity of the surface $\mathrm{d}F_1$ and ω_{2-1} represents the solid angle of $\mathrm{d}F_2$ to $\mathrm{d}F_1$, then energy $\mathrm{d}F_1$ emitting intercepted by $\mathrm{d}F_2$ is

$$\mathrm{d}q_{1\to 2} = I_1 \cos\varphi_1 \mathrm{d}F_1 \, \mathrm{d}\omega_{2-1} \tag{6.18}$$

Because all the radiative energy emitted by $\mathrm{d}F_1$ is $\pi I_1 \mathrm{d}F_1$, so the view factor F_{12} is

$$F_{12} = \frac{\mathrm{d}q_{1\to 2}}{\pi I_1 \mathrm{d}F_1} = \frac{\cos\varphi_1}{\pi} \mathrm{d}\omega_{2-1} \tag{6.19}$$

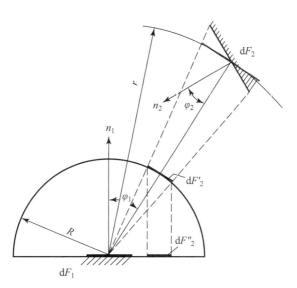

Fig. 6.7 Nusselt sphere view factor technique

First, draw a sphere with radius R around the geometric center (CG) of dF_1, the projection of the sphere on the plane F_1 is a circle with radius R. Next, the perimeters of dF_2 are projected onto the sphere forming dF_2', by drawing lines from the center of the sphere to the perimeters, Obviously, dF_2' and dF_2 have equal solid angle to dF_1, $d\omega_{2-1} = \frac{dF_2 \cos\varphi_2}{r^2} = \frac{dF_2'}{R^2}$, so

$$F_{12} = \frac{\cos\varphi_1}{\pi} d\omega_{2-1} = \frac{dF_2' \cos\varphi_1}{\pi R^2} \qquad (6.20)$$

And $dF_2' \cos\varphi_1$ is just equals to dF_2'', i.e, the projection of dF_2' on the plane dF_1, so the final result is:

$$F_{12} = \frac{dF_2''}{\pi R^2} \qquad (6.21)$$

The above process can be directly used to calculate the view factor between dF_1 and finite surface F_2: Firstly, F_2 is projected onto the sphere surface, this projection is further projected onto the plane dF_1, and the ratio of the final projection area to the area of circle is the view factor between dF_1 and F_2. The view factor between the finite surface F_1 and finite surface F_2 can be obtained similarly by dividing the F_1 into many subsurfaces, repeating the above process and applying the reciprocity and superposition, property of view factor.

If a unit sphere (radius $= 1$) is taken, the area of the circle is equal to π, so it is easier to calculate. In this case, the Nusselt sphere method is often called the unit sphere method.

3. **Hemicube method** [4]

Hemicube method is based on Nusselt sphere method and combines computer graphics techniques to calculate view factor. This method takes the advantages of computer graphics cards supporting Open GL in graphics processing and can calculate view factors faster, especially for large-scale radiation models.

Referring to Fig. 6.8, imaging drawing a cube around surface 1 and keeping a half to obtain a hemicube, the sides of the hemicube are equally divided into small squares measured in pixels. The surface 2 is projected onto the side of the hemicube with the center of the surface 1 as the focus to form the surface 3. Obviously, the view factor F_{12} is completely equal to F_{13} and is also approximately equal to F_{14} in Fig. 6.9. The problem is converted into the solution of the view factor dF_P of the surface to the element ΔA_P in Fig. 6.10.

According to the definition of view factor,

$$dF_P = \frac{\cos\theta_{iP} \cos\theta_P}{\pi r^2} \Delta A_P \qquad (6.22)$$

$$\cos\theta_{iP} = \cos\theta_P = \frac{1}{r}$$

6.4 Radiation Computing

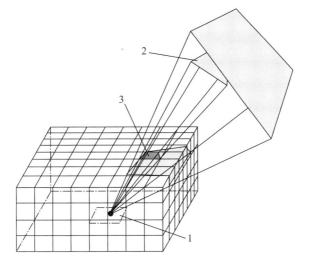

Fig. 6.8 Other surface projections viewed from surface 1

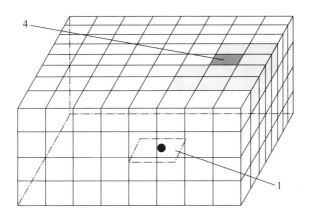

Fig. 6.9 Approximate projection area of surface 2 on hemicube side

Taking local coordinates and making the hemicube a unit height, we get

$$dF_P = \frac{\Delta A_P}{\pi r^4} \qquad (6.23)$$

where $|r| = \sqrt{x_P^2 + y_P^2 + 1}$

Similarly, for the element ΔA_P located on the lateral side in Fig. 6.10,

$$dF_P = \frac{z_P \Delta A_P}{\pi r^4} \qquad (6.24)$$

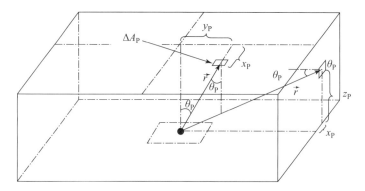

Fig. 6.10 View factor of surface to element on hemicube side

With the center of one surface as the focus, all the surrounding surfaces (receiving surfaces) are projected onto the sides of the hemicube, each receiving surface is assigned a unique color, and then the sides of the semicube are rendered. The information about areas can be obtained through counting the pixel colors. For graphics cards supporting Open GL, the above projection and rendering processes are relatively easy, so the above view factors to element can be easily obtained, and the calculation of view factors between finite areas can be completed through repeated cumulations.

Compared with Nusselt sphere method, hemicube technique only needs one projection process and is based on Cartesian coordinate system, which is simpler than sphere projection. The calculation of projection area can make full use of graphics card's rendering ability. Because the projection area is measured based on pixels, the accuracy is limited by the pixel resolution of the drawing area. It can also be seen from the concept of hemicube method that calculation of view factors for finite surface is based on computing of view factors for elements; hence, it is usually necessary to subdivide the surface to improve the calculation accuracy and the shadow checks. Obviously, the more subdivision used, the higher the accuracy usually achieved (but inherently limited by the resolution of the display) and the more time it takes.

4. **Monte Carlo technique** [7]

Monte Carlo method uses a large number of randomly emitted rays to simulate the radiative effects through statistics. Monte Carlo technique has the advantages of less memory-consuming, and its computational complexity is almost not affected whether surface is regular or not. The technique is widely used in the calculation of view factor, radiative absorption factor and radiative heat.

The calculation steps of Monte Carlo technique are demonstrated by the solution process of the view factor between two parallel rectangles shown in Fig. 6.11, where variables Counts and Hits, respectively, represent amounts of the energy emitting from plane A_1 and the effective energy hitting plane A_2.

6.4 Radiation Computing

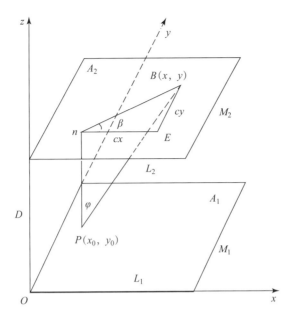

Fig. 6.11 Energy beams of parallel rectangular surfaces

(1) Generating random numbers ξ_1 and ξ_2 over range [0, 1], respectively, to obtain a random point $P(x_0 = \xi_1 L_1, y_0 = \xi_2 M_1)$ on A_1 plane.

(2) Generating random numbers ξ_3 and ξ_4 over range [0, 1], respectively, let $\varphi = \xi_3 \frac{\pi}{2}$, $\beta = \xi_4(2\pi)$, then a random ray PB with zenith angle being φ and azimuth being β is obtained. dA_1 is the cross section of PB intersecting plane A_1, $d\varphi$ and $d\beta$ refer to interval of zenith angle and azimuth angle at the cross section of PB with plane A_2 are. According to Lambert's cosine law, energy per unit solid angle represented by beam PB should be $dA_1 I \cos\varphi$ (I is the full wavelength radiation intensity), and the solid angle in the direction of point $B = \sin\varphi d\varphi d\beta$. Therefore, PB represents that the energy received by surface A_2 is $I\cos\varphi\sin\varphi dA_1 d\varphi d\beta$, and I, dA_1, $d\varphi$, $d\beta$ can all be simultaneously eliminated from the numerator and denominator in the final calculation formula, so they are simply regarded as 1.

(3) Judging whether PB intersects with surface A_2: With D representing the distance between these two planes, the coordinate of point B is

$$x = x_0 + D \tan\varphi \cos\beta$$
$$y = y_0 + D \tan\varphi \sin\beta$$

If $x \in [0, L_2]$ and $y \in [0, M_2]$, the effective energy count is accumulated:

$$\text{Hits} = \text{Hits} + \cos\varphi \sin\varphi$$

(4) Emitting energy accumulation Counts = Counts + $\cos\varphi \sin\varphi$.
(5) Repeating steps (1) to (4).
(6) $F_{12} = \frac{\text{Hits}}{\text{Counts}}$ at the end of loop.

Note that the radiation energy, not the number of beams, should be counted. If the energy counts in steps (3) and (4) above are substituted by simple counting of rays, a great error will occur: Giving the above squares a unit side length, the distance between them is 1, and the accurate result of view factor is 0.2. The above Monte Carlo method gives a result of 0.196,7 if number of emitted rays is 10,000, but if the counting rule in steps (3) and (4) is changed into counting of rays, the result becomes 0.268,1, which is quite different from the exact solution.

6.4.2 Radiative Absorption Factor

In the 1950s and 1960s, Gebhardt B. proposed the concept of radiative absorption factor to calculate the radiation exchange between surfaces. The physical definition of B_{ij} is: the fraction of the emitted or reflected radiant thermal energy leaving the surface i and absorbed by j after multiple reflections by all surfaces. There is no restriction on whether the radiation emission belongs to the infrared spectrum and whether the involved surface is diffuse surface.

Both B_{ij} and F_{ij} refer to the fraction of emitted energy leaving surface i, and the difference is:

(1) Arrival versus absorption: F_{ij} defines the fraction of radiation reaching surface j, while B_{ij} defines that of absorbed by surface j.
(2) First versus multiple radiation: F_{ij} aims at the first radiation leaving surface i and reaching surface j, while radiation absorbed by surface j represented by B_{ij} also includes those via multiple reflections by all surfaces in the closure.

Radiative absorption factors B_{ij} is also called Gebhardt coefficient, Gebhardt absorption coefficient, Gebhardt absorption factor, Gebhardt radiation exchange factor, gray body radiation exchange factor, etc. Some software defines $\varepsilon_i B_{ij}$ as radiation exchange factor or gray body radiation exchange factor. For thermal analysts, it is important to clearly understand the relationship of B_{ij} to radiative heat exchange calculation in thermal analysis software they used.

When the surfaces in the closure are all diffuse surfaces, B_{ij} can be mathematically expressed as

$$B_{ij} = \varepsilon_j F_{ij} + \sum_{k=1}^{N} \rho_k F_{ik} B_{kj} \qquad (6.25)$$

where ρ is reflectivity and $\rho = 1 - \varepsilon$. The first term at the right end of the formula is the fraction of direct emitted energy leaving surface i and absorbed by surface j; the second term is the fraction of radiation reflected by all surfaces and absorbed by surface j.

6.4 Radiation Computing

B_{ij} also has the property of unity and reciprocity:

$$\text{Unity}: \sum_{j=1}^{N} B_{ij} = 1$$

$$\text{Reciprocity}: \varepsilon_i A_i B_{ij} = \varepsilon_j A_j B_{ji}$$

When the surfaces in the closure are all diffuse, that is, B_{ij} can be calculated by Formula (6.25); B_{ij} matrix is linearly related to F_{ij} matrix. B_{ij} can be solved according to Cramer's rule or by matrix manipulation.

In addition, Monte Carlo technique is also widely used to directly solve B_{ij}. It does not require whether the surface is a diffuse surface.

6.4.3 Radiative Heat

The purpose of calculating the view factor or absorption factor is to obtain radiative heat, such as absorbed radiation, radiative heat exchange or net radiation.

1. **Absorption of radiation**

If emission intensity of a surface j is $q_j(\text{W/m}^2)$, the radiative energy emitted is $\varepsilon_{\lambda,j} A_j q_j$. According to the definition of absorption factor, the energy absorbed by the surface i is $\varepsilon_{\lambda,j} A_j B_{ji} q_j$; considering the reciprocity between surfaces is also equal to $\varepsilon_{\lambda,i} A_i B_{ij} q_j$.

Now suppose that surface j represents a celestial radiation source (solar direct radiation, Earth albedo or Earth infrared), E_R is the radiation intensity of corresponding source, obviously the radiative energy absorbed by surface i is $\varepsilon_{\lambda,i} A_i B_{ij} E_R$. In other words, once the calculation of the absorption factor corresponding to a certain celestial body radiation source is solved, the absorbed radiation of the external orbital heating in the thermal network equation is also solved. $\varepsilon_{\lambda,i}$ is emissivity, which is also absorptivity corresponding to the source E_R's spectrum range.

Although it is also possible to solve the absorption factor B_{ji} of the celestial radiation source to the surface i, B_{ij} is preferable to B_{ji} in the calculation of external orbital heating.

2. **Radiation heat exchange**

For the surface represented by nodes in the thermal network equation, if it is a problem of enclosure composed by diffuse surfaces, the energy radiated by surface i itself is $\varepsilon_i A_i \sigma T_i^4$, of which being absorbed by surface j is $\varepsilon_i A_i \sigma B_{ij} T_i^4$, according to the definition of absorption factor. Similarly, the energy emitted from surface j itself and absorbed by surface i is $\varepsilon_j A_j \sigma B_{ji} T_j^4$. According to concept of reciprocity, the radiation heat exchange between surfaces i and j is

$$q_{ij} = \varepsilon_i A_i \sigma B_{ij}\left(T_i^4 - T_j^4\right) \tag{6.26}$$

This means that the radiative conductance in the thermal network equation is

$$G_{ij} = \varepsilon_i A_i \sigma B_{ij} \tag{6.27}$$

3. **Net radiation** [9]

The thermal network equation can also be written as

$$m_i c_i \frac{dT_i}{dt} = Q_i + q_i + \sum_{j=1}^{N} D_{ji}(T_j - T_i) - q_{R,i} \tag{6.28}$$

where $q_{R,i}$ represents the net radiation of surface i.

For an enclosure formed by a diffuse gray surface, using J_i to denote effective radiation.

Since

$$q_{R,i} = \sum_j \frac{J_i - J_j}{(A_i F_{ij})^{-1}}$$

the thermal network equation can be written as

$$m_i c_i \frac{dT_i}{dt} = Q_i + q_i + \sum_{j=1}^{N} D_{ji}(T_j - T_i) - \sum_j \frac{J_i - J_j}{(A_i F_{ij})^{-1}} \tag{6.29}$$

So,

$$\frac{\sigma T_i^4 - J_i}{(1-\varepsilon_i)/\varepsilon_i A_i} = \sum_j \frac{J_i - J_j}{(A_i F_{ij})^{-1}} \tag{6.30}$$

According to Formula (6.30), we can obtain N equations for N surfaces, combined with the thermal network equations, the temperature and effective radiation J_i can be solved. This is called Oppenheim's method. Compared with Gebhardt's method, which uses F_{ij} matrix to calculate B_{ij}, this method requires less storage space and less computations for matrix inversion. For the case that emissivity varies with temperature, only the thermal conductance between surface i and virtual surface, $\frac{\varepsilon_i A_i}{1-\varepsilon_i}$, needs to be updated in iteration, so it is more efficient. For the above reasons, Oppenheim's method is also adopted by some analysis software.

Oppenheim's method can be regarded as creating a virtual Oppenheim node J_i for each surface i, and the real node i is only connected to the virtual node J_i via the radiation conductance $\varepsilon_i A_i/(1-\varepsilon_i)$; the radiation conductance between additional node J_i and J_j is $A_i F_{ij}$.

The thermal conduction of radiation heat exchange between the two surfaces cannot be directly obtained by Oppenheim's method. In this method, the net radiation heat transfer between the surfaces i and j is

$$q_{ij} = A_i F_{ij} (J_i - J_j) \tag{6.31}$$

Only after the effective surface radiation J_i and J_j are determined, can the radiation thermal conductance be determined by the following formula:

$$G_{ij} = \frac{A_i F_{ij} (J_i - J_j)}{T_i^4 - T_j^4} \tag{6.32}$$

6.4.4 Non-diffusive Radiation

Sections 6.4.1–6.4.3 focus on discussion of diffuse surface. In the thermal analysis of spacecraft, it is also possible to encounter a variety of non-diffusive radiation problems, such as the directionally dependent surface properties. More common problems are specular reflection or transmissive radiation, as well as non-diffusive illumination (such as collimated direct solar irradiation).

1. **Specular reflection or transmissive radiation**

Gebhardt's method utilizing Formula (6.25) is no longer applicable for problems involving specular reflection or transmissive radiation. The most popular alternative is to solve B_{ij} directly with Monte Carlo method assisted ray-tracing.

When Monte Carlo method is used to simulate reflection for each incident ray, it is reflected probabilistically according to a random number within [0, 1] which represents the probability of reflection. Only when the random number is not greater than the surface's reflectivity, the reflection is considered to be valid and need to be registered. If the surface is diffusive, the direction of the reflected ray is randomly determined according to equal probability in the hemisphere direction of the surface. However, for a specularly reflective or transmitting surface, incident rays must be deterministically (rather than uniformly) reflected by specular surface, or transmitted by transparent surface. In the sequent interactions with other surfaces, it is necessary to continuously track the reflected or transmitted rays.

Ray-tracing refers to registering the direction (coordinates of intersection points) and strength changes of rays during radiation transfer. It is usually used in combination with Monte Carlo method to obtain accurate results or facilitate error checking.

If the receiver surface j is partially specular/transparent, the recording of energy strength is similar to the fully diffuse problem: The energy ray carried by the rays is absorbed/diffusely reflected/specularly reflected/transmitted proportionally to the absorptivity/diffuse reflectivity/specular reflectivity/transmissivity. But it is different from the fully diffuse problem in terms of the determination of the direction: Only diffusely reflected rays are distributed in all directions uniformly. The rays specularly reflected must reach the mirror paths of the incident rays, or transmitted along the incident direction. The rays are traced through all reflection or transmission until they hit a fully diffuse surface or their energy falls below the level to be discarded.

2. **Collimated light source**

Direct sunlight is a collimated rather than a diffuse light source. The above two methods can also be used to calculate the absorption factor or absorbed heat of the solar irradiation. However, it should be noted that if rays are launched from a surface rather than the sun, for each ray intersecting the surface representing the sun, the intensity of the ray needs to be multiplied by the cosine of the angle between the normal of the surface emitting or reflecting the ray and the sun vector; that is, only the component parallel to the sun vector should be counted as the effective solar irradiation.

6.4.5 Spatial Decomposition Method for Radiation Calculation

Some thermal analysis software use spatial decomposition technology to accelerate radiation analysis. The principle of this technology is: The space occupied by the model is subdivided into several virtual small boxes, when Monte Carlo algorithm is used; if a random launched ray cannot hit the sides of a certain box, all surfaces in this box cannot receive the ray, so it is not necessary to judge it intersects each surface inside the box or not one by one. In Monte Carlo algorithm, it is rather time-consuming to judge the intersections of lines and surfaces. Therefore, the spatial decomposition method can speed up the calculation.

This technology has different name in different software. It is called oct-tree in Thermal Desktop and Voxel in NEVADA. All current software partitions the model space equally into the virtual separated boxes. They are not smart enough to do the partition right along the cabin–cabin section.

6.4.6 Residual Processing

In general, neither F_{ij} nor B_{ij} can reach the theoretical unity in numerical radiation calculation. Therefore, in addition to specifying reasonable closure criteria, amount by which sum of factors depart from unity need to be handled. There are normally three ways of treatment: reallocating to space, to surface itself, or proportionally to other surfaces. There is no priority in the choice of reallocation method, and some particular limits are clear: If it can be confirmed that the space in the model is not visible to all surfaces, then reallocating the residual error off unity to space in invalid, otherwise the result of temperature calculation is often greatly affected (in most cases space represents cold black background or low-temperature heat sink). In most cases, it is recommended to proportionally reallocate the residual to other surfaces.

Due to the rounding errors in computer processing, radiation calculations are likely to yield pseudo-view factors to deep cold space, which should also be discarded.

6.5 Simulation of Specific Problems

6.5.1 Flow and Heat Transfer in Pressurized Cabin

For the three-dimensional forced air flow in the pressurized cabin, it can be assumed that the air density is independent of temperature. If only the velocity field is to be analyzed, the nonslip velocity boundary condition is suitable for the wall, and the wall can be set as a constant temperature or an adiabatic boundary. After the velocity field is obtained, the convective coefficient on the surface of solid (cabin wall and equipment) can be given by empirical formula. If the air temperature is known, it can be used as the third kind of boundary condition for the heat conduction–radiation problem of solid part.

However, the air temperature in the pressurized cabin is usually unknown and uniform and coupled with the solid surface temperatures. Therefore, the coupled analysis of air field and solid field is usually needed. Only the external heat flux on the exposed surface of the spacecraft and the radiation to space are boundary conditions.

The external orbital heat flux can be obtained through separated analysis and then applied. However, as the data is numerous and the surface is mostly irregular, it would be quite difficult to apply heat flux manually. It is better to use software to implement the data mapping automatically.

If the empirical formula is used to calculate the convective coefficient according to the surface flow speed, there exists a problem of determining distance away from the surface to represent the outer edge of the boundary layer, and there is a big ambiguity in manual judgment. Therefore, some software use more reasonable log-law wall function to describe the velocity and temperature distribution near-wall.

In the pressurized cabin, there is almost no high air flow speed. Therefore, the radiation heat exchange between solid surfaces is usually not negligible [12].

6.5.2 Flow and Heat Transfer in Ducts

For the incompressible flow of single-phase fluid in a duct, if the pressure distribution is not to be solved and only heat exchange is concerned, and the following methods can be used to couple the flow heat transfer with the thermal network of the solid part.

The one-way conductance is used to represent the linear conductance between serial nodes of the fluid. One-way conductance $G_{ij} = \dot{m}C_p$, where m is the mass flow rate and C_p is the specific heat of the fluid. One-way conductance means that heat is only transferred from upstream node i to downstream node j, but not from node j to i.

The linear thermal conductance between node i and the corresponding duct wall node is described as hA, where A is the contact area between node i and the corresponding duct wall node. The convective coefficient h is calculated based on Nusselt number, which is obtained according to the corresponding empirical formula. For example, for fully developed laminar flow with the constant heat flux wall, $Nu_D =$ 4.36. The analyst must be aware that aspects of the flow, including flow regime (laminar or turbulent), the geometric feature of cross section, Reynolds number and Prandtl number, are all within the scopes satisfying the empirical formula he used.

This method only requires analysis software to support one-way conductance. No flow solving ability is required. But, if the pressure change in the flow needs to be analyzed, the analysis software supporting the tube flow is needed. These software use Bernoulli equation and automatically select empirical formula to calculate the friction drag along the path, but the complexity lies in the simulation of local head loss. In engineering, there are some empirical formulas of head loss coefficient for simple cross sections, but no such coefficients are known for most actual fluid components such as cold plates and valves. For these components, the head loss should be calibrated in advance, or the component-level flow model depicting detailed geometric characteristics should be established.

For two-phase flow, it is also necessary to set the flow regime (i.e., bubbly, or slug, or annular, or stratified regime) and to consider dryness fraction. At the same time, detailed fluid properties and special simulation models should be specified for components such as accumulator (reservoir) and pump.

6.5.3 Heat Transfer of Heat Pipe

The heat transfer process inside the heat pipe is quite complicated. Generally, only the heat transfer performance externally behaved by the heat pipe needs to be simulated in engineering. For constant or fixed conductance heat pipes, the dependence of heat transfer ability on temperature is generally ignored. Therefore, the following three models are usually used to describe the heat transfer performance of heat pipe:

(1) Constant-conductance heat pipe is regarded as an isothermal body and thus represented by one node. Under this approximation, the heat pipe is equivalent to a thermal superconductor, which gives too ideal performance, so it is rarely used.
(2) Ignoring axial thermal resistance of the adiabatic section, the heat pipe is subdivided into nodes representing evaporating and condensing section in the axial direction. At the same time, the vapor is considered as an isothermal node. Heat conduction between evaporating section node and vapor node is calculated according to evaporating area and evaporation heat transfer coefficient, and heat conduction between condensing section node and vapor node is calculated according to condensation area and condensation heat transfer coefficient. This simulation method is closer to the real heat transfer process of the heat pipe.
(3) The heat pipe is subdivided into nodes axially, and the empirical value is adopted for the heat transfer coefficient between nodes. The empirical values should make the analytical temperature difference at both ends of the heat pipe close to the value measured in previous performance check to the pipe and should not be too high to stall convergence of the numerical computing.

6.5.4 Low Pressure Gas Heat Conduction

Multilayer insulation blankets need to be modeled mostly in vacuum, but sometimes it is necessary to consider whether there is rarefied gas between multiple layers. For example, when the engine is fired (especially in the ground bench test), low-pressure gas heat conduction may occur between the spacing layers of the nearby MLI blankets. During the ascent phase and the short time after the injection into orbit, the spacecraft is depressurized slowly and the MLI blankets are "gas-containing" for a while. In this case, in addition to the inter-layer radiation, the heat transfer of inter-layer gas is also sometimes to be modeled. Low pressure precludes the conventional heat conduction formula. In vacuum realm, pressure is usually classified according to the relative scale of the mean free path λ of gas to the container size d. When $\bar{\lambda} > d$, it is called low pressure, where the heat conduction of gas mainly attributes to free molecular movement. $\bar{\lambda} \approx d$ means intermediate pressure, where temperature transition should be considered for gas heat conduction. Taking air as an example, when the pressure is 1.3 Pa, the mean free path is about 5 mm, scaling by the characteristic dimension of the distance d between the multilayer spacers; it already belongs to the low pressure range of $\bar{\lambda} > d$. For the pressure of 13 Pa, $\bar{\lambda}$ is about 0.5 mm, which roughly belongs to the intermediate pressure range of $\bar{\lambda} \approx d$. These two cases may be encountered in the above situations.

There is no precise formula widely recognized for the theoretical description of gas heat conduction, whether it is low pressure or medium pressure. From the engineering point of view, many theoretical formulas are mostly similar except for slight differences. Here is one of them [13].

Assuming that the solid materials on both sides of the gas layer are the same, the temperature is respectively T_1', T_2', the wall adaptation factor is a, and the gas thermal conductivity is expressed in k_g.

(1) $\bar{\lambda} \approx d$.

Flat plate:

$$Q = k_g \frac{T_2' - T_1'}{d + 2\bar{\lambda}\frac{2-\alpha}{\alpha}} A \qquad (6.33)$$

Concentric tubes:

$$Q = \frac{2\pi k_g L(T_2' - T_1')}{\ln\left(\frac{r_1}{r_2}\right) + \bar{\lambda}\frac{2-\alpha}{\alpha}\left(\frac{1}{r_1} + \frac{1}{r_2}\right)} \qquad (6.34)$$

where A is area of flat plate; r_2 is radius of inner tube; r_1 is radius of outer tube; L is length of tube.

(2) $\bar{\lambda} > d$.

Flat plate:

$$Q = K_0 P f'(T_2' - T_1')A \qquad (6.35)$$

where

$$K_0 = \frac{1}{2}\frac{\gamma_T + 1}{\gamma_T - 1}\sqrt{\frac{R}{2\pi \mu_m T}}$$

$$f' = \frac{\alpha}{2 - \alpha}$$

γ_T is the specific heat ratio; μ_m is the molecular weight; and R is the universal gas constant.

Concentric tubes:

$$Q = K_0 P f'(T_2' - T_1')2\pi r_2 L \qquad (6.36)$$

In the formula

$$f' = \frac{\alpha}{1 + (1 - \alpha)\frac{r_2}{r_1}}$$

6.5.5 Thermal Behavior of Solid–Liquid Phase Change

The detailed internal behavior simulation of solid–liquid phase change process is very complicated. What is discussed here is a system-level engineering simulation method that regards phase change materials as one node. In the typical thermal analysis software, phase change can be directly simulated; that is, it is done just by setting the latent heat of phase change, mass, phase change temperature and specific heat of liquid and solid. However, these functions sometimes might cause non-convergence of the calculation, so enthalpy method is often used in engineering.

Considering that the temperature of the phase change material is almost constant during the phase change, which is equivalent to heat capacity being infinite, the enthalpy method utilizes this characteristic: Assuming that the material does not undergo phase change, but the heat capacity in the phase change zone is quite large, and the thermal energy stored by the heat capacity of small temperature change zone near the phase change point is equal to the latent heat of phase change, as shown in Fig. 6.12.

The resulting temperature error is shown in Fig. 6.13. If isosceles triangle curve is used to define the relationship between heat capacity and temperature in the interval $[T_{pc} - \Delta T/2, T_{pc} + \Delta T/2]$, it is required to satisfy that the pulse area of total heat capacity shown in Fig. 6.12 is equal to latent heat, namely

$$\frac{1}{2}\Delta T\left[C_p\left(T_{pc}\right) - C_p\left(T_{pc} - \Delta T/2\right)\right] = q_{pc} \qquad (6.37)$$

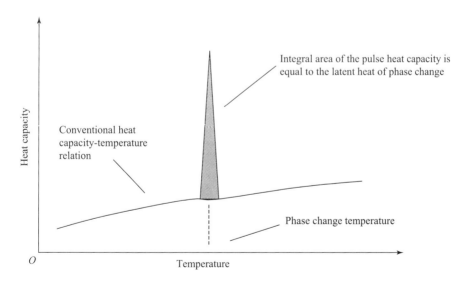

Fig. 6.12 Heat capacity function in enthalpy method

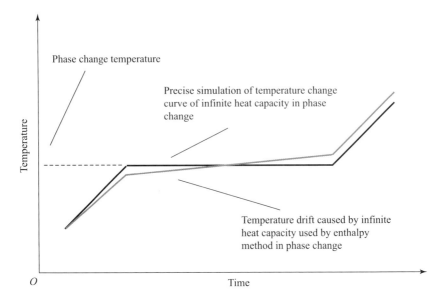

Fig. 6.13 Error of enthalpy method

$$C_p\left(T_{pc} - \Delta T/2\right) = C_p\left(T_{pc} + \Delta T/2\right)$$

where q_{pc} is the latent heat of phase change material. It is believed that the heat capacity $C_p(T_{pc} - \Delta T/2)$ before phase change is equal to the heat capacity $C_p(T_{pc} + \Delta T/2)$ after phase change. By assigning a small interval ΔT, we can obtain

$$C_p\left(T_{pc}\right) = C_p\left(T_{pc} - \Delta T/2\right) + \frac{2q_{pc}}{\Delta T} \tag{6.38}$$

The heat capacities at the three vertices of the isosceles triangle are determined, and the virtual heat capacity–temperature relationship of the material over the phase change interval $[-\Delta T/2, \Delta T/2]$ can be defined consequently.

If it is required to estimate the duration of the phase change process according to the temperature result given by the enthalpy method, the assigned ΔT must be equal to the actual phase change temperature interval.

6.5.6 Thermal Behavior of Semiconductor Cooling

When Peltier cooler works, the heat absorbed at cold end and that dissipated at hot end are both temperature-dependent, but the relationship between heat absorption(or dissipation) and temperature is often unknown. Therefore, an alternative solution is

6.5 Simulation of Specific Problems

to calculate its thermal behavior using electric current and voltage [4]:

$$Q_{cold} = -2N\left[aIT_{cold} - \left(\frac{I^2\rho}{2G}\right) - \lambda\Delta T G\right] \tag{6.39}$$

$$Q_{hot} = 2N\left[aIT_{hot} + \frac{I^2\rho}{2G} - \lambda\Delta T G\right] \tag{6.40}$$

where T_{hot} is the hot end temperature (K), T_{cold} is the cold end temperature (K) and ΔT is $T_{hot} - T_{cold}$ (K); G is thermocouple pin area/length (m); N is the number of thermocouples in device; I is current (A); a is the Seebeck coefficient (V/K); ρ is electrical resistivity ($\Omega \cdot$m); λ is the thermal conductivity (W/(m²•K)).

$$V = 2N\left(\frac{I\rho}{G} + \alpha\Delta T\right)$$

where V is the voltage (V).

It can be seen that the heat absorbed and dissipated are dependent on the electric current and the temperature difference between both ends. Since the current is also influenced by the temperature difference, the heat absorption and dissipation are coupled with cold or hot end temperature, and iteration is required in solving the temperature.

6.5.7 Junction-Case Heat Transfer of Electronic Components

Temperature has a significant influence on the reliability of electronic components. The electronic products in the military or aerospace community are generally required to obey derating requirements including temperature, usually demanding that the junction temperature of components is lower than a certain value, or that the junction temperature should not be higher than the maximum allowable junction temperature subtract a certain value given in the component specifications. Thermal analysis of electronic equipment generally requires the analysis of junction temperature of components.

If heat dissipation of a component is P, junction-case thermal resistance is θ_{jc}, and case (outside surface of a package) temperature is T_c. According to the definition of junction-case thermal resistance:

$$\theta_{jc} = \frac{T_j - T_c}{P} \tag{6.41}$$

Junction temperature can be obtained by:

$$T_j = T_c + P\theta_{jc} \tag{6.42}$$

This calculation method is only applicable under strict conditions: Junction-to-case thermal resistance refers to the thermal resistance between the surface of die/chip and the nearest case surface. It is measured according to top cold plate test: The top surface is isothermal, the rest is adiabatic. In the definition formula, heat dissipation P is used to represent the heat transfer from the junction to the case, and heat dissipation only equals to heat transferred under such a test condition. The actual heat transfer path of most components is not simple. Taking the component illustrated in Fig. 6.14 as an example, there are two paths: junction-case and junction-substrate (or board).

Under the installation condition shown in Fig. 6.14, it is incorrect to calculate junction temperature by using Formula (6.42): Suppose that the component generates heat at 10 W, in which 5 W is transferred to the case and 5 W to the substrate, the case and plate temperature are both 10 °C, if the junction-case resistance is 1 °C/W, according to Formula (6.42), the junction is 20 °C. In fact, the heat transfer through the junction-case thermal resistance path is 5 W, and the junction should actually be 15 °C.

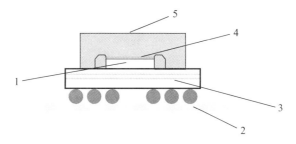

Fig. 6.14 Schematic diagram of component structure. 1—Die; 2—PCB; 3—Substrate; 4—Surface of Junction/Die; 5—Case

In other words, for the case where the heat transfer paths are more than junction-to-case, formula

$$T_j = T_c + q_{jc} \theta_{jc} \tag{6.43}$$

should be used to calculate junction temperature, provided that q_{jc} can be determined accurately.

In most cases, the component has more than one heat transfer path outward, so it is unreasonable to describe multiple complex heat transfer paths by a single thermal resistance. Strictly speaking, the junction-case thermal resistance is only used as a comparative rather than a predictive metric.

Considering that the heat transfer path is not unique in most cases, and the junction temperature cannot be calculated directly by Eq. (6.42); another more accurate simulation method, two resistor model [14], is introduced; that is, both the junction-case and the junction-board thermal resistance θ_{jb} are used at the same time, where the junction-board thermal resistance is specified in JESD51-8, a standard of Joint

Electronic Device Engaging Council (JEDEC), and is to be tested according to a ring cold plate test as shown in Fig. 6.15.

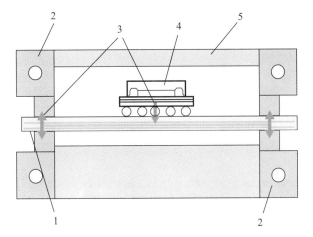

Fig. 6.15 Junction-plate thermal resistance test method. 1—2S2P board; 2—Cold plate; 3—Heat flow; 4—Package; 5—Insulation

Once a two resistor model adopted, if the analysis model can give the case temperature T_c and the board temperature T_b, the junction temperature can be solved by the following formula:

$$\frac{T_j - T_c}{\theta_{jc}} + \frac{T_j - T_b}{\theta_{jb}} = P \qquad (6.44)$$

In fact, in most cases, the actual heat transfer path of components is more complicated, and the temperatures of case surfaces are not uniform; even there may exist temperature gradient on the same surface. Therefore, a more accurate models are needed, such as the so-called DELPHI model, or a more sophisticated model that can express the detailed heat transfer path of components shall be established, which, generally speaking, can only be provided by component suppliers.

In short, when calculating the junction temperature of a component, if the its outer surface temperature and the junction-surface thermal resistance(s) are known, and if the additional known condition is only the heat dissipation of the component, attention should be paid to whether the application environment is the same as the condition under which resistance be tested, only when the answer is "yes" can we regard heat dissipation as heat transfer through a path to estimate the junction temperature. In most cases, they differ from each other, and the heat transfer along one path is unknown. Therefore, heat dissipation should be applied onto the junction and the heat transfer path shall be simulated with the known thermal resistance. In this case, the accuracy of junction temperature calculation depends on the level of depicted details of heat flow path.

6.6 Equivalent Transformation of Radiation Term of Thermal Network

A surface in a heat conductance–radiation system is usually radiated by multiple surfaces. It is favorable to convert these multiple radiation into a single thermal factor, such as attached electric heating or a virtual equivalent sink, which is conducive to the simplification of boundaries to be implemented in thermal balance test, and cooperation of several analysts in the thermal analysis of complex spacecraft.

6.6.1 Equivalent Heating [15]

The radiation heat exchange in the thermal network equation can be regarded as the gap between the radiation absorbed and the emitted. Using (6.27) to express the radiative conductance, and considering the reciprocity and unity property of radiative absorption factor, the thermal network Eq. (6.1) can also be expressed as

$$m_i c_i \frac{dT_i}{dt} = Q_i + q_i + \sum_{j=1}^{N} D_{ji}(T_j - T_i) + \sum_{j=1}^{N} B_{ij} A_i \varepsilon_i \sigma T_j^4 - \varepsilon_i A_i \sigma T_i^4 \quad (6.45)$$

where Q_i is the external heat flux in-orbit or simulated in test.

Grouping the radiative heating (item 4 of the right hands in Formula (6.45)) nodes into two: groups $(1, 2, \ldots, M)$ and $(M+1, M+2, \ldots, N)$, and Formula (6.45) becomes

$$m_i c_i \frac{dT_i}{dt} = Q_i + q_i + \sum_{j=1}^{N} D_{ji}(T_j - T_i) + \sum_{j=1}^{M} B_{ij} A_i \varepsilon_i \sigma T_j^4 +$$
$$\sum_{j=M+1}^{N} B_{ij} A_i \varepsilon_i \sigma T_j^4 - \varepsilon_i A_i \sigma T_i^4 \quad (6.46)$$

Let $Q_i' = Q_i + \sum_{j=M+1}^{N} B_{ij} A_i \varepsilon_i \sigma T$, then Eq. (6.46) can be converted into

$$m_i c_i \frac{dT_i}{dt} = Q_i' + q_i + \sum_{j=1}^{N} D_{ji}(T_j - T_i) +$$
$$\sum_{j=1}^{M} B_{ij} A_i \varepsilon_i \sigma T_j^4 - \varepsilon_i A_i \sigma T_i^4 \quad (6.47)$$

Obviously, Eq. (6.47) is equivalent to Eq. (6.45).

6.6 Equivalent Transformation of Radiation Term of Thermal Network

The above process proves the rationale of converting radiation into equivalent heating. If certain parts (like solar array or large-scale antenna) of a vehicle cannot participate in the thermal balance test, this method can be used to reflect the radiation effects of the absent parts, just like they still exist.

Considering that the node group $(M + 1, M + 2, \ldots, N)$ represents parts absenting the thermal balance test, assuming that the thermal conductance between this part and those participating in the test can still be simulated (usually not a difficult thing), we have the following steps to make the test still reflect the radiation effect of the part $(M + 1, M + 2, \ldots, N)$:

(1) Building a complete thermal model including all parts, and obtain the radiative absorption factor B_{ij} and the temperature result.
(2) Building the radiation model without part $(M + 1, M + 2, \ldots, N)$, then get the radiative absorption factor B'_{ij} $(i, j = 1, 2, \ldots, M)$ under the test state.
(3) Calculate the compensating heating P_i and modified external heat flux Q'_i by using B_{ij}, B'_{ij} and the temperature obtained in (1).

$$P_i = \sum_{j=1}^{N} B_{ij} A_i \varepsilon_i \sigma T_j^4 - \sum_{j=1}^{M} B'_{ij} A_i \varepsilon_i \sigma T_j^4 \tag{6.48}$$

$$Q'_i = Q_i + P_i \tag{6.49}$$

(4) In the test, the external heat flux is simulated according to Q'_i.

If the deviation of the sink from the cold black background is to be considered, the above process is also valid, as long as the sink radiation heating is also deducted from the compensating heating.

6.6.2 Equivalent Heat Sink

The rationale of equivalent heat sink is similar to that of equivalent heating; that is, the radiation effects of multiple nodes are alternated by a virtual heat sink radiation.

The nodes of the thermal network equation are divided into two groups $(1, 2, \ldots, M)$ and $(M + 1, M + 2, \ldots, N)$. The thermal network Eq. (6.1) can be expressed as

$$m_i c_i \frac{dT_i}{dt} = Q_i + q_i - \sum_{j=1}^{N} D_{ij}(T_i - T_j) - \sum_{j=1}^{M} G_{ij}(T_i^4 - T_j^4) - \sum_{j=M+1}^{N} G_{ij}(T_i^4 - T_j^4) \tag{6.50}$$

Taking Definitions [16]:

$$G_{s,i} = \sum_{j=M+1}^{N} G_{ij} \tag{6.51}$$

$$T_{s,i} = \left(\frac{\sum_{j=M+1}^{N} G_{ij} T_j^4}{G_{s,i}} \right)^{1/4} \tag{6.52}$$

then Eq. (6.50) can be rewritten as

$$m_i c_i \frac{dT_i}{dt} = Q_i + q_i - \sum_{j=1}^{N} D_{ij}(T_i - T_j) - \sum_{j=1}^{M} G_{ij}(T_i^4 - T_j^4) - G_{s,i}(T_i^4 - T_{s,i}^4) \tag{6.53}$$

That is, the multiple radiation relations between node i and node group ($M + 1$, $M + 2$, $\cdots$, N) are exactly equivalent to one radiation exchange with a virtual surface, which is the virtual heat sink of node i.

Spacecraft development is a system engineering. The thermal control system is responsible for the interface temperature at the unit fixation points, and the thermal control of the unit is under the responsibility of the unit supplier. The thermal control system analyst does not know the details of heat transfer inside the units, while the unit analyst cannot build the entire thermal model of the spacecraft. Therefore, it is often necessary to construct several models with different objects of interest with different details. The equivalent heat sink facilitates the exchange of radiation boundaries between multiple models [17].

The equivalent heat sink method is used to reduce the radiation boundary conditions of a node in the full network to 1. On this basis, each lower-level partial model can be arbitrarily inserted or deleted of nodes and heat transfer paths independent of this equivalent sink, eliminating recalculation of the same radiative relations (if exist) in every partial model. As a result, the relationships of multiple partial models are thus decoupled, which enable parallel computing of multiple models. External heat flux can be also included into the equivalent sink, then it no longer need to be computed repeatedly, and data to be exchanged between models is further reduced. The equivalent heat sink method is also applicable to transient problems, in which equivalent sink might be generated at every instants, when variational radiation couplings (e.g., displacement or rotation of articulator) exist or time-varying external orbital heating is to be included in the sink. Data to be exchanged for these transient problems is numerous.

6.7 Thermal Model Correlation

Thermal model correlation is the process of modifying uncertain parameters of a thermal mathematical model to fit/match the data from thermal balance or flight test, so as to improve the quality of the model, further providing more accurate predication or exploration in scenarios that may not have been tested or even untestable. Most of the test data used in correlation is temperature. Although there are always errors in the temperatures measured in the test, unless the obvious measurement errors or deviations are recognized, the temperature measurement errors are usually ignored in the correction. In other words, the correction is based on the supposition that temperature readings are accurate.

Thermal model-to-test correlation mainly includes two steps: parameter analysis and optimization.

In order to carry out thermal model correlation, one should first understand the scope of parameters to be correlated and the correlation success criteria and then correctly update the status of the thermal model.

6.7.1 Basic Knowledge of Thermal Model Correlation

1. **Parameters to be correlated**

The errors of thermal analysis mainly arise from model error, observation error, truncation error and rounding error. For example, if the thermal conductivity of a material at temperature $T = 0$ is k_0, using k_T to represent the calculated thermal conductivity of it at temperature T, and the following model is established.

$$k_T = k_0(1 + \alpha T)$$

If research shows that this linear function cannot accurately represent the relationship between the actual thermal conductivity of the material and T, using k'_T to represent the actual thermal conductivity at the temperature T, then $k'_T - k_T$ is the "model error." a is the coefficient by experiment. If $a = (0.000{,}23 \pm 0.000{,}002)/°C$, then $0.000{,}002/°C$ is the "observation error" of a; that is, the error of observation data contained in the model. Due to the difficulty of obtaining with the analytical solution, approximate numerical computing is generally used. The discrepancy between the accurate analytical and numerical solution is "truncation error" (also known as "method error"). The fourth error is caused by limited digits for values in calculation, which is called "rounding error."

In general, thermal model correlation cannot eliminate the "model error" and "rounding error." Sometimes, a coarse model may be refined to reduce the "truncation error" to some extent. But in most cases, the correlation of thermal model if only "observation error"-oriented.

As for the thermal model, the correlation adjusts the known parameters with uncertainty in the model, such as thermal conductance (linear conductance, effective emissivity of MLI), heat source terms (heat dissipation, heating power and external heat flux). Linear thermal conductance is the most widely modified parameter. No matter what parameter is to be modified, the adjustment should be reasonable; i.e., the value of the parameter should be alternated within the physically possible range. To help to do this judgment, parameters are better be measurable. For instance, the radiative conductance is hard to be directly measured, yet the emissivity contained therein can be calibrated. Therefore, the modification of emissivity is more practical and appropriate than the radiative conductance. From this perspective, it is more difficult to correlate radiation terms directly using integrated radiative factor method, proposed by [18], which create a virtual radiative factor to represent all radiative conductances of a node.

It is worth mentioning that the correlation of thermal environment parameters is related to the external heat flux. Thermal environment parameters such as albedo and Earth infrared are seldom adjusted in most thermal model correlation. However, the Earth albedo and the Earth infrared intensity vary with the latitude, longitude, and solar elevation angle of the field viewed by spacecraft. For exposed units whose temperatures are sensitive to the values of thermal environment parameters, it may be necessary to consider the correlation of these parameters with a wider range scatter instead of fixed values [19].

Another important factor affecting external heat flux absorbed and radiative heat rejection ability is the thermo-optical properties of the surface coating, among which the relevant issues are also worthy of attention. The degradation/evolution of coatings is quite complicated and diversiform. For example, ultraviolet and charged particles have a great influence on the degradation of the white paint. OSR features good endurance to ultraviolet and charged particles but is vulnerable to outgassing. Degradation of a same kind of material located in different sides of the vehicle differs due to different exposure and particle impinging conditions. However, measurements about thermal coating's degradation in flight are not adequate. The ground space environment simulation test can provide only limited reference, because it is not sure how close the values of items (ultraviolet, atomic oxygen, etc.) simulated in these tests are to the real situation. In particular, it cannot be guaranteed that these simulation conditions can make sound representation for each location at all. Therefore, it is very difficult to adjust the thermo-optical properties of surfaces on different sides according to respective degradation laws. Generally, the same coating on different locations can only be treated according to the same degradation law. Besides, the change of coating during long-time storage on the ground is also difficult to evaluate. All these problems affect the correlation of the model and increase the difficulty of thermal model correlation.

2. **Correlation Success Criteria**

Even best efforts is paid out, a thermal model cannot be correlated up to 100% accurate. The error of thermal analysis cannot be precisely evaluated because the evaluation is always based on certain assumptions rather than definitely identified

6.7 Thermal Model Correlation

scatter profile of parameters. Therefore, no organization can declare its analysis accuracy before the real flight. Even the same analyst cannot guarantee that each analysis he did can always achieve a certain accuracy. However, if there is no requirement, the accuracy of the thermal analysis results will be unknown at all, so it is absolutely necessary to put forward certain requirements on the accuracy of the thermal model in engineering. Fortunately, if test data are available, it is possible to evaluate the error of a thermal model in a certain extent. Therefore, the correlation success criteria are exactly the accuracy requirement for thermal model to establish the adequacy of the thermal model correlation.

The test (including ground or flight test) data used to evaluate/correlate the thermal model is always limited. Therefore, even if the thermal model meets the criteria after correlation, strictly speaking, it can only prove that the analysis is close to the test data at a limited number of points, and it cannot guarantee that the accuracy of predicted temperature at any point meets the criteria. In other words, satisfaction of the correlation criteria is not a sufficient condition for the accuracy of the model, and it only means that the model satisfying the criteria can be deemed to have higher confidence and lower risk. In this sense, the criteria act as a "threshold" measure of correlation between the analysis and known test data.

The model correlation criteria refer to the allowable temperature discrepancies between the analytic predications and the test results. The three commonly used criteria are defined as follows:

$$\text{Absolute deviation}: \delta_i = T_{\text{m},i} - T_{\text{P},i} \tag{6.54}$$

$$\text{Mean deviation}: \Delta = \frac{\sum_{i=1}^{N} \left(T_{\text{m},i} - T_{\text{P},i}\right)}{N} \tag{6.55}$$

$$\text{Standard deviation}: \sigma = \sqrt{\frac{\sum_{i=1}^{N} (\delta_i - \Delta)^2}{N-1}} \tag{6.56}$$

where T_{P} and T_{m} represent predicted and measured temperature, respectively.

Criteria based on absolute deviation are the most commonly used, with slight difference among countries or organizations in values of absolute deviation and percentage of temperature points to be satisfied. The mean deviation trends to give an overvalued matching degree of analysis and test due to offset of the positive and negative values, so it is rarely used solely.

In order to ensure the final flight temperature range of a unit does not exceed the temperature range required by the thermal control system, accounting for inherent uncertainties in the thermal analysis, thermal uncertainty margin ought to be added to the model temperature predication (calculated temperature) to obtain the predicated or design temperature, which must be recognized and accepted in the thermal control system design. The compliance of thermal uncertainty margin does not imply that the actual final flight temperature meets the requirements completely. The well-known ±11 °C margin is deemed to provide a 95% confidence level that the flight temperature will be within ±11 °C of preflight temperature predictions from a thermal model correlated to thermal balance test data. This recommendation is originally based on standard deviations and assumed near-zero mean deviation, biasing of data (nonzero mean) has a significant impact on the required thermal uncertainty margin [20]. The residual deviation of the model correlated with test is another additional error besides uncertain parameters for predicted temperature. This will reduce the confidence level. Assuming that a thermal model meets the correlation criteria for standard deviation of 3 °C, in the worst case, the correlation error (±3 °C) will reduce the margin of 11–8 °C, and the probability of the actual flight temperature does not violate the required limits will be reduced from 95 to 85% [21]. In a word, thermal model correlation criteria should be stated in combination of thermal uncertainty margin. More relax correlation criteria require more stringent thermal uncertainty margin.

3. **Thermal model status**

As a measurement of correlation between analysis and test, the thermal model correlation should be conducted on the premise that the real model and the test model have identical status. For example, if the primary unit is set to operate in thermal model according to the default mission while the redundant is actually powered on in the test, the assignment of heat source should be updated correspondingly in the model to be correlated; if one heat pipe does not work due to the gravity in the test, the model should be modified accordingly. In addition, supplementaries due to heat flux simulation or fixation in ground test need to be evaluated intentionally. If accurate heat flux measuring is unavailable for the complex shaped test object, heat flux simulation facilities might have to be incorporated in the model.

4. **Summary**

The thermal model can only be correlated for partial error sources. Temperature readings are always limited, so there is no unique solution for this inverse problem. If the qualified/success criteria for correlation are a discrepancy between analysis and test below certain value prestated, the correlated model can only be cited to meet the criteria on the limited points, and the correlated parameters herein obtained are only one set of possible solutions. Even if the least square algorithm is used, the error source other than the observation errors are ignored, aiming at only those selected parameters, so some potential candidates may be missed. Therefore, its solution of uncertain parameters is not actually capable of the unique solution.

6.7.2 Parameter Analysis

Obviously, it is unlikely to adjust all the input parameters with uncertainties when the model is correlated. In practice, only those parameters sensitive to thermal model

6.7 Thermal Model Correlation

need to be modified, so the thermal model correlation begins with the screening of parameter. It is impossible to find out parameters to be modified just by a casual glance or snapshot. Blind trials are inefficient [22]. This task requires interaction of good insight and trial analysis.

Parameter analysis consists of two parts: determining the rule of alternating parameters, that is, determining the parameter scan/sweeping/sampling policy; identifying the sensitivity of the model to parameters according to the response of the model to the variation of parameters.

1. **Parameter sampling**

Early parameter analysis is usually a trial-and-error method-based experience. It is a one-factor-at-a-time method. It is inefficient and blind and often needs a lot of repeated analyses.

In fact, all sensitive parameters affect the results simultaneously. There are lots of more reasonable and efficient alternative approaches. The following three methods are briefly introduced: orthogonal analysis, full-space random sampling and Latin hypercube sampling.

1) Orthogonal analysis

Orthogonal analysis can be classified by full factorial and the partial factorial methods. Illustrating with a problem of three design parameters A, B and C, when the full factorial method is adopted, combinations of parameters are calculated one by one according to the eight combinations in Table 6.3, where each row represents the unique combination of values for all the design parameters, -1 symbolizes the low limit of a particular design parameter, $+1$ represents the upper high limit. The partial factorial method uses a reduced orthogonal array configuration for parameter values. Table 6.4 is a combination configuration of parameters according to Taguchi method, which is one of the most popular partial factorial methods.

2) Full-space random sampling method

Orthogonal analysis is costly, and its coverage is poor; only extreme values (lower or upper limit) are covered. For large-scale and multi-parameter models, Monte Carlo

Table 6.3 Parameter combinations of full factorial method

A	B	C
−1	−1	−1
+1	−1	−1
−1	+1	−1
+1	+1	−1
−1	−1	+1
+1	−1	+1
−1	+1	+1
+1	+1	+1

Table 6.4 Parameter combinations of Taguchi method

A	B	C
−1	−1	−1
−1	+1	+1
+1	−1	+1
+1	+1	−1

stochastical simulation method has comparative advantage. It randomly generates a set of parameters within the specified variable range of the selected uncertain parameters. By repeating the sampling-solving process, the response of the model to the parameters can be evaluated. This method can evaluate the quantitative dependency of model on the parameters more adequately and can also conduct the analysis and evaluation about the thermal design robustness (i.e., giving the probability that the temperature falls within a certain region when multiple parameters are considered to be variable) [23].

This method needs to specify the distribution of parameters according to the actual situation, such as uniform distribution and normal distribution.

3) Latin hypercube sampling

The Latin hypercube descriptive sampling algorithm regards parameters as random variables and is based on the assumption that "probability distribution function" of parameters is uniform from the lower to upper limit. Its basic idea is: For random parameter $x_i \in X$, dividing its space θ into m regions of equal probability, then only one value from each region is sampled. When there are n random parameters, the sampling space consists a $n \times m$-dimension matrix, and each sampling value is equal probably distributed in the $n \times m$ sampling space. Latin hypercube method can reflect the characteristics of the whole design space with fewer sample points and becomes an effective samples reduction technique.

Referring to Fig. 6.16, for illustration, consider a case with two variables A and B, varying between A_L and A_H, B_L and B_H, respectively. Assuming that the sampling space is equally divided into 5, variables A and B can be sampled with the following formula:

$$A_N = A_L + \frac{2^N - 1}{2^5}(A_H - A_L), N = \text{random}\{1, 2, \ldots, 5\} \quad (6.57)$$

$$B_N = B_L + \frac{2^N - 1}{2^5}(B_H - B_L), N = \text{random}\{1, 2, \ldots, 5\} \quad (6.58)$$

where the two random variables N are independent of each other. Fig. 6.16 depicts two possible samplings of the above case.

Latin hypercube sampling is efficient, but it is more likely to miss the best starting point than is a full scan of every possibility, especially when the parameter space is discretized too sparsely, so full factorial scan is sometimes adopted, which checks

6.7 Thermal Model Correlation

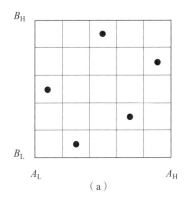
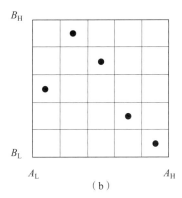

Fig. 6.16 Schematic diagram of Latin hypercube sampling. (a) Random sampling distribution 1; (b) Random sampling distribution 2

every point in the discretized design space: a dot will appear in every box in Fig. 6.16. The efficiency of full factorial scan is much lower.

2. **Parametric sensitivity analysis**

Assuming a model has a total of m variable parameters $P_1, P_2, \ldots, P_m$, and the temperatures to be inspected are $T_1, T_2, \ldots, T_n$; according to the selected sampling method, one temperature result will be obtained every time the parameters are changed, and the corresponding relation of the ith simulation is

$$\left(P_1^i, P_2^i, \ldots, P_n^i\right) \rightarrow \left(T_1^i, T_2^i, \ldots, T_n^i\right)$$

A total of N random simulations are carried out. The rank of the ith value of any parameter $P_k (k = 1, 2, \ldots, m)$ is denoted as $R^i(P_k)$, and the rank of the ith calculated temperature $T_j (j = 1, 2, \ldots, n)$ is denoted as $R^i(T_j)$, so the Spearman rank correlation coefficient between P_k and T_j is

$$r(P_k, T_j) = 1 - \frac{6\sum_{i=1}^{N}\left[R^i(P_k) - R^i(T_j)\right]^2}{n(n^2 - 1)} \tag{6.59}$$

The correlation coefficient between parameter P_k and temperature T_j is thus obtained, matrix $|r(P_k, T_j)|$ indicates the sensitivity to the parameters, but the specific values criteria of high, medium and low sensitivity are determined by the analyst.

According to the correlation between the objective (such as temperature) and the parameters, the design robustness can be further evaluated; that is, the confidence level that a temperature falls in a certain range (or the temperature range satisfying a certain confidence level) can be quantitatively identified. Table 6.5 illustrates the parametric sensitivity analysis for a Chinese GEO satellite [24]. As revealed, the

Table 6.5 Correlation between TWT temperature and thermal parameters of a Chinese GEO satellite

Correlation coefficient	Solar absorptivity of OSR	Emissivity of OSR	Effective emissivity of MLI	Contact heat transfer coefficient	Heat dissipation of TWT
TWT temperature (in equinox)	0.141	0.396	0.102	0.982	0.359
TWT temperature (in solstice)	0.761	0.486	0.146	0.653	0.411

sensitivity of temperatures to the same parameter varies with thermal environment and other conditions.

6.7.3 Correlation Method

After the sensitive parameters are determined, the thermal model correlation, to seek the parameters best fitting to test data, can be regarded as an optimization problem: The first step is to determine the optimization object defined by objective function and the goal represents the success criteria the objective should satisfy; the second step is to find the parameters that satisfy the optimization goal.

1. **Objective function**

The purpose of correlation is generally min (OBJECT), where OBJECT is a measurement of the relative error between the test data T_i^t and predicted value T_i^p. The commonly adopted definitions of *OBJECT* are [16]:

Least Squares Error: OBJECT $= \sum (T_i^p - T_i^t)^2$

Least Average Error: OBJECT $= \sum |T_i^p - T_i^t|$

Least Cubic Error: OBJECT $= \sum |(T_i^p - T_i^t)^3|$

Minimized Maximum Error: OBJECT $= \max(|T_i^p - T_i^t|)$

The above definitions are based on steady-state or the end of transient run. If one needs to compare the predictions with the test data over the course of a transient event, or if a single correlation is to be performed on the basis of multiple tests (and therefore multiple comparisons), the objective and/or the constraints must be cumulative. In case of modification for transient process or repeated steady-state test, the objective function needs to be accumulated (summed again) at different times or for multiple cases. For the same model, an objective function is different from other objective functions in terms of converging speed, stability of convergence and fault tolerance of test data.

6.7 Thermal Model Correlation

2. **Optimization method**

In a sense (measured by objective function), a correlation can be regarded as an inverse problem solving optimization aiming at minimizing the deviation between analysis and test. No matter which expression the objective function is defined by, it is always a function composed of the temperatures network equations generated and the test temperatures. The purpose of correlation is to find the minimum value of the objective function under given constraints. The commonly used optimization methods in industry can roughly fall into sampling-based and gradient-based methods.

The sampling-based method is similar to enumeration method; that is, all parameters are randomly sampled within their variation limits, and then the appropriateness of any set of correlation parameters to a given set of test data is evaluated. When the number of randomly selected samples is large enough, all parameters are approximately swept. The parameter sampling can be completely random, full factorial scanning, Latin hypercube sampling, etc., as aforementioned. Although the process of this random simulation method can be automated, it is still very time-consuming.

The correlation of thermal model can be regarded as a single-valued optimization problem. Once the deviations are merged to a single objective function, the steepest descent method can be used. It calculates the gradient of the objective function at an initial solution, then moves a relevant step toward the negative gradient direction, and then repeats the process until the objective function is no longer responsive to changes in parameters. Steepest descent method converges slowly when approaching the final solution. Furthermore, a key issue involved is how far the step should go. In view of this problem, many algorithms combining conjugate gradient method are derived, including Fletcher–Reeves (FR) and BFGS (proposed by Broyden, Fletcher, Goldfarb and Shanno) method. The gradient-based method relies on the user-supplied initial value. The possible solution is only local rather than global optimum, and thus, multiple runs with different initial conditions are helpful.

Based on the characteristics of popular optimization algorithm and samplings, the steps of effective correlation process in engineering are generally as follows [25]:

(1) Conducting parameter analysis for a large range of uncertain parameters, discarding insensitive parameters to determine a narrower range of parameters that to be modified.
(2) Using Latin hypercube sampling to search parameters and obtaining approximate optimal solution.
(3) Using FR or BFGS method to carry out the ultimate optimization solution near the approximate optimum.

References

1. Min Guirong. Research on approximate simulation method of space heat flux. J. Astronaut. (4), (1981)
2. Z. Qi, G. Xiaoming, J. Hai, et al. Numerical analysis of the differences between transient, quasi-transient and steady-state thermal network models. Spacecraft Eng 13(2), 10–15 (2003)
3. Zhong Qi, Investigation of deviation between quasi-transient/steady state and transient model. Chin. Space Sci. Technol. 27(1), 21–26 (2007)
4. Maya Heat Transfer Technologies Ltd. NX Thermal Solver TMG Reference Manual for NX 6.0 (2008)
5. Maya Heat Transfer Technologies Ltd. I- DEAS ESC Electronic System Cooling Reference Manual, ESCref10a. (2003)
6. G.E. Schneider, M.J. Raw, Control- volume finite element method for heat transfer and fluid flow using Co- located Variables- 1. Numer. Heat Transf. 11, 363–390 (1987)
7. G. Kuanliang, K. Xiangqian, C. Shannian, *Computational heat transfer* (University of Science and Technology of China Press, Hefei, 1988)
8. J. Appelbaum, G.A. Landis, I. Sherman, Sloar radiation on mars-update 1991. Sol. Energy 50(1), 35–51 (1993)
9. F.P. Incropera, D.P. Dewitt, *Fundamentals of heat and mass transfer*, 2nd edn. (Wiley, New York, 1985)
10. W. Baoguo, L. Shuyan, W. Xinquan et al., *Heat transfer* (China Machine Press, Beijing, 2009)
11. Y. Shiming, *Heat transfer*, 2nd edn. (Higher Education Press, Beijing, 1989)
12. Z. Qi, *Gas flow and thermal analysis in a pressurized cabin of spacecraft* (China Acadamy of Space Technology, Beijing, 2001)
13. G. Benhui, C. Suyan, *Vacuum physics* (Science Press, Beijing, 1983)
14. Flomerics Corporation. Modeling integrated circuit packages using FLOPACK (2006)
15. Z. Qi, Thermal radiation equivalent test method of complex or large structures. Chin. Space Sci. Technol. 30(2), 31–36 (2010)
16. Cullimore & Ring Technologies Inc. SINDA/FLUINT User's Manual, Version 4.8 (2005)
17. Z. Qi, P. Wei, Y. Hua, et al. Modular thermal analysis method for heat conduction-radiation system, in *The 11th Space Thermophysics Society*, Beijing (2013)
18. W. Jianhua, *Theoretical and experimental study of spacecraft thermal network* (China Academy of Space Technology, Beijing, 1995)
19. Z. Qi, W. Yaopu, L. Guoqiang, Influences of near-Earth thermal environment parameters on spacecraft temperature: a first review. Spacecraft Eng. 16(3), 74–77 (2007)
20. J.W. Welch, Comparison of recent satellite flight temperatures with thermal model predictions, in *Proceedings of the 36th International Conference on Environmental Systems*, Norfolk, Virginia (2006)
21. J.W. Welch, Assessment of thermal balance test criteria requirements on test objectives and thermal design, in *46th International Conference on Environmental Systems*, Vienna, Austria (2016)
22. Z. Qi, L. Wei, M. Huitao, A brief review on thermal model correlation, in *The 11th Space Thermophysics Conference*, Chengdu (2005)
23. M. Molina, A. Ercolifinzi, Montecarlo techniques for thermal analysis of space vehicles: practical examples of robustness determinination in preliminary design, in *XVIII Congresso Nazionale AIDAA* (2005)
24. Y. Huning, Z. Qi, Montecarlo method for thermal model of spacecraft parametric analysis and robustness determination. Chin. Space Sci. Technol. 29(5), 21–27 (2009)
25. C. Wenlong, L. Na, Z. Qi et al., Study on parameters correction method of steady-state thermal model for spacecraft. J. Astronaut. 31(1), 270–275 (2010)

Chapter 7
Spacecraft Thermal Testing

7.1 Overview

Spacecraft generally operates in vacuum and deep cold environment. Some special planetary surface detectors, such as Mars surface detectors may operate in atmospheric environment. Spacecraft receives orbital heat flux in such forms as solar radiation, planetary (lunar) albedo radiation and planetary (lunar) infrared radiation and dissipate heat through the spacecraft heat rejection surface to the deep cold space. This process finally leads to the dynamic balance of temperature.

Spacecraft thermal tests generally include thermal balance test, atmospheric pressure test and low-pressure test. Most tests are thermal balance tests (TBT). During the TBT, the spacecraft is installed in the space environment simulator and operate in orbit. Orbital heat flux and other environments such as vacuum and deep cold background are provided to obtain the temperature of each equipment or structure. These results are utilized to verify the correctness of the thermal control design and thermal analysis model.

Firstly, this chapter introduces the simulation method of space thermal environment and orbital heat flux. Secondly, the spacecraft thermal balance test methods are introduced in detail. Finally, some studies on atmospheric pressure thermal test and low-pressure test are given.

7.2 Simulation Methods for Space Thermal Environment

The major factors to be simulated for space thermal environment include vacuum, low temperature and dark background and orbital heat flux. It is extremely difficult to strictly and accurately simulate these space thermal environmental conditions at the same time on the ground. Therefore, the approximate simulation method is commonly applied in space thermal environment simulations.

7.2.1 Vacuum

The orbits of general spacecraft are outside of the atmosphere. With increase of the orbit attitude, the pressure will drop from 10^{-5} Pa to below 10^{-14} Pa. Currently, it is technically difficult to maintain such a low pressure in a large-scale space environment simulator for a long time. Since the high vacuum environment in space eliminates heat convection, this pressure is acceptable as long as the convective heat transfer effect caused by the residual air in the space environment simulator is negligible.

According to the relationship between heat transfer capacity and gas pressure, the heat transfer capacity of gas at 10^{-1} Pa is only 0.01% of that at atmospheric pressure. The calculation also shows that at the pressure of 10^{-3} Pa, if the accommodation coefficient of gas is 1.0, the heat transferred by hydrogen molecules with relatively high thermal conductivity is extremely small and is approximately equal to 0.33% of the total energy emitted by a satellite with hemispherical emissivity of 0.1 and surface temperature of 300 K. In fact, the emissivity of the satellite surface is much larger than 0.1, and the heat transfer capability of other gases is lower than that of hydrogen. Therefore, from the perspective of engineering application, the impact of simulating the high vacuum space environment with the vacuum of 10^{-3} Pa in TBT is completely negligible.

In some tests, it is necessary to simulate the rapid changing process from atmospheric pressure to high vacuum, namely, the depressurization process. This often occurs in the fairing during the ascent phase of a launch vehicle. For example, the pressure variation curve inside the fairing in the CZ-5 ascent phase is shown in Fig. 7.1. The maximum pressure drop rate is about 6.9 kPa/s.

The simulation of depressurization process is typically performed using the combination of a small vacuum chamber and a very large one. Usually, it is not economic or acceptable to directly release air in the small chamber to the large one so as to obtain a minimum pressure of 10 Pa. The reason is that the volume ratio of the large chamber to the small chamber is required to be as large as about 1,000, according to the ideal state equation $PV = nRT$.

A feasible engineering practice is that, at a relatively high-pressure stage, due to the relatively low-pressure relief rate, the rough vacuum mechanical pump of the small chamber is generally utilized to satisfy the requirement. When the pressure decreases to the point where the mechanical pump of the small chamber cannot meet the requirement, the pressure relief valve to the large chamber is turned on and the residual air is directly released into the large chamber to obtain a larger depressurization rate. Besides, the mechanical pumps for the two chambers work at constant speed until the pressure in the small chamber meets the specified depressurization rate and required pressure level.

7.2 Simulation Methods for Space Thermal Environment

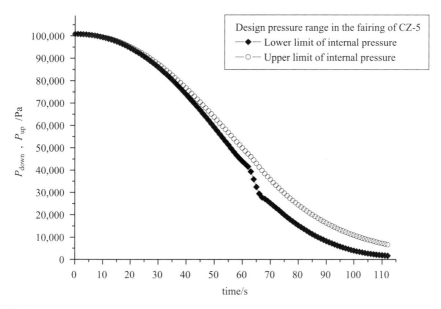

Fig. 7.1 Pressure variation curve in the fairing of CZ-5 launch vehicle (design pressure range in the fairing of CZ-5, upper limit of internal pressure, lower limit of internal pressure)

7.2.2 Cold and Dark Background

The radiation energy emitted from the deep cold space is approximately equal to that of a 4 K black body, which can almost be ignored. To simulate this, the heat sink temperature of space environment simulator should be as close to 4 K as possible. The emissivity of the heat sink should be as close to 1 as possible. However, it is impossible to achieve this in engineering. Therefore, the infrared heat flux from space environment simulator to spacecraft, which is usually referred to as the background heat flux, is always higher than that of 4 K deep cold space. Provided that the background heat flux can be accurately measured, and regarded as part of the simulated orbital heat flux, the deviation between the background heat flux and the heat flux from 4 K cold space will be completely eliminated.

If the background heat flux is larger than the simulated orbital heat flux, the influence must be analyzed and evaluated (provided that the background heat flux can be accurately measured). Or, the background heat flux should be reduced to an acceptable level as far as possible. Since emissivity of heat sink surface is always less than 1.0, the radiative heat transfer between spacecraft and heat sink is related to the surface area ratio. Hence, the intensity of the background heat flux is function of heat sink temperature, emissivity and the surface area ratio of heat sink to spacecraft. For a certain local surface, the background heat flux is also affected by the relative orientation from the surface to heat sink. Thus, it is still very difficult to accurately analyze and evaluate the influence of background heat flux. In order to simplify the problem as much as possible, it is generally required that the heat sink temperature be as low as possible, the emissivity of heat sink surface as high as possible, and the ratio of surface area to spacecraft surface area as large as possible.

Usually, liquid nitrogen is used to simulate the low-temperature space environment. In this circumstance, heat sink temperature can only be as low as 100 K. Coatings with emissivity of about 0.95 are applied on the surface of heat sink. The surface area ratio of heat sink to spacecraft is required to be no less than 2. Under these conditions, the overall deviation between the real cold dark space and the simulated environment is acceptable. However, it is still needed to assess their impact in some cases, especially in the worst cold case.

7.2.3 Orbital Heat Flux

In general, the space orbital heat fluxes to be simulated include the solar radiation, planetary (lunar) albedo and planetary (lunar) infrared radiation. According to the spectral characteristics of the simulated heat flux, the simulation methods of space orbital heat flux can be divided into two categories: the incident flux method and the absorbed flux method. The incident flux method is to simulate the intensity, spectral distribution and direction of radiation heat flux incident on the spacecraft surface. The absorbed flux method is to simulate only the intensity of radiant heat flux absorbed by the spacecraft surface.

Some references proposed the equivalent heat sink temperature method paralleling to the incident flux method and the absorbed flux method. The equivalent heat sink temperature method can be implemented in two ways.

One way is to calculate the surface equivalent temperature given zero internal heat rate. During the test, this temperature is measured under closed-loop control. The temperature can be measured by either a thermistor or a radiometer whose sensitive surfaces is made of the same material as that of spacecraft surface. In either way, it should be ensured that the temperature is only affected by external heat flux.

Another way is to use an infrared plate. In the test, the spacecraft surface is only directly visible only to the infrared plate. Besides, all the orbital heat fluxes to be simulated are converted into the infrared radiation of the infrared plate. Then, according to Stefan–Boltzman's law, the radiation of the infrared plate is further converted to temperature, which is controlled during the test.

In principle, the first method is identical to the absorbed flux method. As a result, if the equivalent heat sink temperature method is regarded as independent to the absorbed heat flux method, the classification crossing and overlapping will occur. The equivalent heat sink temperature method can be completely classified as the absorbed flux method if the classification criterion is whether the spectrum of the radiant heat flux is simulated, and the classification crossing will not occur. Therefore, the classification into incident flux method and absorbed flux method is more reasonable. This classification method is currently adopted by both Chinese and European space communities and is consistent with international standards.

7.3 Environmental Heat Flux Simulator and Heat Flux Measurement

7.3.1 Environmental Heat Flux Simulator

The environment heat flux simulators in common use include solar simulator, infrared cage, infrared lamp array, infrared rod, infrared plate and surface contact-type electric heater are commonly used. Among them, the solar simulator uses the incident flux method, while other devices use the absorbed flux method.

1. **Solar simulator**

The solar simulator consists of light source, optical system, cooling system and control system. Structurally, solar simulators can be divided into coaxial solar simulator and off-axis solar simulator. The performance indexes of solar simulator include irradiance, irradiation non-uniformity, irradiation instability, collimation angle and spectral matching.

The thermal balance test with solar simulator has certain advantages, since it can directly evaluate the thermal optical properties of spacecraft coatings and obtain the effect of thermal coupling between the various parts of the spacecraft. However, only solar radiation heat flux can be simulated by solar simulator, while the planetary (lunar) albedo radiation and the planetary (lunar) infrared radiation heat flux must be simulated by other measures.

The spacecraft is constantly changing in position relative to the sun, so the relative position of the sunlight and the spacecraft must be adjusted during the test. The spacecraft motion is usually achieved by a motion simulator, while the light is fixed. If a moving light source is essential, the solar lamp array should be equipped with longitudinal, lateral and pitch motion mechanisms. The solar lamp array is mounted on a gantry crane. Its longitudinal position can be adjusted through the chain, its lateral position can be adjusted through the casters, and its tilt angle can be adjusted through a pitching mechanism.

2. **Infrared heater**

The infrared heater includes infrared cage, infrared lamp array, infrared heating rod and infrared plate.

The infrared cage simulates the heat flux in the far-infrared spectrum. Usually, it is made up of stainless steel straps, as shown in Fig. 7.2. It is designed into various shapes such as a plate, a cylinder or a cone, depending on the shape of the heated surface. The strap can be divided into different heating loops as needed. In order to increase the heating efficiency, the strap facing the spacecraft is painted with high-emissivity coating, such as black paint. The other side is polished or with a low emissivity. Sometimes the outside surface of the strap is also painted with high-emissivity coating, when very low background flux is simulated. The thermal inertia of the infrared cage is relatively large. Thus, it is not suitable to simulate strong transient effect.

Fig. 7.2 Infrared cage

Usually, the infrared lamp arrays are fixed on a bracket. In order to improve efficiency and uniformity, reflectors with high reflectivity are mounted on the back side of the infrared lamp arrays, while stainless steel baffles are mounted around the array. The shapes of infrared lamp array can be flat, circular or of other relatively regular shapes. The light spectrum generated by the heat flux of infrared lamps is related to the input voltage and thus is partly in the visible spectrum. The thermal inertia of lamp is small, so it can be used to simulate transient heat flux.

Infrared rod is an armored heating rod fixed on the bracket, as shown in Fig. 7.3. The heating rod is a metal tube filled with heating wires and the void filled with powder of MgO, which has good thermal conductivity and electric insulation, as shown in Fig. 7.4. In order to improve heat efficiency and heat flux uniformity, reflectors with high reflectivity are mounted on the back side of the infrared heating rods and stainless steel baffles are mounted around the infrared heating rods. The spectrum of infrared heating rod is far infrared and its heat flux is similar to that of infrared cages. The heat capacity is larger than that of infrared lamps. Thus, it is not suitable for transient cases.

3. **Contact-type electric heater**

The contact-type electric heater is directly attached to the surface of the test object. Damaging the object surface is not allowed when attaching or removing the heater. In general, the emissivity of the heater and that of object surface will be significantly different. Thus, the heater shall be painted with the alternative coating having the same emissivity as that of spacecraft. In addition, the background flux, if existing on the heated surface, shall be subtracted from the heat flux of the heater.

7.3 Environmental Heat Flux Simulator and Heat Flux Measurement

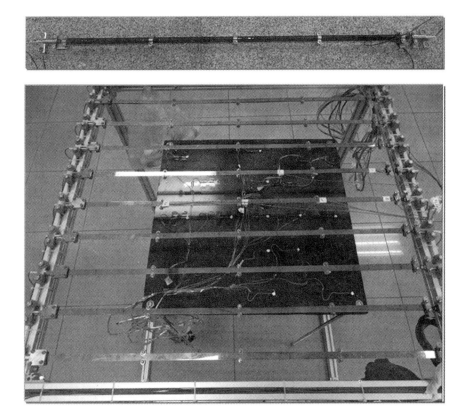

Fig. 7.3 Infrared heating rod

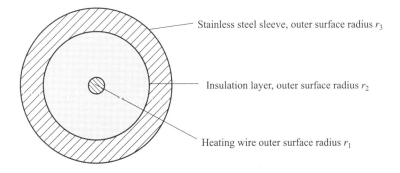

Fig. 7.4 Schematic diagram of infrared heating rod structure

7.3.2 Environmental Heat Flux Measurement

In thermal balance test, the heat flux arriving at the surface of the tested spacecraft should be measured in all the simulation methods except heat flux method and absorbed flux method (surface contact-type electric heater or infrared plate. Radiometer is mostly used in the spacecraft heat flux simulation. Insulation radiometer is often used and its structure is shown in Fig. 7.5. The sensitive slice that absorbs heat flux is a round metal flake sprayed with a high absorptivity coating. The radiometer is mounted to a predetermined location on the spacecraft surface and the absorbed heat flux can be obtained according to the temperature measurement of sensitive slice surface and the Stefan–Boltzman's law.

The coating properties and the spectral distributions of the sensitive slice are very important to heat flux measurement. The change of simulated heat source spectrum has no influence on the measurement, when the coating of sensitive slice is the same as that of spacecraft surface. However, the result will have discrepancy when the two coatings are different. When the back and side faces of radiometer are adiabatic to the spacecraft and the simulated heat source is infrared spectrum, the flux can be corrected according to the emissivity of the two different coatings. The radiometer with the same coating as that of spacecraft surface must be used when using infrared lamp, because the spectrum of infrared lamps is not in far-infrared band. When the infrared cage is used, the coating on the radiometer surface can be different from that on spacecraft surface. Usually, the sensitive slice surface is sprayed with black paint and the emissivity is close to 1.0.

The insulation between the sensitive surface and its back has a great influence on the measurement accuracy. In order to improve the radiometer measurement accuracy, usually, the heat transfer between the sensitive slice and the back must be reduced. A multi-stage insulation radiometer is shown in Figs. 7.6 and 7.7. The sensitive slice is mounted on the inner cup through a bracket with low thermal conductivity. The inner cup (accumulator of the insulation radiometer) is mounted on the outer cup through a bracket with low thermal conductivity. The outer cup is

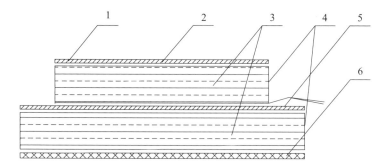

Fig. 7.5 Schematic diagram of insulation radiometer structure. 1—Sensitive slice; 2—Thermocouple; 3—Multi-layer insulation blankets; 4—Protective film; 5—Compensating slice; 6—Bottom

7.3 Environmental Heat Flux Simulator and Heat Flux Measurement

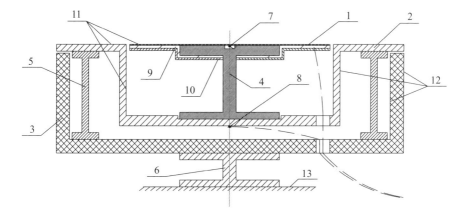

Fig. 7.6 Multi-stage thermal insulation radiometer. 1—Sensitive slice; 2—Inner cup; 3—Outer cup; 4—Sensitive slice bracket; 5—Inner cup bracket; 6—Outer cup bracket; 7—Sensitive slice thermocouple; 8—Inner cup thermocouple; 10—Multi-layer insulation blanket; 11—Black paint; 12—Aluminized polymide film; 13—Radiometer bracket

Fig. 7.7 Multi-stage thermal insulation radiometer (real object)

mounted on the test object surface through a bracket with low thermal conductivity. The heat exchange between the sensitive slice and the inner cup is well controlled to a very low level when their temperature is close to each other, because the insulation method is used.

With the development of deep space exploration, more and more thermal balance tests need to simulate and measure transient heat flux. The dynamic response characteristic of radiometer is very important to ensure test success. Figure 7.8 is a transient radiometer. The sensitive slice is 0.2 mm pure copper slice. The heat leak is reduced by two annular slices, so good dynamic response characteristics and measurement accuracy can be obtained. The radiometer measuring range is from 10 to 1,400 W/m^2.

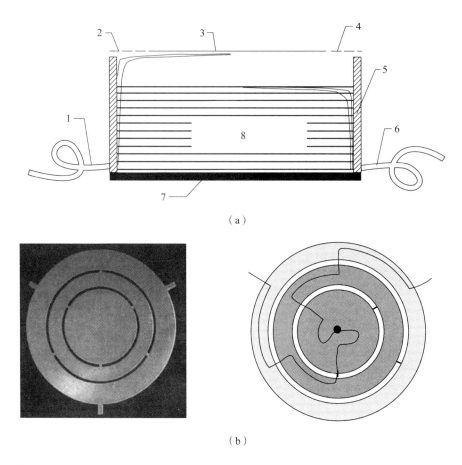

Fig. 7.8 Transient radiometer with two annular slices. (a) Sectional view of radiometer; (b) Structure of sensitive slice (the left is positive, the right is back)

The relative mean deviation is less than 0.9% when the heat flux is greater than 60 W/m^2. The absolute mean deviation is less than 0.8 W/m^2 and the response time is less than 10 s when the heat flux is less than 60 W/m^2.

Figure 7.9 is another radiometer with a thermocouple stack. Its response time is less than 2 s. The maximal relative deviation is less than 10%. For most cases, the deviation is less than 4%.

Fig. 7.9 Transient radiometer with thermal couple stack

7.4 Thermal Balance Test

7.4.1 Thermal Test Model

Besides thermal environment simulator, a proper spacecraft thermal test model is also required in the thermal balance test. Depending on the development stage, the thermal test model is either engineering model or flight model. According to the scale of the model, it may be assembly-level model, module-level model or the system-level model.

To ensure the validity of results, heating effect of thermal test model should be consistent with that of the design state. That is, the model configuration, structural materials and connection pattern, unit layout, shape and surface coating should be the same as those in the design state. The unit heat dissipation in the thermal test model can be simulated by the resistance heater. The effect of natural convection under test conditions should be evaluated if the thermal test model has gas inside, as in the manned spaceship. If the effect is not negligible, the natural convection should be suppressed by the corresponding measures or corrected by the experimental thermal analysis model.

7.4.2 Determination of Test Cases

The selection of test cases and their quantity can affect the development schedule and cost. So it is necessary to analyze the test cases and reduce the test cases as much as possible while satisfying the purpose of the test.

The test case refers to a combination of the environmental heat flux and the internal heat source of the thermal test model. It simulates a certain working state during the orbital operation of the spacecraft. The test cases can be classified into steady-state cases, quasi-steady-state cases, periodic transient cases and transient cases according to the features of internal heat source, simulated environmental heat flux and other thermal boundaries in the thermal test model.

(1) Steady-state cases: Internal heat flux, simulated environmental heat flux and other thermal boundaries are constant.
(2) Quasi-steady-state cases: Internal heat source is periodically changing, while the simulated environmental heat flux and other thermal boundaries are constant.
(3) Periodic transient cases: Internal heat source, environment heat flux and other thermal boundaries are periodically changing.
(4) Transient cases: One of the thermal boundaries, either internal heat source, environmental heat flux or others, changes non-periodically within the given time.

For some spacecraft, their in-orbit working mode is not cyclic for each orbital period but for multiple orbital periods. The quasi-steady-state cases or the periodic transient cases should be repeated for multiple continuous orbital periods until the working condition is stable. The thermal test cycle should correspond to the spacecraft duty cycle.

According to the temperature of spacecraft, the thermal balance test cases usually include worst hot case and worst cold case. Worst hot case refers to the predicted highest temperature in the mission period obtained by the flight thermal analysis model. It is generally the largest combination of the absorbed environmental heat flux and the heat dissipation of equipment. Worst cold case refers to the indicated lowest temperature in the mission period obtained by the flight thermal analysis model. It is generally the smallest combination of the absorbed environmental heat flux and the heat consumption of equipment.

In general, the thermal balance test aims at verifying the correctness of the spacecraft thermal control. For this purpose, the test conditions can only include worst hot case and worst cold case. However, for Mars surface detector, the major purpose of thermal tests is to verify the thermal analysis model, because the ground test cannot simulate the real environment of Mars surface. When thermal analysis model is corrected via test, it is then used to verify the spacecraft thermal control. For this thermal balance test, the test conditions are determined by the need of correcting the thermal analysis model. In short, the thermal balance test type, quantity and scale of the test conditions should be determined according to the purpose of the test. The general principles for determining the thermal balance test conditions are as follows:

(1) Key parameters should be obtained to support thermal analysis model validation and flight mission prediction.
(2) Generally, when the ground test equipment is difficult to achieve high-temperature condition and low-temperature condition, the test conditions should be designed in accordance with the requirement of (1).
(3) According to the thermal analysis model correction requirements, quasi-steady-state conditions, periodic transient conditions or transient conditions are needed for thermal analysis model correction.
(4) Generally, consider verifying the power requirement and control capabilities of the heating circuit.
(5) Generally, consider the fault conditions that may occur and have a significant impact on the function of the spacecraft.
(6) Depending on the demand, the calibration cases for determining the relationship between the heat flux and the current applied by the heat flux simulator may be added.
(7) The engineering model should consider the full verification of the thermal control. The flight model should consider the overall verification requirements, the limitations of the thermal test model and the feasibility of the test on the ground.

7.4.3 Test Process and Method

The thermal balance test process is related to the characteristics of the test object, and the test process can be roughly divided into three stages: pretest preparation, test and post-test check.

Pretest preparation includes the preparation and inspection of the test model, inspection of ground equipment and test equipment. For the test model, the temperature measurement circuit, the heating circuit, the heat flux simulation circuit and the radiometer are mainly checked. After the inspection, cables are connected, vacuum chamber is closed, and the vacuum system is started. When a certain degree of vacuum is reached, liquid nitrogen is added to the heat sink system. Then, according to the test program, test equipment is charged and heat flux is provided. After all tests are completed, the vacuum chamber can be heated and repressurized. During the heating and repressurizing process, temperature of each part must be maintained higher than that of heat sink, so as to prevent condensation of contaminants on the surface of the test model. When the environment simulator pressure returns to normal and the heat sink temperature rises to room temperature, environmental simulator is opened, test model is taken out, and the spacecraft surface is checked.

After the test, the energy balance of the test model should be analyzed and the absorbed and radiant energy of the test model should be compared to confirm the correctness and validity of the test results. According to the difference of thermal test model and the heat flux simulation method, the ratio of the difference between the absorption and radiant energy of the thermal test model is generally within $\pm 10\%$.

If the ratio is too large, further analysis should be made to find out the reasons and propose the treatment suggestion.

The above thermal balance test process is for spacecraft operating in a vacuum environment. For the Mars surface detector operating in a low-pressure environment, the thermal balance process also involves the simulation of gas pressure control, gas temperature control and heat sink temperature control. See Sect. 7.5 for details.

7.5 Atmospheric Thermal Test

Convection is a tool for heat transfer. Reasonable flow field is the premise that the heat rejection of sealed cabin (including dehumidification for the manned spacecraft sealed cabin) meets the requirements. The atmospheric test is to verify whether the convection design is reasonable from the perspective of flow, grasp the characteristics of the ventilation system, verify the matching of flow field design and the fans system and provide the basis for the optimal design of the ventilation system.

The factors affecting the flow field and heat transfer performance of the sealed cabin include the gas flow rate around the equipment and the flow field distribution in the sealed cabin, as well as the related air flow rate and noise. The measurement of air flow rate and noise is difficult to be carried out in the thermal balance test, and easier to be measured under atmospheric pressure. For the test of gas flow rate and flow field, it is necessary to control the influence of natural convection on the ground. It is hoped that the test will be carried out under near-isothermal conditions. The "shutdown state" where the spacecraft is not powered on the ground is more consistent with the test requirements. Therefore, spacecraft convection tests are generally performed separately on the ground.

Noise test is generally performed using a hand-held phonometer. The surrounding environment should be kept quiet during the test and the background noise should be recorded. The wind speed test around the equipment or the air outlet is generally carried out by a hand-held anemometer. The other equipment remains unpowered except that the fan works during the test. For the ventilation pipe that can be connected to the flowmeter with little influence on the supply air flow, the flowmeter can be directly used for the supply air flow test. In other locations, the flow rate is generally measured by a hand-held anemometer, and then the supply air flow is calculated based on the flow rate.

The flow field distribution in the sealed cabin refers to the flow velocity distribution in the large space. Since it is impossible to arrange too many wind speed sensors in a large space, the test time with the hand-held anemometer is too long and the positioning is not accurate. Meanwhile, the human body would interfere with the flow field. Generally, the sensors are arranged in a plane array, and the plane-by-plane scanning method is used to obtain the spatial distribution of the speed. Fig. 7.10 shows the schematic diagram of sensor array for measuring the flow field in the sealed cabin of the Tiangong-1 manned spacecraft. 16 TSI8475 hot-ball wind speed sensors are installed on the movable test fixture-coordinate frame to form a test array. During

7.5 Atmospheric Thermal Test

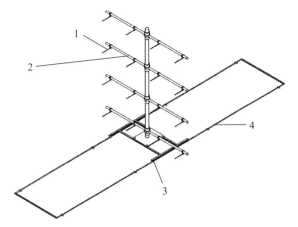

Fig. 7.10 Schematic diagram of coordinate frame for wind speed sensors. 1—Air speed sensor; 2—Bracket; 3—Wheel; 4—Guide rail

the test, the wind speed test array is installed in the cabin, and the movement of the coordinate frame is controlled outside the cabin to prevent the tester from entering the cabin to interfere with the flow field. The measurement point's arrangement of the wind speed test array in the cabin is shown in Fig. 7.11. Since the front end of the hot-ball wind speed sensor has a protective cover which has a great influence on the wind speed measurement parallel to the axis of the sensor, it is necessary to perform two scanning tests on the wind speed at the same point in the test: The first probe is

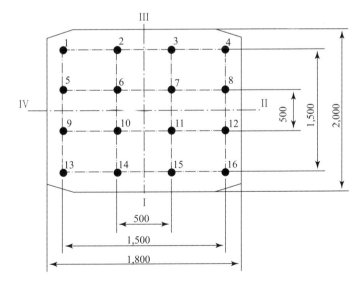

Fig. 7.11 Layout of air speed sensor measuring points

in the horizontal direction, and the second probe is in the vertical direction. The two measurements are compared and the maximum value is taken as the wind speed of the measurement point.

7.6 Low-Pressure Test

7.6.1 Overview

The low-pressure test discussed in this section refers to a type of thermal test conducted in the environment that simulates low temperature and low pressure (fluctuation range of 150–1,400 Pa) on the surface of Mars. On the surface of Mars, there is an atmosphere with an average pressure of about 700 Pa, including the main gas component is CO_2, which accounts for 95.3%, and a small amount of N_2, Ar, O, H_2O, etc. The wind speed on the surface of Mars is about 2–15 m/s, and the maximum wind speed can reach 150 m/s during dust storms. The temperature range of the Mars surface is about -123 °C to $+27$ °C. The low-temperature, low-pressure environment and flow field on the surface of Mars have a significant impact on the thermal control of the Mars surface detector and its devices, so it is generally necessary to carry out special design and test verification.

7.6.2 Selection of Test Gas

The main component of the atmosphere on the surface of Mars is CO_2, thus, CO_2 is theoretically the best choice for conducting tests. However, under the pressure condition of Mars surface, when the temperature drops to -123 °C, CO_2 will condense on the wall surface of the heat sink. As a result, the internal pressure fluctuations and heat sink inefficiency of space environment simulation container will be caused, and the test cannot be carried out. Therefore, selecting an appropriate test gas and controlling the condensation of the gas at low temperature is the keys to conducting the low-pressure test smoothly.

In the early development of Mars probes in the USA, gases such as CO_2, N_2 and Ar were used. In the subsequent development of rover opportunity, courage and curiosity, N_2 was used as the test gas. With reference to the experience of the Mars probes tests that have been carried out abroad, and in accordance with the existing conditions of China's test equipments, the measures that can be taken by China are as follows: ① The tests with Ar gas instead of CO_2 gas under low-temperature conditions, and the tests with CO_2 gas under high-temperature conditions; and ② the tests with N_2 gas under both high- and low-temperature conditions.

For the test using Ar gas instead of CO_2 gas, the analysis is as follows:

7.6 Low-Pressure Test

(1) The use of Ar gas will make the test results of convection effect more positive and the verification insufficient. Under the same pressure and temperature conditions, the experiment with Ar will lead to smaller convective heat transfer coefficient, and more positive verification result (the convective heat transfer coefficient is 11% lower at -40 °C and 17% lower at 0 °C).

(2) The use of Ar gas will lead to more negative verification results of gas insulation layer effect at low temperature and more positive at high temperature. The thermal conductivity of Ar gas is greater than or equal to the thermal conductivity of CO_2 below 55 °C and is less than the thermal conductivity of CO_2 above 55 °C.

(3) Using Ar gas is dangerous. Ar gas is an inert gas with a higher density than air. Ar gas itself does not do harm to the human body directly. But if it is industrially used, the generated exhaust gas will be very harmful to the human body and may cause silicosis, eye damage and so on. There is a risk of suffocation when the Ar gas content in the air exceeds 33%. When it exceeds 50%, severe symptoms may occur. When it exceeds 75%, people will die within a few minutes.

(4) The American Viking Detector used Ar gas in the test, and the subsequent Mars Pathfinder, MER, MSL and other detectors were all tested with N_2.

For the test using N_2 instead of CO_2, the analysis is as follows:

(1) The use of N_2 will lead to more negative verification results of gas thermal conduction. In the case where there is only gas thermal conduction in the cabin, the thermal conductivity coefficient of N_2 is about 40%–84% higher than that of CO_2 at the temperature of 220–350 K. The higher the temperature, the smaller the difference.

(2) The use of N_2 will make the test results of convection effect tend to be more negative. In the external convective heat transfer, the average convective heat transfer coefficient of N_2 is larger than that of CO_2 gas. The deviation is between 2 and 28% at the temperature of 220–350 K, and the deviation is relatively larger at low temperature.

(3) The use of N_2 is safe. N_2 is the main component of the atmosphere and is relatively safe to use and easy to obtain.

It can be seen from the above analysis results that, whether Ar or N_2 is used, the direct verification of product performance cannot be achieved. The thermal analysis model can only be corrected by using the test results, and then the thermal analysis model is used for analysis and verification. Considering the thermal performance, safety, availability and the like, N_2 is preferably used as the test gas in the low-pressure test. In addition, for the situation where the test temperature is higher than the solidification temperature of CO_2 gas under the test pressure, it is recommended to use CO_2 gas for direct test verification.

7.6.3 Gas Temperature Simulation

The gas temperature is controlled by a temperature-adjustable heat sink (thermostatic heat sink) to simulate the indoor gas temperature. According to the required gas temperature control curve, by adjusting the temperature changing rate of the temperature-adjustable heat sink, the temperature changing rate of the simulated indoor gas is consistent with or close to the required gas temperature control requirements. At present, China has the ability to simulate low-pressure and low-temperature situations and can control the temperature of gas between −110°C and +25 °C.

7.6.4 Flow Field Simulation

In the low-pressure test, the flow field should be simulated at the same time in the space environment simulator with a temperature-adjustable heat sink. This is achieved by the direct inspiration wind tunnel. As shown in Fig. 7.12, the entire flow field simulation system consists of the import turning section, import rectifying section (anti-separation net, cellular device, turbulence suppression net), test section, contraction section, fan section, exit turning section and contraction backflow channel, among which the fan section adopts the low-noise DC fan.

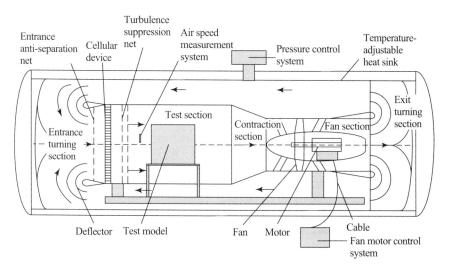

Fig. 7.12 Flow field simulation system

7.6.5 *Measurement*

7.6.5.1 Wind Speed Measurement

The wind speed of low-pressure wind field is generally measured by using a thermal anemometer. There are two types of thermal anemometers: hot wire anemometry and hot film anemometry. The working principle is to measure the heat exchange between the hot wire and the hot film at different wind speeds and obtain the wind speed according to the relationship between heat exchange and wind speed. Therefore, the premise of measuring the wind speed with thermal anemometer is to calibrate the anemometer to obtain the correspondence between the heat exchange and the wind speed.

The rotation method is a more traditional method of wind speed calibration, and the calibration equipment is simple and low-cost. As shown in Fig. 7.13, the basic principle of the rotary calibration is to fill a sealed container with the gas at specified temperature and pressure. The rotation of the rotating arm drives the anemometer through the fluid, which does not move. The relative motion between the anemometer and the fluid medium causes convective heat transfer, thereby obtaining the relationship between the rotational speed of the anemometer and the heat transfer between the anemometer and the gas. In the NASA's rotary calibration method, the calibration equipment has a 2-m-diameter rotating arm and a 3-m-diameter vacuum container, and its maximum measuring speed is 10 m/s.

The main drawback of rotary anemometer calibration is that it depends on the assumption that the gas is stationary. The rotating arm is similar to a paddle, which causes a circulation of the gas flow field and an error in the wind speed system. At the same time, the turbulent vortex will be generated in the container, thus, causing the wind speed to oscillate and measurement errors to occur.

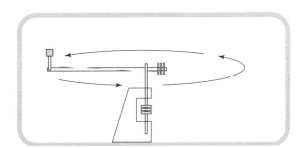

Fig. 7.13 Rotary calibration method for anemometer

7.6.5.2 Gas Temperature Measurement

Thermocouples are still usually used to measure gas temperature in low-pressure tests. However, in the low-pressure test, the thermocouple that measures the gas temperature makes not only convective heat transfer (forced convection, natural convection) with the gas but also radiative heat transfer with the temperature-adjustable heat sink, causing a deviation between the temperature measured by thermocouple and the gas temperature. In order to eliminate this deviation, the measures similar to those as shown in Fig. 7.14 are generally taken:

(1) The thermal junction of thermocouple is coated with a gold-plated film to reduce the emissivity. Then, the thermocouple is surrounded by a multi-layer cylinder (which should not affect the gas flow) to reduce the radiation heat exchange with the heat sink.
(2) Starting from the thermal junction of thermocouple, pasting aluminized film of a certain length along the direction of the thermocouple wire to reduce heat leakage along the wire direction.

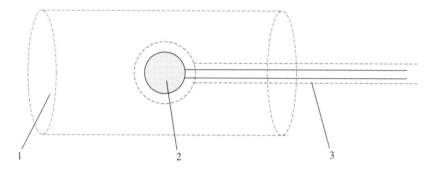

Fig. 7.14 Measurement of gas temperature by thermocouple. 1—Cylindrical cylinder; 2—Thermocouple; 3—Gold-plated film